I0061764

Petrogenic Polycyclic Aromatic Hydrocarbons in the Aquatic Environment:
Analysis, Synthesis, Toxicity and Environmental Impact

Edited by

Daniela M. Pampanin

International Research Institute of Stavanger
University of Stavanger, Norway

&

Magne O. Sydnes

University of Stavanger, Norway

Petrogenic Polycyclic Aromatic Hydrocarbons in the Aquatic Environment: Analysis, Synthesis, Toxicity and Environmental Impact

Editors: Daniela M. Pampanin and Magne O. Sydnes

eISBN (Online): 978-1-68108-427-5

ISBN (Print): 978-1-68108-428-2

©2017, Bentham eBooks imprint.

Published by Bentham Science Publishers – Sharjah, UAE. All Rights Reserved.

First published in 2017.

BENTHAM SCIENCE PUBLISHERS LTD.
End User License Agreement (for non-institutional, personal use)

This is an agreement between you and Bentham Science Publishers Ltd. Please read this License Agreement carefully before using the ebook/echapter/ejournal (**"Work"**). Your use of the Work constitutes your agreement to the terms and conditions set forth in this License Agreement. If you do not agree to these terms and conditions then you should not use the Work.

Bentham Science Publishers agrees to grant you a non-exclusive, non-transferable limited license to use the Work subject to and in accordance with the following terms and conditions. This License Agreement is for non-library, personal use only. For a library / institutional / multi user license in respect of the Work, please contact: permission@benthamscience.org.

Usage Rules:

1. All rights reserved: The Work is the subject of copyright and Bentham Science Publishers either owns the Work (and the copyright in it) or is licensed to distribute the Work. You shall not copy, reproduce, modify, remove, delete, augment, add to, publish, transmit, sell, resell, create derivative works from, or in any way exploit the Work or make the Work available for others to do any of the same, in any form or by any means, in whole or in part, in each case without the prior written permission of Bentham Science Publishers, unless stated otherwise in this License Agreement.

2. You may download a copy of the Work on one occasion to one personal computer (including tablet, laptop, desktop, or other such devices). You may make one back-up copy of the Work to avoid losing it. The following DRM (Digital Rights Management) policy may also be applicable to the Work at Bentham Science Publishers' election, acting in its sole discretion:

- 25 'copy' commands can be executed every 7 days in respect of the Work. The text selected for copying cannot extend to more than a single page. Each time a text 'copy' command is executed, irrespective of whether the text selection is made from within one page or from separate pages, it will be considered as a separate / individual 'copy' command.
- 25 pages only from the Work can be printed every 7 days.

3. The unauthorised use or distribution of copyrighted or other proprietary content is illegal and could subject you to liability for substantial money damages. You will be liable for any damage resulting from your misuse of the Work or any violation of this License Agreement, including any infringement by you of copyrights or proprietary rights.

Disclaimer:

Bentham Science Publishers does not guarantee that the information in the Work is error-free, or warrant that it will meet your requirements or that access to the Work will be uninterrupted or error-free. The Work is provided "as is" without warranty of any kind, either express or implied or statutory, including, without limitation, implied warranties of merchantability and fitness for a particular purpose. The entire risk as to the results and performance of the Work is assumed by you. No responsibility is assumed by Bentham Science Publishers, its staff, editors and/or authors for any injury and/or damage to persons or property as a matter of products liability, negligence or otherwise, or from any use or operation of any methods, products instruction, advertisements or ideas contained in the Work.

Limitation of Liability:

In no event will Bentham Science Publishers, its staff, editors and/or authors, be liable for any damages, including, without limitation, special, incidental and/or consequential damages and/or damages for lost data and/or profits arising out of (whether directly or indirectly) the use or inability to use the Work. The entire liability of Bentham Science Publishers shall be limited to the amount actually paid by you for the Work.

General:

1. Any dispute or claim arising out of or in connection with this License Agreement or the Work (including non-contractual disputes or claims) will be governed by and construed in accordance with the laws of the U.A.E. as applied in the Emirate of Dubai. Each party agrees that the courts of the Emirate of Dubai shall have exclusive jurisdiction to settle any dispute or claim arising out of or in connection with this License Agreement or the Work (including non-contractual disputes or claims).

2. Your rights under this License Agreement will automatically terminate without notice and without the need for a court order if at any point you breach any terms of this License Agreement. In no event will any delay or failure by Bentham Science Publishers in enforcing your compliance with this License Agreement constitute a waiver of any of its rights.

3. You acknowledge that you have read this License Agreement, and agree to be bound by its terms and conditions. To the extent that any other terms and conditions presented on any website of Bentham Science Publishers conflict with, or are inconsistent with, the terms and conditions set out in this License Agreement, you acknowledge that the terms and conditions set out in this License Agreement shall prevail.

Bentham Science Publishers Ltd.
Executive Suite Y - 2
PO Box 7917, Saif Zone
Sharjah, U.A.E.
Email: subscriptions@benthamscience.org

BENTHAM SCIENCE

CONTENTS

FOREWORD

Research about polyaromatic hydrocarbons (PAHs) as environmental stressors is a long term topic with multiple aspects. Its importance implies that we will not see any final papers within the area for a long time. As the research unfolds, both extensive global reviews and more limited reviews addressing recent or expected advances are needed. As new investigative methods emerge it is useful to sum up recent findings and the possible directions being opened to us. This is the purpose of this book.

The analysts now refer to the oil fingerprinting field as "petroleomics", indicating how the field is advancing based on more powerful tools. The analytical advances are not only regarding the parent PAH compounds, but also the metabolized PAHs, known for their high hazardousness. High resolution mass spectrometry will probably soon lead to the detection of a wider range of metabolized and oxidised PAHs too, which will need follow up in different directions.

The analysts ability to synthesize reference PAH metabolites is one of them, and new research challenges will be posed regarding their potential environmental effects. Other logical following steps are revisits of the PAH compounds bioavailability, biodegradability and carcinogenicity. This is the reason these themes are taken up in this book, accompanied by other advances within each of their respective topics. Finally, suggestions are given to how the advances will open for new monitoring methods and strategies leading to one of our important goals: an improved control of a challenged marine environment for which there is an increasingly growing concern.

May this book be an important drop in the ocean that diminishes the PAH part of that concern and help fellow researchers and ourselves to be on top of this complex issue!

Steinar Sanni
International Research Institute of Stavanger
University of Stavanger
Norway

PREFACE

Crude oil and consequently polycyclic aromatic hydrocarbons (PAHs) have been released to the marine environment long time before humans discovered oil through natural oil seeps. Nature has quite efficient systems for cleaning up these releases of contaminants. Since the first human oil excavation started at sea, many oil spills of small and large sizes, have been taking place (chapter 1, 2 and 3). These types of incidents represent an overdose to the natural processes.

The natural defense system for oil degradation is micro organisms, mainly bacteria, which can use the components in oil as an energy source (chapter 7). However, these bacteria preferably use the readily available straight chain hydrocarbons as their first choice of nutrition. This results in a very slow natural removal of larger components such as PAHs, which requires more processing in order to utilize the energy found in the molecules. With great abundance of easily accessible energy sources available, *e.g.* alkanes, which is the case in an oil spill, PAHs will accumulate in the environment. The micro organisms will only start breaking down these compounds in order to get energy once other energy sources are used up. In the meantime PAHs, which have been pointed out as the most toxic organic compounds found in oil, will impact marine life.

Research has shown that PAHs cause severe effects on fish at all life stages, with the most severe effects found on fish embryos. The book will convey an up to date overview of the current state of knowledge regarding the negative health effects of PAHs on marine life (chapter 1 and 4). Most of the health problems related to exposure to PAHs are caused by the oxidized metabolites that organisms generate *in vivo* in order to excrete the unwanted compounds (chapter 5). The metabolites are much more reactive and therefore also more toxic than their mother compounds (chapter 4 and 5). These secondary compounds are prone to react with DNA and proteins to form adducts.

Central in research directed towards understanding the mode of action for PAH metabolites *in vivo* has been the preparation of synthetic material (chapter 6). This material has been utilized as standards for analysis and starting point for studying further metabolism *in vivo*. Moreover, due to their presence in the marine environment and their genotoxicity, there is a standing requirement for oil and gas operators to monitor their concentration and influence on marine life through environmental monitoring studies (*e.g.* the Water Column monitoring) (chapter 2).

Although a lot is known about the influence of PAHs on the environment, and the marine environment in particular, there are still many unresolved questions that are awaiting answers. It is our intention that this book will give a solid basis facilitating the pursue of these unanswered questions.

Daniela M. Pampanin
International Research Institute of Stavanger
University of Stavanger
Norway

&

Magne O. Sydnes
University of Stavanger
Norway

List of Contributors

Andrea Bagi	Faculty of Science and Technology, Department of Mathematics and Natural Science, University of Stavanger, Stavanger, Norway
Anna Marqueño	Environmental Chemistry Department, IDAEA-CSIC, 08034 Barcelona, Spain
Cinta Porte	Environmental Chemistry Department, IDAEA-CSIC, 08034 Barcelona, Spain
Daniela M. Pampanin	International Research Institute of Stavanger, Stavanger, Norway Faculty of Science and Technology, Department of Mathematics and Natural Science, University of Stavanger, Stavanger, Norway
Denise Fernandes	Environmental Chemistry Department, IDAEA-CSIC, 08034 Barcelona, Spain
Emil Lindbäck	Faculty of Science and Technology, Department of Mathematics and Natural Science, University of Stavanger, Stavanger, Norway
Jérémie Le Goff	ADn'tox SAS, Centre de Lutte contre le Cancer François Baclesse, Caen, France
Magne O. Sydnes	Faculty of Science and Technology, Department of Mathematics and Natural Science, University of Stavanger, Stavanger, Norway
Montserrat Solé	Marine Science Institute (ICM-CSIC), 08003 Barcelona, Spain

Petrogenic Polycyclic Aromatic Hydrocarbons
in the Aquatic Environment:
Analysis, Synthesis, Toxicity and Environmental Impact

Introduction to Petrogenic Polycyclic Aromatic Hydrocarbons (PAHs) in the Aquatic Environment

Daniela M. Pampanin[*]

International Research Institute of Stavanger, Stavanger, Norway

Faculty of Science and Technology, Department of Mathematics and Natural Science, University of Stavanger, Stavanger, Norway

Abstract: This chapter serves as an introduction to the book. It contains key points and background information that will be followed up and expanded in the coming chapters. Polycyclic aromatic hydrocarbons (PAHs) are a constituent of crude oil and PAHs from petroleum are called petrogenic PAHs. This book focuses on petrogenic PAHs and research results related to them, including sources of contamination in the aquatic environment, analytical methods utilized in order to detect and quantify their presence in various matrixes, biological effects of PAHs and their metabolites, and microbial degradation with special focus on the situation in the Arctic. The tracking of sources of petrogenic contamination is of significance and it is a focus of this book. Both current knowledge and future challenges related to petrogenic PAHs are introduced and discussed.

Keywords: Biological effects, Biota, Contamination, Crude oil, PAHs.

PETROGENIC PAHS

Polycyclic aromatic hydrocarbons (PAHs) are a constituent of crude oil and PAHs from petroleum are called petrogenic PAHs. These groups of compounds have been of high concern due to their toxic effect and in particular their carcinogenic potential [1]. PAHs are known for causing adverse effects in aquatic organisms, however, their toxicity is not directly due to the parent compounds, but predomi-

[*] **Corresponding author Daniela M. Pampanin:** International Research Institute of Stavanger, Stavanger, Norway; Tel/Fax: +47 51875502; E-mail: dmp@iris.no

Daniela M. Pampanin and Magne O. Sydnes (Eds.)
All rights reserved-© 2017 Bentham Science Publishers

nantly due to the oxidation products generated *in vivo* [2, 3]. The oxidation products are formed during the process of making the compounds more water soluble so that they can be more easily excreted. There has been a great research effort in order to unravel the harm caused by PAHs on biota. Hylland reviewed the ecotoxicology knowledge related to PAHs in the marine ecosystem, underlining the link between these compounds and their adverse effects on biota and recognizing the importance of the determination of adverse effects using biological variables (*i.e.* biomarkers) [4].

Chapter 2 will give an overview of the possible sources of contamination.

It is important to provide information about the analytical methods utilized to identify the source of contamination (see Chapter 3) and the methods used for their quantification in biota.

Significant improvements in analytical chemistry methods are increasing the possibility to analyse a large fraction of PAH compounds found in oil in a single analysis. Descriptions of these methods, especially new approaches using mass spectrometry, and how they are and can be used in environmental investigations are reported in Chapter 3.

More than 600 aromatic hydrocarbons have been listed, from the monocycle benzene (molecular weight (MW) = 78) to the nine ring compounds (MW up to 478). They are classified according to their physical and chemical properties, *i.e.* the temperature of compound formation and the origin. They can be classified in: 1) natural, *i.e.* biogenic or diagenetic origin; 2) pyrogenic, *i.e.* originated from pyrolysis substrates; 3) petrogenic, *i.e.* originated from petroleum sources. Characterisation of PAHs has been reviewed and physical and chemical properties of the most common PAH contaminants (*e.g.* MW, aqueous solubility, vapor pressure, the octanol-water partition coefficient, boiling point) are available online.

The characterization and quantification of PAHs in the sediment compartment have received growing attention since the 1970s (*i.e.* lakes, rivers, estuaries and seas). Most studies during this time reported a predominance of pyrogenic PAHs *versus* petrogenic PAHs, except in specific cases of oil spills or oil related activities. Petrogenic PAHs in general can result in more bioavailability since they tend to bind strongly to sediment particles.

The identification of PAH type is essential for evaluating the risk to biota. The ratio between alkylated PAHs and the parent compound is commonly used to distinguish between PAH types [5 - 7]. In particular, petrogenic PAH composition is dominated by alkyl constituents.

Petrogenic PAHs primarily consists of 2- and 3-ring compounds and alkyltated forms. The alkyltated forms can make up to 90% of the PAHs found in crude oil. It is important to note that alkylated PAHs have higher toxicity than other forms of PAHs.

The concern over the toxicity of PAHs resulted in U.S. EPA (Environmental Protection Agency) establishing the narcosis model for protecting the benthic community [8]. These guidelines require the measurement of the so-called 34 PAHs in sediment sample (*i.e.* 18 parent PAHs and 16 alkyl PAH derivatives) to evaluate the impact of PAH contamination on benthic species [9].

Knowledge and regulations have been improved since the introduction of the 16 EPA PAH concept in the 1970s and more research is focusing on toxicity, environmental and chemical analysis of other polycyclic aromatic compounds, *e.g.* alkylate PAHs, amino-PAHs, cyno-PAHs [10 - 12].

BIOACCUMULATION OF PAHS IN AQUATIC ORGANISMS

Petrogenic PAHs are bioavailable compounds in the aquatic environment and their presence in biota represents a problem. The estimation of the total PAH load entering the aquatic environment is clearly quite difficult and the value of 0.5 million tons/year suggested in the 1980s is now at least one order of magnitude greater.

Most studies related to the measurement of PAH concentration in aquatic organisms are linked to the risk they pose on human health [13 - 17]. Therefore, the management of aquatic resources have been traditionally based only on the quantification of PAHs in biota. PAHs are included in the often referred to as the Hygiene Package from the European regulation, which reports the limits for contaminants in shellfish as foodstuff. Table **1** reports the maximum level of benzo[*a*]pyrene (B[*a*]P) allowed in food (EC regulation No 1881/2006). A recent review from Guéguen *et al.* summarises the chemical hazard of shellfish (*e.g.* oysters, mussels, scallops) collected along French coasts [14]. The authors concluded that the chemical monitoring of contaminants is important for evaluating the risk to human health from the consumption of contaminated food, however it is not sufficient to estimate the real risk. Moreover, they highlight the necessity to monitor PAH compounds.

Table 1. Maximum level of benzo[*a*]pyrene allowed in foodstuff (EC regulation No 1881/2006).

	µg/kg wet weight
Bivalves	10
Crustaceans, cephalopods	5
Fish meat	2

Bioavailability is a key element when discussing aquatic environment contaminations. Adverse biological effects need to be considered and possibly predicted [18]. The bioavailability of a chemical depends on biogeochemical and physiological processes and determining the amount of pollutants that are capable of entering the target (*i.e.* organism).

The degree of bioavailability also dependents on the possibility that PAHs make a stable complex with dissolved organic matter in the water and in the sediment (*i.e.* pore water). This affinity is known to increase with the PAH MW and the hydrophobicity of the compounds.

It is well-known that a prediction of PAH bioavailability can be done by taking into account key factors such as partitioning of the compound between sediment, water and tissue compartments (which is controlled by lipid and organic carbon) and the octanol-water partition coefficient [19]. The organic carbon associated with the sediment or dissolved in water influences the bioavailability (Fig. **1**). Since this is the crucial step for both exposure and effects of PAH on biota, scientific knowledge regarding the bioavailability of PAH in aquatic organisms is regularly reviewed [20 - 22].

Bioavailability

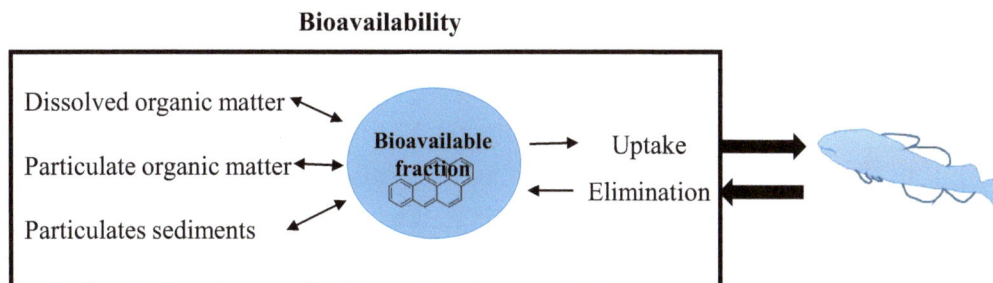

Fig. (1). PAH uptake into aquatic organisms depends on the fraction of bioavailable PAHs. Schematic representation of factors that influence bioavailability.

If PAHs are bioavailable in the environment, there will be an uptake of PAHs into the organism (*i.e.* bioaccumulation). Bioaccumulation occurs in all the aquatic organisms, and the level of exposure is highly dependent to the capability of the target species to metabolise PAHs [20]. Being hydrophobic compounds, they tend to accumulate in lipid rich tissues. The lipid reservoir also makes the maximum level of accumulation into an organism (*i.e.* maximum body burden).

In fish *e.g.*, PAHs and their metabolites can be found in most tissues, however higher levels are normally found in the liver (metabolic site) and the bile (excretion site).

In invertebrates, the highest concentrations of PAHs are found in the hepatopancreas or the digestive gland.

In general, benthic species have been found to have higher levels of PAHs in their tissues compared to pelagic species, due to the ingestion of contaminated particles.

Various studies have shown that PAH bioaccumulation follows a clear seasonal variation [23 - 27]. This seasonal variability is related to different factors, including the lipid content of organisms that varies with reproduction cycles [28].

Bioaccumulation models have been developed, resulting in a great improvement in the modelling of bioaccumulation of PAHs [29, 30]. Their application is still not accurate enough to predict the risk since many factors can influence the performance of the model (*e.g.* exposure duration, biotransformation, abiotic factors) [30, 31]. Therefore, their use should be considered in combination with experimental data.

Aquatic organisms are capable of metabolising PAHs and different species use different metabolic systems, consequently having different efficiency. It is well known that fish have very efficient metabolic pathways to transform PAH and up to 99% of the uptaken compounds are metabolised within 24 hours. Nevertheless, these processes can also bring to the creation of more toxic forms capable of damaging the organisms metabolism, tissue or even physiological processes. For an accurate estimation of PAH exposure, it is therefore desirable to determine the presence of both PAHs and their metabolites. This clearly represents a challenge in the determination of PAHs in biota.

PAH concentration also increases due to bioaccumulation at the different trophic levels (*i.e.* biomagnification). Biomagnification is still not a fully understood process and research efforts are directed to fill the knowledge gaps [32]. A recent study regarding the biomagnification of PAH included *Daphnia magna*, zebrafish and cichlids as test organisms. Results showed that predation of contaminated *D. magna* increased the uptake and elimination rates of PAHs in fish. However, predation did not change the bioaccumulation equilibrium, which in fish species depended on the freely dissolved PAHs in water. The authors concluded that biomagnification occurred due to an increased uptake (caused by predation) if the bioaccumulation equilibrium was not reached [32].

Once more, the risk posed by the biomagnification has a high priority because of the risk to human health through consumption of aquatic organisms and various knowledge gaps have been identified [33]. A critical review has been recently published regarding the potential sources, risk and effects of PAHs in marine food [17].

QUANTIFICATION OF PAHS IN BIOTA

Quantification of PAHs in biota has been carried out for many years in both laboratory exposure and monitoring surveys.

Studies have been conducted all over the world, however remote areas like the

Arctic and sub-Arctic still have a need for baseline studies. With the opening of sailing routes (as a consequence of global warming) and new oil field explorations, these previously pristine regions are coming under increasing pressure from contamination and the scope for disaster only increases with increasing activities. Researchers have therefore started to collect information regarding background level of PAHs in these areas [34 - 36].

The Biota Sediment Accumulation Factor (BSAF) is commonly used to measure the bioaccumulation of PAHs [37], allowing the comparison of sediment contaminations in different locations of the world. A recent publication from Szczybelski *et al.* is providing BSAF values from the Arctic [38], where petrogenic PAHs are of concern [39].

An overview of the available techniques for chemical evaluation of PAHs in biota (body burden quantification) includes:

- gas chromatography–mass spectrometry (GC–MS) operated in selected ion mode;
- high-performance liquid chromatography (HPLC);
- gas chromatographic tandem mass spectrometric method (GC–MS/MS) (developed by Kasiotis *et al.* [40];
- HPLC with fluorometric detection, and atomic absorption spectrophotometry [41].

The European Commission has recently published, under the Water Framework Directive, guidelines regarding analytical methods for the determination of PAHs in biota.

BIOLOGICAL EFFECTS OF PAHS

Once entering the organism, these compounds are metabolised and partly excreted *via* the bile as PAH metabolites. Methods to detect and quantify PAH metabolites in biota are therefore presented in detail in this book, especially in relation to their endocrine disruption potential (see Chapter 5).

Different species metabolise PAHs differently giving rise to various products. The diverse oxidation products also have different toxicity. These events result in a range of exposure molecules in the organism and large variation of toxicity for a given PAH between species. This represents significant information for establishing concentration limits, since different parts of a geographical region can tolerate different levels of PAH contamination. Biomarker analyses are available and commonly used in monitoring programs to help understand the effects of PAH contamination. A comprehensive review of the ecotoxicological effects of PAHs was written by Hylland [4].

A search for existing literature quickly shows that biomarkers have been used since the 1980s to evaluate PAH contamination and the list of analysed parameters has been adjusted to fit environmental cases. The following list of biomarkers has been successfully used in the evaluation of petrogenic PAH effects [21, 42]:

- Metabolites occurrence:
    PAH metabolites in bile;
- Biomarkers of xenobiotic transformation:
    cytochrome P4501A (CYP1A), ethoxyresorufin-O-deethylase (EROD), glutathione-S-transferase (GST), gene expression related to the aryl hydrocarbons receptor (AHR) activation;
- Antioxidant defences:
    glutathione-S-transferase (GST), superoxide dismutase (SOD), catalase (CAT), glutathione peroxidase (GPx), glutathione reductase (GR), glutathione (GSH), NADPH-dependent isocitrate dehydrogenase (IDP), pyruvate kinase (PK), phosphoenolpyruvate carboxykinase (PEPCK), TOSC (total oxyradical scavenging capacity);
- Oxidative damages:
    malondialdehyde (MDA) production, lipid peroxidation (TBARS);
- Genotoxicity
    DNA adducts, comet assay, micronucleus assay;
- Immuno responses:
    total hemocyte count (THC) (including percentage of different cell types); hemocyte mortality (apoptosis), phagocytosis, parasitic infection;
- Enzyme activity:
    Acetylcholine esterase (AChE) (*i.e.* in relation to neurotoxic effect), acyl CoA

oxidase (AOX);
- Histological evaluation of target tissues:
  alteration of internal tissues;
- Physiological measurements:
  Swimming performance, metabolic rate, escape capacity (*e.g.* valve closing/opening in scallops, growth).

Some emerging approaches are related to omics techniques (genomics, transcriptomics, metabolomics, proteomics, adductomics). As an example, Deng *et al.* published a research article related to the reproductive toxicity of benzo[*a*]pyrene in scallops using digital gene expression analysis. Results showed a clear response in gene expression related to the exposure and consolidate the idea to use this approach in relation to other PAH compounds and in other aquatic species [43]. Similarly, the transcriptional responses of zebrafish due to PAH exposure has been reported. In this case, the omics method was suggested as useful for classifying the PAH toxicity [44].

The omics approach seems to be particularly useful when the contamination is due to a mixture of PAH compounds, which is what organisms are generally exposed to in the natural aquatic environment. Our research group has also been developing a method for using the bile proteome as biomarker of exposure to PAHs [1, 45] (detailed information can be found in Chapter 2).

A major advantage in using biomarkers is the possibility to have a time-integrated response of the organism, even at very low PAH contamination level (concentration below detection limit for quantification in the water and/or in the biota). The correlation between body burden and biological effects is definitively not straight forward [46]. Therefore the chemical quantification of PAH in an organism does not provide enough information about biological effects or the risk for the organism. Being sub-lethal measurements, biomarkers are good early warning system for monitoring the ecosystem and are capable of highlighting the long term effects of contamination.

Factors as seasonal variations can cloud results and cause confusion. These factors have been studied and the conclusion is that they are within acceptable limits in environmental monitoring, as far as the origins are known (*e.g.* reproduction state, temperature) [47, 48].

A summary of the effects of PAHs on marine organisms can be found in the review from Mearns *et al.*, where the authors reviewed more than 2000 scientific contributions [49]. This study aimed to give an overview of studies conducted both in the laboratory and the field on various species. In Mearns *et al.* review, a special attention was dedicated to the effects of oil spill events and PAHs in form of dispersed crude oil.

CONCLUDING REMARKS

Considerable research has been done to evaluate the presence and effects of petrogenic PAHs in the aquatic environment as confirmed by the large number of scientific publications (*e.g.* scientific articles, books, reviews). Tools are available for monitoring petrogenic PAHs and they include both chemical and biological measurements. Most environmental monitoring activities nowadays include both approaches, when aiming to provide information for decision makers. It is far more common now for legislation and regulation to take a broader more comprehensive approach to preserve the health of both humans and ecosystems.

Nevertheless, the research effort needs to continue towards the identification and quantification of PAHs and their metabolites in aquatic organisms, in order to fill knowledge gaps regarding the toxic effects and long term consequences of PAH contamination.

The tracking of sources of PAH contamination is a topic of great interest both for regulating the input into the aquatic environment and for predicting/estimating the biological effects of the pollution.

In general, the current knowledge starts to allow us to predict the adverse effects of petrogenic PAH contamination both in short term (acute event like an oil spill) and long term (chronic contamination due to anthropogenic activities).

CONFLICT OF INTEREST

The author confirms that the author has no conflict of interest to declare for this publication.

ACKNOWLEDGEMENTS

Declared none.

REFERENCES

[1]   Pampanin, D.M.; Sydnes, M.O. Polycyclic aromatic hydrocarbons a constituent of petroleum: Presence and influence in the aquatic environment. In: *Hydrocarbons*; Vladimir, K.; Kolesnikov, A., Eds.; InTech: Rijeka, **2013**; pp. 83-118.

[2]   Dasgupta, S.; Cao, A.; Mauer, B.; Yan, B.; Uno, S.; McElroy, A. Genotoxicity of oxy-PAHs to Japanese medaka (*Oryzias latipes*) embryos assessed using the comet assay. *Environ. Sci. Pollut. Res. Int.,* **2014**, *21*(24), 13867-13876.
[http://dx.doi.org/10.1007/s11356-014-2586-4] [PMID: 24510601]

[3]   Lundstedt, S.; White, P.A.; Lemieux, C.L.; Lynes, K.D.; Lambert, I.B.; Öberg, L.; Haglund, P.; Tysklind, M. Sources, fate, and toxic hazards of oxygenated polycyclic aromatic hydrocarbons (PAHs) at PAH-contaminated sites. *Ambio,* **2007**, *36*(6), 475-485.
[http://dx.doi.org/10.1579/0044-7447(2007)36[475:SFATHO]2.0.CO;2] [PMID: 17985702]

[4]   Hylland, K. Polycyclic aromatic hydrocarbon (PAH) ecotoxicology in marine ecosystems. *J. Toxicol. Environ. Health A,* **2006**, *69*(1-2), 109-123.
[http://dx.doi.org/10.1080/15287390500259327] [PMID: 16291565]

[5]   Blumer, M. Polycyclic aromatic compounds in nature. *Sci. Am.,* **1976**, *234*(3), 35-45.
[http://dx.doi.org/10.1038/scientificamerican0376-34] [PMID: 1251182]

[6]   Garrigues, P.; Budzinski, H.; Manits, M.P.; Wise, S.A. Pyrolitic and petrogenic inputs in recent sediments: a definitive signature through phenanthrene and chrysene compounds distribution. *Pol. Arom. Comp,* **1995**, *7*, 275-284.
[http://dx.doi.org/10.1080/10406639508009630]

[7]   Saha, M.; Togo, A.; Mizukawa, K.; Murakami, M.; Takada, H.; Zakaria, M.P.; Chiem, N.H.; Tuyen, B.C.; Prudente, M.; Boonyatumanond, R.; Sarkar, S.K.; Bhattacharya, B.; Mishra, P.; Tana, T.S. Sources of sedimentary PAHs in tropical Asian waters: differentiation between pyrogenic and petrogenic sources by alkyl homolog abundance. *Mar. Pollut. Bull.,* **2009**, *58*(2), 189-200.
[http://dx.doi.org/10.1016/j.marpolbul.2008.04.049] [PMID: 19117577]

[8]   Berry, J.A.; Wells, P.G. Integrated fate modeling for exposure assessment of produced water on the Sable Island Bank (Scotian shelf, Canada). *Environ. Toxicol. Chem.,* **2004**, *23*(10), 2483-2493.
[http://dx.doi.org/10.1897/03-458] [PMID: 15511109]

[9]   Hawthorne, S.B.; Miller, D.J.; Kreitinger, J.P. Measurement of total polycyclic aromatic hydrocarbon concentrations in sediments and toxic units used for estimating risk to benthic invertebrates at

manufactured gas plant sites. *Environ. Toxicol. Chem.,* **2006**, *25*(1), 287-296.
[http://dx.doi.org/10.1897/05-111R.1] [PMID: 16494254]

[10]   Andersson, J.T.; Achten, C. Time to Say Goodbye to the 16 EPA PAHs? Toward an Up-to-Date Use of PACs for Environmental Purposes. *Polycycl. Aromat. Compd.,* **2015**, *35*(2-4), 330-354.
[http://dx.doi.org/10.1080/10406638.2014.991042] [PMID: 26823645]

[11]   Hong, W-J.; Jia, H.; Li, Y-F.; Sun, Y.; Liu, X.; Wang, L. Polycyclic aromatic hydrocarbons (PAHs) and alkylated PAHs in the coastal seawater, surface sediment and oyster from Dalian, Northeast China. *Ecotoxicol. Environ. Saf.,* **2016**, *128*, 11-20.
[http://dx.doi.org/10.1016/j.ecoenv.2016.02.003] [PMID: 26874984]

[12]   Sørensen, L.; Meier, S.; Mjøs, S.A. Application of gas chromatography/tandem mass spectrometry to determine a wide range of petrogenic alkylated polycyclic aromatic hydrocarbons in biotic samples. *Rapid Commun. Mass Spectrom.,* **2016**, *30*(18), 2052-2058.
[http://dx.doi.org/10.1002/rcm.7688] [PMID: 27470186]

[13]   Wang, H.S.; Man, Y.B.; Wu, F.Y.; Zhao, Y.G.; Wong, C.K.; Wong, M.H. Oral bioaccessibility of polycyclic aromatic hydrocarbons (PAHs) through fish consumption, based on an *in vitro* digestion model. *J. Agric. Food Chem.,* **2010**, *58*(21), 11517-11524.
[http://dx.doi.org/10.1021/jf102242m] [PMID: 20929255]

[14]   Guéguen, M.; Amiard, J.C.; Arnich, N.; Badot, P.M.; Claisse, D.; Guérin, T.; Vernoux, J.P. Shellfish and residual chemical contaminants: hazards, monitoring, and health risk assessment along French coasts. *Rev. Environ. Contam. Toxicol.,* **2011**, *213*, 55-111.
[http://dx.doi.org/10.1007/978-1-4419-9860-6_3] [PMID: 21541848]

[15]   Crowe, K.M.; Newton, J.C.; Kaltenboeck, B.; Johnson, C. Oxidative stress responses of gulf killifish exposed to hydrocarbons from the Deepwater Horizon oil spill: Potential implications for aquatic food resources. *Environ. Toxicol. Chem.,* **2014**, *33*(2), 370-374.
[http://dx.doi.org/10.1002/etc.2427] [PMID: 24122941]

[16]   Barhoumi, B.; El Megdiche, Y.; Clerandeau, C.; Ameur, W.B.; Mekni, S.; Bouadhallah, S.; Derouiche, A.; Touil, S.; Cachot, J.; Driss, M.R. Occurrence of polycyclic aromatic hydrocarbons (PAHs) in mussel (*Mytilus galloprovincialis*) and eel (*Anguilla anguilla*) from Bizerte lagoon, Tunisia, and associated human health risk assessment. *Cont. Shelf Res.,* **2016**, *124*, 104-116.
[http://dx.doi.org/10.1016/j.csr.2016.05.012]

[17]   Balcıoğlu, E.B. Potential effects of polycyclic aromatic hydrocarbons (PAHs) in marine foods on human health: a critical review. *Toxin Rev.,* **2016**, *35*, 98-105.
[http://dx.doi.org/10.1080/15569543.2016.1201513]

[18]   Fent, K. Ecotoxicological problems associated with contaminated sites. *Toxicol. Lett.,* **2003**, *140-141*, 353-365.
[http://dx.doi.org/10.1016/S0378-4274(03)00032-8] [PMID: 12676484]

[19]   Walker, C.H.; Sibly, R.M.; Hopkin, S.P.; Peakall, D.B. *Principles of ecotoxicology*; CRC Press, Taylor and Francis Group, **2012**, p. 360.

[20]   Meador, J.P.; Stein, J.E.; Reichert, W.L.; Varanasi, U. Bioaccumulation of polycyclic aromatic hydrocarbons by marine organisms. *Rev. Environ. Contam. Toxicol.,* **1995**, *143*, 79-165.
[http://dx.doi.org/10.1007/978-1-4612-2542-3_4] [PMID: 7501868]

[21]   van der Oost, R.; Beyer, J.; Vermeulen, N.P. Fish bioaccumulation and biomarkers in environmental risk assessment: a review. *Environ. Toxicol. Pharmacol.,* **2003**, *13*(2), 57-149.
[http://dx.doi.org/10.1016/S1382-6689(02)00126-6] [PMID: 21782649]

[22]   Gauthier, P.T.; Norwood, W.P.; Prepas, E.E.; Pyle, G.G. Metal-PAH mixtures in the aquatic environment: a review of co-toxic mechanisms leading to more-than-additive outcomes. *Aquat. Toxicol.,* **2014**, *154*, 253-269.
[http://dx.doi.org/10.1016/j.aquatox.2014.05.026] [PMID: 24929353]

[23]   Baumard, P.; Budzinski, H.; Garrigues, P.; Dizer, H.; Hansen, P.D. Polycyclic aromatic hydrocarbons in recent sediments and mussels (*Mytilus edulis*) from the Western Baltic Sea: occurrence, bioavailability and seasonal variations. *Mar. Environ. Res.,* **1999**, *47*, 17-47.
[http://dx.doi.org/10.1016/S0141-1136(98)00105-6]

[24]   De Lange, H.J.; Peeters, E.T.; Harmsen, J.; Maas, H.; De Jonge, J. Seasonal variation of total and biochemically available concentrations of PAHs in a floodplain lake sediment has no effect on the benthic invertebrate community. *Chemosphere,* **2009**, *75*(3), 319-326.
[http://dx.doi.org/10.1016/j.chemosphere.2008.12.046] [PMID: 19167023]

[25]   Copat, C.; Brundo, M.V.; Arena, G.; Grasso, A.; Oliveri Conti, G.; Ledda, C.; Fallico, R.; Sciacca, S.; Ferrante, M. Seasonal variation of bioaccumulation in *Engraulis encrasicolus* (Linneaus, 1758) and related biomarkers of exposure. *Ecotoxicol. Environ. Saf.,* **2012**, *86*, 31-37.
[http://dx.doi.org/10.1016/j.ecoenv.2012.09.006] [PMID: 23020986]

[26]   Michel, C.; Bourgeault, A.; Gourlay-Francé, C.; Palais, F.; Geffard, A.; Vincent-Hubert, F. Seasonal and PAH impact on DNA strand-break levels in gills of transplanted zebra mussels. *Ecotoxicol. Environ. Saf.,* **2013**, *92*, 18-26.
[http://dx.doi.org/10.1016/j.ecoenv.2013.01.018] [PMID: 23490194]

[27]   Asker, N.; Albertsson, E.; Wijkmark, E.; Bergek, S.; Parkkonen, J.; Kammann, U.; Holmqvist, I.; Kristiansson, E.; Strand, J.; Gercken, J.; Förlin, L. Biomarker responses in eelpouts from four coastal areas in Sweden, Denmark and Germany. *Mar. Environ. Res.,* **2016**, *120*, 32-43.
[http://dx.doi.org/10.1016/j.marenvres.2016.07.002] [PMID: 27423807]

[28]   González-Fernández, C.; Albentosa, M.; Campillo, J.A.; Viñas, L.; Franco, A.; Bellas, J. Effect of mussel reproductive status on biomarker responses to PAHs: Implications for large-scale monitoring programs. *Aquat. Toxicol.,* **2016**, *177*, 380-394.
[http://dx.doi.org/10.1016/j.aquatox.2016.06.012] [PMID: 27379756]

[29]   Arnot, J.A.; Gobas, F.A. A food web bioaccumulation model for organic chemicals in aquatic ecosystems. *Environ. Toxicol. Chem.,* **2004**, *23*(10), 2343-2355.
[http://dx.doi.org/10.1897/03-438] [PMID: 15511097]

[30]   De Hoop, L.; Huijbregts, M.A.; Schipper, A.M.; Veltman, K.; De Laender, F.; Viaene, K.P.; Klok, C.; Hendriks, A.J. Modelling bioaccumulation of oil constituents in aquatic species. *Mar. Pollut. Bull.,* **2013**, *76*(1-2), 178-186.
[http://dx.doi.org/10.1016/j.marpolbul.2013.09.006] [PMID: 24064372]

[31]   Sappington, K.G.; Bridges, T.S.; Bradbury, S.P.; Erickson, R.J.; Hendriks, A.J.; Lanno, R.P.; Meador, J.P.; Mount, D.R.; Salazar, M.H.; Spry, D.J. Application of the tissue residue approach in ecological risk assessment. *Integr. Environ. Assess. Manag.,* **2011**, *7*(1), 116-140.
[http://dx.doi.org/10.1002/ieam.116] [PMID: 21184572]

[32] Xia, X.; Li, H.; Yang, Z.; Zhang, X.; Wang, H. How does predation affect the bioaccumulation of hydrophobic organic compounds in aquatic organisms? *Environ. Sci. Technol.,* **2015,** *49*(8), 4911-4920.
[http://dx.doi.org/10.1021/acs.est.5b00071] [PMID: 25794043]

[33] Borgå, K.; Kidd, K.A.; Muir, D.C.; Berglund, O.; Conder, J.M.; Gobas, F.A.; Kucklick, J.; Malm, O.; Powell, D.E. Trophic magnification factors: considerations of ecology, ecosystems, and study design. *Integr. Environ. Assess. Manag.,* **2012,** *8*(1), 64-84.
[http://dx.doi.org/10.1002/ieam.244] [PMID: 21674770]

[34] Jörundsdóttir, H.O.; Jensen, S.; Hylland, K.; Holth, T.F.; Gunnlaugsdóttir, H.; Svavarsson, J.; Olafsdóttir, Á.; El-Taliawy, H.; Rigét, F.; Strand, J.; Nyberg, E.; Bignert, A.; Hoydal, K.S.; Halldórsson, H.P. Pristine Arctic: background mapping of PAHs, PAH metabolites and inorganic trace elements in the North-Atlantic Arctic and sub-Arctic coastal environment. *Sci. Total Environ.,* **2014,** *493*, 719-728.
[http://dx.doi.org/10.1016/j.scitotenv.2014.06.030] [PMID: 24995638]

[35] Geraudie, P.; Bakkemo, R.; Milinkovitch, T.; Thomas-Guyon, H. First evidence of marine diesel effects on biomarker responses in the Icelandic scallops, *Chlamys islandica. Environ. Sci. Pollut. Res. Int.,* **2016,** *23*(16), 16504-16512.
[http://dx.doi.org/10.1007/s11356-016-6572-x] [PMID: 27169408]

[36] Grotti, M.; Pizzini, S.; Abelmoschi, M.L.; Cozzi, G.; Piazza, R.; Soggia, F. Retrospective biomonitoring of chemical contamination in the marine coastal environment of Terra Nova Bay (Ross Sea, Antarctica) by environmental specimen banking. *Chemosphere,* **2016,** *165*, 418-426.
[http://dx.doi.org/10.1016/j.chemosphere.2016.09.049] [PMID: 27668719]

[37] Klosterhaus, S.L.; Ferguson, P.L.; Chandlert, G.T. Polycyclic aromatic hydrocarbon bioaccumulation by meiobenthic copepods inhabiting a superfund site: techniques for micromass body burden and total lipid analysis. *Environ. Toxicol. Chem.,* **2002,** *21*(11), 2331-2337.
[http://dx.doi.org/10.1002/etc.5620211111] [PMID: 12389911]

[38] Szczybelski, A.S.; van den Heuvel-Greve, M.J.; Kampen, T.; Wang, C.; van den Brink, N.W.; Koelmans, A.A. Bioaccumulation of polycyclic aromatic hydrocarbons, polychlorinated biphenyls and hexachlorobenzene by three Arctic benthic species from Kongsfjorden (Svalbard, Norway). *Mar. Pollut. Bull.,* **2016,** *112*(1-2), 65-74.
[http://dx.doi.org/10.1016/j.marpolbul.2016.08.041] [PMID: 27575395]

[39] Laender, F.D.; Hammer, J.; Hendriks, A.J.; Soetaert, K.; Janssen, C.R. Combining monitoring data and modeling identifies PAHs as emerging contaminants in the arctic. *Environ. Sci. Technol.,* **2011,** *45*(20), 9024-9029.
[http://dx.doi.org/10.1021/es202423f] [PMID: 21888330]

[40] Kasiotis, K.M.; Emmanouil, C.; Anastasiadou, P.; Papadi-Psyllou, A.; Papadopoulos, A.; Okay, O.; Machera, K. Organic pollution and its effects in the marine mussel *Mytilus galloprovincialis* in Eastern Mediterranean coasts. *Chemosphere,* **2015,** *119* Suppl., S145-S152.
[http://dx.doi.org/10.1016/j.chemosphere.2014.05.078] [PMID: 24953521]

[41] Benedetti, M.; Gorbi, S.; Fattorini, D.; DErrico, G.; Piva, F.; Pacitti, D.; Regoli, F. Environmental

hazards from natural hydrocarbons seepage: integrated classification of risk from sediment chemistry, bioavailability and biomarkers responses in sentinel species. *Environ. Pollut.,* **2014**, *185*, 116-126.
[http://dx.doi.org/10.1016/j.envpol.2013.10.023] [PMID: 24246782]

[42]   Davies, I.M.; Vethaak, A.D. Integrated marine environmental monitoring of chemicals and their effects. *ICES Coop. Res. Rep.,* **2012**, *315*, 277.

[43]   Deng, X.; Pan, L.; Miao, J.; Cai, Y.; Hu, F. Digital gene expression analysis of reproductive toxicity of benzo[a]pyrene in male scallop *chlamys farreri. Ecotoxicol. Environ. Saf.,* **2014**, *110*, 190-196.
[http://dx.doi.org/10.1016/j.ecoenv.2014.09.002] [PMID: 25244687]

[44]   Goodale, B.C.; Tilton, S.C.; Corvi, M.M.; Wilson, G.R.; Janszen, D.B.; Anderson, K.A.; Waters, K.M.; Tanguay, R.L. Structurally distinct polycyclic aromatic hydrocarbons induce differential transcriptional responses in developing zebrafish. *Toxicol. Appl. Pharmacol.,* **2013**, *272*(3), 656-670.
[http://dx.doi.org/10.1016/j.taap.2013.04.024] [PMID: 23656968]

[45]   Pampanin, D.M.; Larssen, E.; Øysæd, K.B.; Sundt, R.C.; Sydnes, M.O. Study of the bile proteome of Atlantic cod (*Gadus morhua*): Multi-biological markers of exposure to polycyclic aromatic hydrocarbons. *Mar. Environ. Res.,* **2014**, *101*, 161-168.
[http://dx.doi.org/10.1016/j.marenvres.2014.10.002] [PMID: 25440786]

[46]   Baussant, T.; Bechmann, R.K.; Taban, I.C.; Larsen, B.K.; Tandberg, A.H.; Bjørnstad, A.; Torgrimsen, S.; Naevdal, A.; Øysaed, K.B.; Jonsson, G.; Sanni, S. Enzymatic and cellular responses in relation to body burden of PAHs in *Bivalve molluscs*: a case study with chronic levels of North Sea and Barents Sea dispersed oil. *Mar. Pollut. Bull.,* **2009**, *58*(12), 1796-1807.
[http://dx.doi.org/10.1016/j.marpolbul.2009.08.007] [PMID: 19732912]

[47]   Nahrgang, J.; Brooks, S.J.; Evenset, A.; Camus, L.; Jonsson, M.; Smith, T.J.; Lukina, J.; Frantzen, M.; Giarratano, E.; Renaud, P.E. Seasonal variation in biomarkers in blue mussel (*Mytilus edulis*), Icelandic scallop (*Chlamys islandica*) and Atlantic cod (*Gadus morhua*): implications for environmental monitoring in the Barents Sea. *Aquat. Toxicol.,* **2013**, *127*, 21-35.
[http://dx.doi.org/10.1016/j.aquatox.2012.01.009] [PMID: 22310169]

[48]   Farcy, E.; Burgeot, T.; Haberkorn, H.; Auffret, M.; Lagadic, L.; Allenou, J-P.; Budzinski, H.; Mazzella, N.; Pete, R.; Heydorff, M.; Menard, D.; Mondeguer, F.; Caquet, T. An integrated environmental approach to investigate biomarker fluctuations in the blue mussel *Mytilus edulis* L. in the Vilaine estuary, France. *Environ. Sci. Pollut. Res. Int.,* **2013**, *20*(2), 630-650.
[http://dx.doi.org/10.1007/s11356-012-1316-z] [PMID: 23247516]

[49]   Mearns, A.J.; Reish, D.J.; Oshida, P.S.; Ginn, T.; Rempel-Hester, M.A.; Arthur, C.; Rutherford, N. Effects of pollution on marine organisms. *Water Environ. Res.,* **2013**, *85*, 1828-1933.
[http://dx.doi.org/10.2175/106143013X13698672322949]

The Presence of Petrogenic PAHs in the Aquatic Environment, a Focus on Monitoring Studies

Daniela M. Pampanin*

International Research Institute of Stavanger, Stavanger, Norway

Faculty of Science and Technology, Department of Mathematics and Natural Science, University of Stavanger, Stavanger, Norway

Abstract: Due to their toxic properties and their carcinogenic potential, petrogenic PAHs are routinely monitored in environmental surveys and following accidental events such as oil spills, boat accidents, and accidental discharges. Measurements of PAH concentrations in water and sediment can give an estimation of their abundance, but this does not reveal the real environmental risk. Bioaccumulation in biota and further biological effects of these compounds on various organisms needs to be evaluated to establish the risk posed to the aquatic environment by their presence. In this chapter, monitoring activities of petrogenic PAHs in the aquatic environment are described, including specific cases as examples. Due to the increased knowledge and technological improvements in recent years, new monitoring strategies are also proposed.

Keywords: Bioaccumulation, Biomarkers, Biomonitoring, Invertebrates, Monitoring, PAHs, Sources of contamination, Vertebrates.

INTRODUCTION

Monitoring of the aquatic environment is extremely important for providing concrete information for the preservation of ecosystems against the adverse effect of anthropogenic sources of contamination. Great attention is focused nowadays on the cocktail of chemicals present in the aquatic environment and their potential

---

* **Corresponding author Daniela M. Pampanin:** International Research Institute of Stavanger, Stavanger, Norway; Tel/Fax: +47 51875502; E-mail: dmp@iris.no

Daniela M. Pampanin and Magne O. Sydnes (Eds.)
All rights reserved-© 2017 Bentham Science Publishers

detrimental influence [1]. Within the cocktail of contaminants, the presence of polycyclic aromatic hydrocarbon (PAH) compounds are of particular concern due to their proven carcinogenic properties [2, 3]. Some refer to PAHs as PBT substances: Persistent, Bioaccumulative and Toxic substances. Sensitive and solid source of contamination trackers are still under development. This is particularly relevant for legal issues and chronic contaminations.

Due to the large number of publications related to techniques and approaches for monitoring the presence of PAHs in the aquatic environment [4], this book chapter will focus on giving an overview of the most recent successfully applied strategies in sediment, water and biota. In particular, the chapter will highlight new promising methodologies for tracking PAH contamination sources in biota that concurrently provide information about the real effect of PAH contamination on the aquatic ecosystem and about the source of contamination.

Causes of the Presence of PAHs in the Aquatic Environment

Petrogenic PAHs are present in the marine environment in significant concentrations [5]. They are naturally present in crude oil and coal. In coastal areas, they enter the water primarily from sewage, runoff from roads, smelter industries and oil spills while the presence of PAHs offshore is mostly related to oil seeps, oil spills and produced water discharge from oil and gas installations [3] (Fig. **1**).

Evaluation of PAH sources and their effects has been reported since the '80s, as shown for example in the National Research Council (US) Committee on pyrene and selected analogues [6]. Examples of sources of PAHs in the aquatic environment are reported herein and a selection of relevant references is proposed.

Natural Occurrence

PAHs are naturally found in aquatic plants, in bottom sediments, in fresh and marine waters, and in emissions from volcanoes and forest fires. The natural occurrence of PAHs in the aquatic environment does not represent an environmental risk *per se*. A good example is the study of Olivella *et al.* [7]; the authors reported the evaluation of exceptional leakage of ashes from a forest fire into riverine waters in Catalonia (Spain) in 1994. Even if the studied areas were clearly affected by extensive forest fires (and consequently exceptional emission of PAHs), levels of PAHs in sampled drinking water were always below the limit established by the European Community. This confirms that naturally occurring PAHs are most likely not harmful to the aquatic environment or to human health.

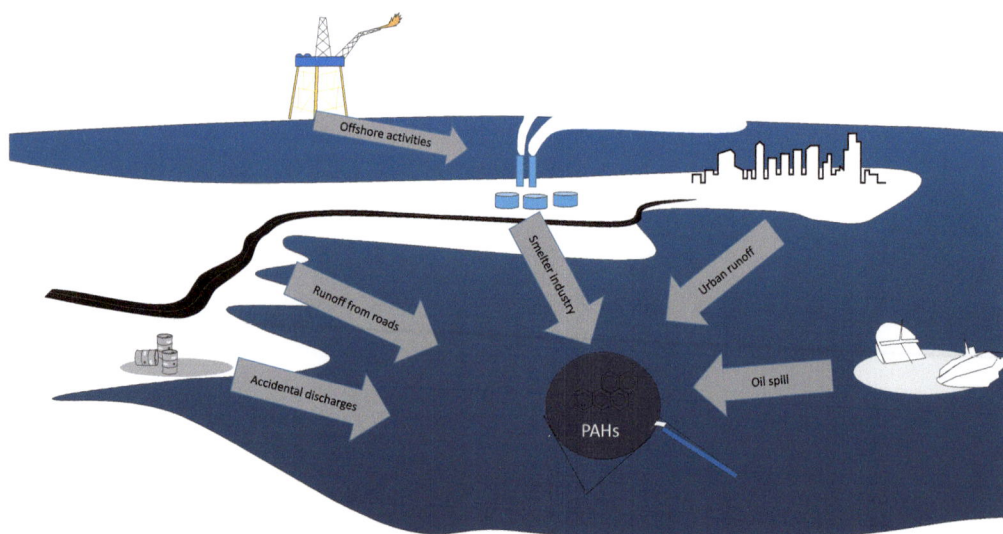

Fig. (1). Anthropogenic sources of PAH contamination in the Sea (figure courtesy of J. Beyer).

Natural processes are also in place to preserve a functional aquatic environment (*i.e.* biodegradation processes, see Chapter 7 of this book and [8]). Nevertheless, the knowledge gained studying the natural presence of PAH represents a good base for the establishment of monitoring tools [9, 10]. Developments in molecular and analytical chemistry have been improving the understanding of the microbial metabolic networks, providing valuable tools to both verify and assess the presence of PAHs [11].

It is important to note that bioremediation, which exploits the natural microbial degradation of organic compounds, is also considered the most cost-effective and sustainable cleaning technology related to PAHs [12].

Oil Exploration and Production

Discharges from oil and gas exploration and production have been subjected to intense PAH monitoring studies in the last three decades worldwide [13, 14], and in particular in the Norwegian Continental Shelf [15 - 19], to control and eventually minimize the environmental impact of these activities. Organisms living in waters and sediments around oil and gas production facilities are potentially exposed to various chemicals, especially through the discharge of produced water (PW) [20]. Data from offshore oil production platforms in the North Sea have shown that PAHs are one of the major organic components in PW [20, 21].

Special attention has been given to the extraction of petroleum hydrocarbon resources in vulnerable areas like the Arctic. A review by country of the Arctic region's oil and gas potential activities, including information about environmental regulation related to controlling, monitoring and mitigating the impact of these activities, reported severe local and regional environmental risk related to the operational and accidental release of contaminants and in particular PAHs [22].

It is important to note that the ability to track the source of PAH contaminations is key. During the development of the offshore oil exploration activities in the Alaskan Beaufort Sea, the US authority sponsored a multiyear monitoring program to evaluate the chemical and biological characteristics of the Arctic marine environment. It was reported that the hydrocarbons found in amphipod tissues collected between 1999 and 2006, throughout the oil development area, were a mix of natural sources (coastal erosion of natural diagenic and fossil materials including seep oil) and other contamination sources (diet, river runoff), but no link was found with the offshore oil production [23].

Accidental Releases

Large amounts of PAHs can be released by leakages or accidents during extraction, transportation or refinery of petroleum. We have experienced dramatic cases of environmental contaminations since the '60s [24]. Some of the most widely known accidental releases of oil have been brought to the attention of the scientific community and the public alike due to extensive media coverage. Two examples among many are: 1) in 1989, the Exxon Valdez oil tanker ran aground releasing about 42 million L of oil into a pristine area of Alaska [25]; 2) in 2010 about 600000 tons of oil were spilled into the Gulf of Mexico following the disaster at Deepwater Horizon [26, 27] (Fig. **2**). In both cases, a large part of the marine ecosystem was surveyed and monitored for years, proving the long term effects of such accidents. Authorities are well aware of the risks posed by accidental PAH contamination, and collaborate to provide common knowledge and response/restoration efforts (see for example the EPA website containing information in response to the Deepwater Horizon oil spill: https://archive.epa.gov/bpspill/web/html/). Massive marine oil spills are usually followed by an increase in PAH research.

Fig. (2). Deepwater Horizon oil spill accident, Gulf of Mexico 2010 (from Wikipedia By Unknown - US Coast Guard - 100421-G-XXXXL- Deepwater Horizon fire, https://commons.wikimedia.org/w/index.php?curid=10089914).

Other Anthropogenic Sources

As oil combustion is still one of the major sources of energy production, the presence of PAHs in the environment is linked to many anthropogenic activities. Sediment contamination in harbor areas is recognized worldwide and well documented [28 - 31]. Certain key elements are worthy to mention. The 16-PAHs (US Environmental Protection Agency (EPA) pollutants) are largely found in harbor sediments. Their availability varies, being linked to resuspensions, sediment sizes, and other abiotic factors. Sedimentary organic matter with different origins and maturities also have different PAH sorption characteristics, which will influence the PAH distribution and the long term environmental impact [28]. Harbors are both highly impacted environments and critical areas with strategic economic value, and therefore water quality assessment criteria, especially regarding important classes of contaminants such as PAHs, are important. In this case, sediment chemistry, bioaccumulation, biomarker evaluation and bioassays are recognized as fundamental, and only their integrated information can reveal the environmental impact of PAHs [31 - 33].

In general, PAH contamination from land due to various sources/activities is to be expected all around the world. Leaks of crankcase oil from poorly maintained automobiles and/or illegal dumping of waste crankcase oil have been for example documented in Malaysia [34]. Monitoring activity can help in recognizing the presence of PAH contamination, and also contribute in identifying the source of the contamination.

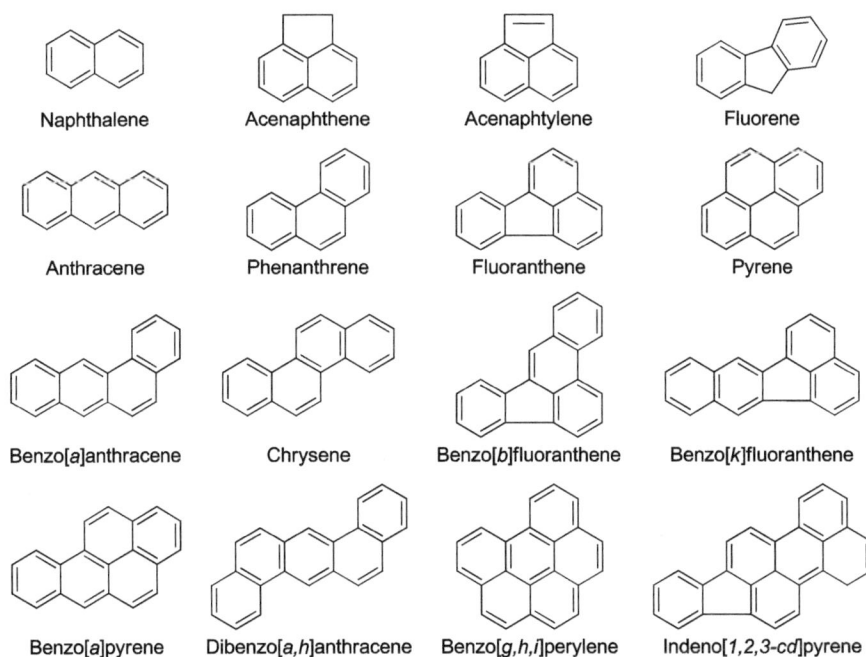

Fig. (3). Structure of the U.S. EPA 16 PAHs listed as priority pollutants.

Monitoring of PAHs in the Aquatic Environment

In 1977, the U.S. EPA added 16 PAHs to the list of priority pollutants in the Clean Water Act [35] (Fig. **3**). Since then, the presence of PAHs in the aquatic environment has been subject to monitoring activity. In the last 20 years, increasing concern about the environmental impact of these compounds on both the ecosystem [36] and human health [37] has brought focus not only on assessing and quantifying their presence, but also on evaluating their biological effects [38].

Table 1. - Physical and chemical properties that influence bioavailability for aquatic organisms for the 16 PAHs of the EPA priority list (modified from Nagpal *et al.* 1993) [42].

PAH compound	Molecular weight	Solubility at 25 °C (µg/L)	Log K_{ow}
Naphthalene	128.2	12500 to 34000	3.37
Acenaphthene	154.2	-	3.98
Acenaphtylene	152.2	3420	4.07
Fluorene	166.2	800	4.18
Anthracene	178.2	59	4.5
Phenanthrene	178.2	435	4.46
Fluoranthene	202.3	260	4.90
Pyrene	202.1	133	4.88
Benzo[*a*]anthracene	228.3	11	5.63
Chrysene	228.3	1.9	5.63
Benzo[*b*]fluoranthene	252.3	2.4	6.04
Benzo[*k*]fluoranthene	-	-	-
Benzo[*a*]pyrene	252.3	3.8	6.06
Dibenzo[*a,h*]anthracene	278.3	0.4	6.86
Benzo[*g,h,i*]perylene	276.4	0.3	6.78
Indeno[*1,2,3-cd*]pyrene	276.3	-	6.58

Indeed, the evaluation of PAH presence does not fully answer the questions about the environmental risk. In this context, bioavailability becomes a key element for the ecotoxicological evaluation of PAH contamination. The same concept applies to organisms, where the first step to cause a harmful effect is to be bioaccessible in the aquatic environment [39]. In general two main factors, lipid and organic carbon, control the partitioning behavior of PAHs in sediment, water and biota;

and these parameters represent the best predictor to estimate PAH behavior and bioavailability in the aquatic environment (Table **1**). PAHs are commonly divided into two groups, according to physical and chemical characteristics: low molecular weight (including 2- and 3-ring compounds) and high molecular weight (including from 4- to 7-ring compounds) PAHs.

Bioaccumulation of PAHs in various marine organisms has been extensively reviewed and biota-sediment accumulation factors (BSAFs) of PAHs have been documented [5, 40].

Moreover, the establishment of PAH background levels is a precursor to detection of "hot spots". An example is the study of the sediment contamination performed by Boitsov *et al.* [41]. The geochemical analysis of sediments in large areas of the Norwegian Continental Shelf has shown clear differences in the composition and distribution of PAHs. Petrogenic inputs contributed less to PAH levels than pyrogenic ones. As a conclusion, a natural PAH background was recognized in the studied areas. Despite the presence of these background levels, anthropogenic PAH contamination was also revealed.

Chemical Monitoring

Chemical monitoring is the exposure assessment determined by measuring levels of a selected set of well-known contaminants in abiotic environmental compartments [5]. Evaluation of the chemical presence of PAHs in sediment, water and biota has been carried out in most monitoring activities since the 80's [3]. Their quantification by chemical analysis in aquatic sediment and water requires the use of highly sensitive equipment, and most of the time environmental concentrations are below the current detection limits.

Sediments are the sink for particle-adsorbed contaminants in aquatic systems and can serve as a reservoir of toxic contaminants that continually threaten the health and viability of aquatic biota [43]. When PAHs are incorporated into sediments, they are somehow immobile because their non-polar structures inhibit them from dissolving in water. Nevertheless, PAHs are not completely insoluble, particularly the lower molecular weight PAHs. Therefore, small amounts of PAHs do dissolve and become bioavailable in the pore water. PAH sediment contamination has been the focus of monitoring activities for a very long time and analytical methods have been developed for this purpose.

Reviews are available regarding the presence of PAHs in surface sediments, including distribution, potential sources and fate of these compounds especially in coastal areas [4, 44].

Interestingly, a number of studies have used PAHs as geochemical markers for evaluating anthropogenic influences since they are strongly linked to local socio-economical activities [44, 45].

PAH patterns and ratios are also helpful in determining the historical contamination of an area. Sediment cores from lakes have the ability to provide reliable environmental archives that can offer a detailed time correlation. PAHs are several orders of magnitude more selective and sensitive than other contamination tracers (*e.g.* heavy metals) due to the absence of a significant natural background [46].

Determining PAH source is still one of the major problems when dealing with environmental contamination in the water, mainly due to the very low environmental concentrations. The occurrence of PAHs in industrial and municipal waters, as well as in ground and surface waters, have been pointed out since the '70s. Due to their ubiquity in the aquatic environment, their surveillance has been considered important since that time, in addition to seeking their reduction when possible [47]. In general, higher concentrations of PAHs are found in the sea-surface microlayer [48]. Different methodologies exist for measuring PAH concentration in water samples. Gas chromatography (GC), mass spectrometry and fluorescence spectroscopy have been providing reliable results in many cases [49]. Although the traditional GC methods are recognized as highly informative and sensitive [50], fluorescence methods for detection of petrogenic PAHs are capable of generating fast and cost-effective results [51]. Fluorescence spectroscopy has been employed extensively worldwide as a means of measuring levels of oil contamination in the field. The unique advantages of flow-through-type fluorimeters include: instrumentation constructed specifically for field use, ease of operation, low detection limits, and no sample preparation is required [52].

Limitations also exist, and more specific information from references about how these instruments are used, including set up and calibration procedures, the oil and dispersant measured, the approximate concentration range of the oil in the water column, and how the real-time data compared to traditional laboratory techniques are reported in the review of Lambert [52].

Effective environmental regulation can only be accomplished if the contamination presence and distribution can be unequivocally linked to a source. Following the progress in analytical chemistry, individual compound-specific stable isotope analysis has been developed for the determination of a single PAH in a complex mixture, using gas chromatography/isotope ratio mass spectrometry (GC/IRMS). This approach has been used to develop protocols for fingerprinting PAHs in contaminated coastal areas [53]. These types of analytical methods are necessary to track sources of contamination and monitor the long term fate of PAHs, and thus distinguish oil spill pollution from any other potential source. Detail on oil fingerprint determination for identification of crude oil source of contamination is available in Chapter 3 of this e-book.

In addition, an overview of the available techniques for chemical evaluation of PAHs in biota (bioaccumulation) is given in Chapter 1 of this e-book. In a typical marine or estuarine monitoring activity, the 16-PAHs are chemically determined in sediment and organisms (*e.g.* fish, crustacean, bivalves) [54]. In general, bioaccumulation of PAHs may include two routes: bioconcentration from water *via* gills, and ingestion of contaminated food or sediment. Clearly this shows that the PAH content of biota is highly dependent on the organism's way of living and feeding. In most cases, demersal fish have higher concentrations of PAHs in their body compared to pelagic fish, even when the concentration is expressed on a lipid weight basis [54]. Usually, hydrophobic organic compounds like PAHs tend to adsorb onto particles and sink and accumulate in the sediment, and the benthic community (*e.g.* fish, crabs) may be more exposed. Influencing factors need to be considered when interpreting bioaccumulation data (*e.g.* fish lipid content). According to specific case studies, different sentinel organisms are utilized, including early life stage. Regarding invertebrates, mussel species are most commonly used. It is important to note that the choice of model used to calculate the uptake of contaminants by an organism depends on the pollution sources to which organisms are exposed and needs to be adapted to the case encountered. If the pollution is located in the sediment, BSAFs can be used. If the pollution is dissolved in the water, bioconcentration factors (ratio of the mussel PAH concentration *vs.* water PAH concentration), calculating the uptake of contaminants from the water, must be used [55].

Other species have proven to be good sentinel organisms for PAH related pollution. As an example, results obtained in recent work from Luna-Acosta *et al.* suggest the use of the juvenile stage of the Pacific oyster (*Crassostrea gigas*) as a sentinel organism for short term assessment of PAH contamination [56].

Clearly, a step forward has been the estimation of PAH bioavailability and their potential in eco-risk assessment. PAHs are not fully available for organisms, and in fact the bioavailability declines as PAHs persist or age within a heterogeneous sediment matrix. Therefore, assessment based on bioavailability is considered to be a critical tool in a risk based approach for both management and remediation of PAH contaminated areas [57]. Increasing emphasis has been placed on the use of bioaccumulation and biomarker responses as monitoring tools for assessment of risks and hazards of PAH presence [58].

Biological Monitoring

In this context, the term biological monitoring includes [5]:

- the bioaccumulation monitoring (BAM), *i.e.* the exposure assessment by measuring contaminant levels in biota or determining the critical dose at a critical site (bioaccumulation);
- the biological effect monitoring (BEM), *i.e.* the exposure and effect assessment by determining the early adverse alterations that are partly or fully reversible (biomarkers);
- the health monitoring (HM), *i.e.* the effect assessment by examining the occurrence of irreversible diseases or tissue damage in organisms; and
- the ecosystem monitoring (EM), *i.e.* the assessment of the integrity of an ecosystem by making an inventory of, for instance, species composition, density and diversity.

Biological effect monitoring of PAHs is part of many international monitoring programmes, among them the OSPAR Joint Assessment and Monitoring Programme [59], the MEDPOL Program [60, 61], and other initiatives within ICES [62, 63].

The MEDPOL Program, designed initially as the environmental assessment component of the Mediterranean Action Plan, has been operational since 1975. During MEDPOL Phase III (1996-2005), biological effects monitoring techniques were included in the monitoring program to test these methods to be used as an early warning tool to detect any negative effects of contaminants, including PAHs, to the organisms [61].

The use of biological effects methods (*i.e.* bioassays and biomarkers) as tools for evaluating the environmental impact of petrogenic PAHs has been demonstrated clearly during the last two decades [64]. Guidelines are available for utilizing this approach [65], including in oil spill pollution monitoring for the NE Atlantic coasts and the NW Mediterranean Sea. In this case, the emphasis has been on fish and invertebrates and on methods at lower levels of organization (*in vitro*, sub-organismal, and individual) [63]. A range of standardized biological effects methods is available that can provide information on the degree of exposure to PAHs and their effects upon the individual organisms in a population [5, 63, 66].

The toxicity of PAHs is the result of the uptake of dissolved hydrocarbons by aquatic organisms, and it can lead to a wide variety of biological responses. The overall picture is often complex since many processes, such as biodegradation, bioaccumulation and biotransformation, determine first the bioavailability and then toxic potency of PAHs in the aquatic environment. Moreover, marine organisms, even from the same taxonomic group, may vary greatly in their sensitivity to the same compound [5].

Biological indicators or markers (biomarkers) have been developed to measure the biological response related to exposure to, or the toxic effect of, an environmental chemical [67]. Common to all of the methods is the capability of performing a time-integrated response assessment over extended periods of time, which is highly relevant in environmental monitoring (Fig. **4**). Most of these methods are highly sensitive, and responses occur at lower concentrations and/or earlier than the appearance of more adverse effects at a higher organization level. This makes these methods convenient early-warning tools for assessing the potential for long term (ecological) effects. Many studies have been published on the biological effects and impact of petroleum-related hydrocarbons, such as PAHs, in the marine environment [68]. Lists are available for both organisms recommended as target monitoring species and biological effects methods for PAH contamination [63, 65, 69].

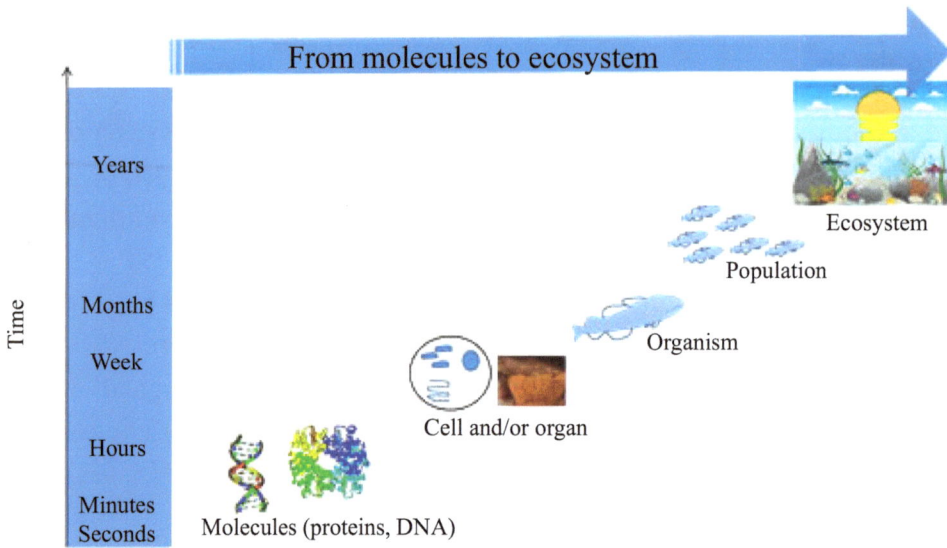

Fig. (4). Time integrated response of biomarkers as early warning system for biological effect of contaminants.

In recent years, a combination of laboratory and field validation of the different biological effects based methods has greatly improved the knowledge of the potential and limitations of these methods. This has made it possible to link responses of biomarker signals to the potential for more adverse effects at the ecological level [70, 71].

Biological monitoring focusses on biological effects and takes the presence of cocktail of contaminants directly into account. As an example, Gauthier *et al.* published an exhaustive review on the co-toxic effects of metal-PAH mixtures, underlying the "more-than-add" effect [72].

Biomonitoring of Sediment Contamination

There are many studies available regarding the evaluation of PAH contamination in sediments using different approaches [73], but providing a complete list of them is not helpful, since specific requirements are usually formulated by authorities and guidelines are available. The most frequently employed approach for evaluating the quality of aquatic sediments is based on data from several lines of evidence [32], including:

1. physical characteristics such as grain size distribution;
2. chemical factors, analysis of nutrients and contaminants in specific phases (*i.e.* water in the sediment–water interface (SWI), interstitial water, and whole sediments);
3. ecotoxicologically based through toxicity tests or biomarker analyses in key organisms;
4. ecological factors through benthic communities structure and function analysis;
5. the assessment of bioaccumulation of contaminants in living organisms.

The last two are part of the so-called biological monitoring of sediments.

An integrated approach based on different lines of evidence is an important tool in the decision making process and for the management of contaminated sediments [74].

It is important to mention a comprehensive review of published studies that investigated the genotoxicity of sediments in rivers, lakes and marine habitats [43]. The Salmonella mutagenicity test is the most frequently used assay and is capable of revealing mutagenic potency values for sediment extracts. Analyses of the Salmonella data (n = 510) showed significant differences between rural, urban/industrial, and heavily contaminated (*e.g.*, dump) sites. Other commonly used assays are the SOS Chromotest, the Mutatox assays and test based on *Vibrio* sp.

A variety of other *in vitro* approaches, employing cultured fish and mammalian cells, have been used to investigate sediment genotoxic activity [75]. Investigated biological measurements include sister chromatid exchange frequency, micronucleus frequency, chromosome aberration frequency, gene mutation, unscheduled DNA synthesis, DNA adduct frequency and DNA strand break frequency. An overview of the genotoxic and carcinogenic evaluation of petrogenic PAH is reported in chapter 4 of this e-book.

PAH contamination can be assessed by measuring biomarkers in sentinel or keystone species, including early life stage. The juvenile stage of the Pacific oyster (*Crassostrea gigas*) has been proposed as a sentinel organism for short term assessment of PAH contamination using biomarkers [76].

Biomonitoring of the Water Column

A comprehensive review on the mutagenicity/genotoxicity of surface waters, including bioassays related to PAH contaminated water samples, was published about 10 years ago [77]. In the last decades, many studies have been dedicated to the biomonitoring of petrogenic PAHs, providing at present a vast amount of material. Homogeneous guidelines are also in the process of being established for the evaluation of this class of contaminants [65, 78]. Later on in this chapter, an informative case study of a biomonitoring program that has been going on since the beginning of this century is reported to add specific information about the monitoring of the water column.

An interesting aspect about long term biomonitoring, albeit difficult to evaluate, is the capacity to address the adaptation of the organisms over multiple generations. This raises the concept of pollution-driven genetic adaptation, which is the potential of contaminants to act as selection pressures potentially driving evolutionary factors. This phenomenon, named evolutionary ecotoxicology, is important to consider in biomonitoring studies, since the risk assessment based on analysis of the effects of contaminants on a population or community that were subjected to adaptation may lead to underestimation of the real environmental risk. An interesting review was recently published about the impacts of PAHs on fish from the Elizabeth River estuary (Virginia, USA) [79].

Regarding biomonitoring approach including biomarker evaluations, there is extensive evidence linking PAHs to induction of phase-I enzymes, development of DNA adducts, and eventually neoplastic lesions in fish. However, most studies have focused on high-molecular-weight, carcinogenic PAHs, such as benzo[*a*]pyrene. It is less clear how 2- and 3-ring PAHs affect organisms like fish. There are experimental evidences indicating that these chemicals may inhibit some components of the phase I system rather than produce induction. Therefore, there is still a need for research to clarify biological effects of 2- and 3-ring PAHs, PAH mixtures, and adaptation processes in marine ecosystems [68].

Confounding Factors

Environmental factors need to be considered wisely, when biomonitoring results are discussed. Many authors have already pointed this out [5, 80, 81]. The season when the biomonitoring activity takes place needs to be considered, the main factors influencing the data are: 1) abiotic factors such as temperature and salinity are known to influence some of the biological markers; 2) stratification of the water column, which can have significant effects on the distribution of the contaminants and thereby affect exposure; 3) algal blooms potentially influence the bioavailability and uptake pathways of contaminants in the studied organisms.

Moreover, the physiological status of the sentinel organisms or bioindicators will have an influence on the biomarker data and need to be taken into account. In mussels, biomarker responses and general health status are known to be affected by the reproductive stage. Recently spawned or spawning mussels should therefore be avoided when monitoring petrogenic PAHs.

The Water Column Monitoring Program in Norway: A Case Study

As mentioned before, PW is a complex mixture of compounds, including PAHs. Discharge of PW from oil and gas activities represents a known source of petrogenic PAHs in the marine environment. Several PW related biomonitoring studies have been conducted globally. A range of different designs have been applied both with respect to type of exposure and to the selection of biological and chemical markers. The most important areas where PW related field biomonitoring investigations based on wild or caged organisms have been carried out are indicated in Fig. (5).

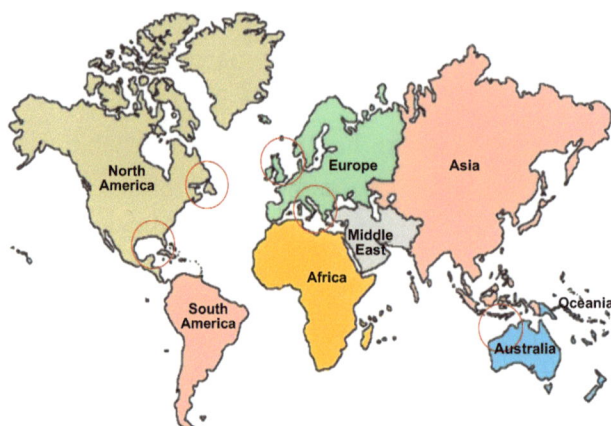

Fig. (5). Circles indicate the main areas where produced water related field biomonitoring, based on wild or caged organisms, has been carried out.

An overview of PW related biomonitoring studies based on caged (transplanted) and wild caught organisms is given by Sundt *et al.* [69], where a compilation of previous experience and suggestions for future survey design for the water column monitoring (WCM) of offshore produced water discharges in Norwegian waters are reported in detail. The evaluation of PW discharge impact in wild organisms has an additional relevance, revealing the risk to wild populations of ecological or environmental significance in the field [82].

The investigations conducted in the North Sea consisting of the BECPELAG [15] and subsequent WCM programs [16 - 19] dominate the picture. Outside the North Sea, caging studies have been conducted in South Europe (Adriatic Sea), in North America (California and Gulf of Mexico) and Australia (Australian Northwest shelf). The most commonly used organisms are bivalves (mainly mussel species). Results from only one study outside Norway using caged fish are reported (Stripy Sea Perch, Australia) [83]. Most of the biological parameters that showed effects were correlated with distance from the PW discharge. Most of the biomarkers tested in caged organisms were well known and they were able to identify effects from PAHs, in particular the genotoxicity biological markers. The use of organisms in optimal conditions and presence of a clean reference station are considered crucial. Most of the biomonitoring studies have focused on individual based parameters, nevertheless some activities were performed on population based parameters, such as density of infaunal organisms and microbial community studies [84].

Accumulation of PAH in zooplankton has also been evaluated in this type of biomonitoring. Zooplankton can easily be collected, and represent an integrated average over a considerable effective sampling volume [85]. Carls *et al.* demonstrated the approach by applying PAH measurements in the copepod Neocalanus collected in Port Valdez, Alaska. Areas with strong current advection (transport of water from one region to another) of planktonic organisms through an affected area present an obvious challenge [86].

The evaluation of the microbial communities in the water column has also been used for monitoring PW discharges. PW contains compounds (*e.g.* organic acids) that may cause organic enrichment increasing the microbial activity in the sea (saprobisation). In a Canadian study the influence of PW on the microbial ecology in the water column was investigated. As for the zooplankton approach above, advection of microorganisms through an affected area is an obvious challenge.

The WCM effect study in Norway is a program conducted by Oil and Gas operators with the objective to determine the potential biological effects of PW on the marine environment [87]. This monitoring is mandatory for operators with PW discharges and new guidelines including specific requirements for the biological monitoring have been recently published [78]. The general design of this part of the WCM program is to measure a suite of biomarkers and chemical exposure concentrations in caged (mussels) and wild (fish) organisms. Previous survey designs have used a grid of stations (between 6 and 12) at increasing distances from the discharge point(s) (*i.e.* 200-300 m to 2 km), and reference stations (1 or 2) away from the studied platform in a 'clean' area. The transplantation period of caged mussels is approximately 6 weeks, which is considered to be sufficient time for significant effects in histological endpoints to occur. Biomarkers measured include markers for the evaluation of PAH contamination, including genotoxicity, metabolic enzyme activities and general stress parameters [17 - 19]. Pre-exposure sampling of caged organisms is also performed to determine the chemical concentrations and biomarker levels in animals prior to the biomonitoring. Additional parameters are measured in order to provide supporting information about the physiochemical conditions. These include measurements of temperature, salinity, current direction and strength. The use of a UV-fluorimeter for the detection of the PW plume and the potential exposure profiles to the caged organisms over the exposure duration has also been investigated. As mentioned before, UV-fluorimeters have the potential to detect hydrocarbons within the water column, and one of the main benefits is to provide real time exposure information throughout the entire duration of the field deployment. The list of requested parameters is reported in the new guidelines [78], this includes specific biological markers for PAH effects, such as: PAH content in biota, genotoxicity biomarkers (DNA adducts, micronucleus and comet assays), gene expression parameters, cytochrome P4501A activity and PAH bile metabolite presence.

The Way Forward in Biological Effect Measurement for PAH Contaminations

In parallel with continuing success in methods and approaches for evaluating petrogenic PAHs in the aquatic environment, scientists are continuously working towards: 1) providing fast and efficient specific methods for the evaluation of exposure; 2) developing *in situ*, on-line and real-time monitoring tools for monitoring purposes; 3) providing highly sensitive methods for tracking the source of PAH contamination. While the first point is limited to development of existing methods, the other two require a more extensive presentation.

In situ, Online and Real-Time Monitoring Tools for PAH Monitoring Purposes

Various sensors have been developed in the recent years, and some are still under testing, for both chemical determination and biological effect monitoring of PAHs in the aquatic environment. These tools have the main advantages of being fast in response, transportable, and in some cases aiming to provide real time information.

Traditional laboratory methods for detection and quantification of low-level PAHs are time-consuming, labor intensive, and costly. Therefore, development of alternative technologies is considered a priority for environmental monitoring [88, 89]. Newly developed field-portable technologies such as thin layer chromatography, synchronous luminescence spectrofluorimetry and portable GC-mass spectrometry are available. However, requirements for sample solvent extraction and concentration are still limiting their common use. Immunoassays developed for PAH quantitative analysis in water samples are currently being developed and include enzyme-linked immunosorbent assays [90, 91], amperometric biosensors [92], piezoelectric biosensors [93], surface plasmon resonance biosensors [94, 95], and polarized fluorescence biosensors [96]. These assays are recognized as affordable high-resolution technologies, promoting their wider use in environmental monitoring. Certainly, they also have limitations and in particular optimal operating conditions. Understanding what the antibody recognizes determines the value of the assay and how the results should be interpreted. The primary advantages are sensitivity, speed, and portability, as well as the small volumes required and limited sample manipulations. Moreover, a good correlation to laboratory-based conventional analytical methods has been demonstrated [89].

Bacterial whole-cell bioreporters are practical and reliable analytical tools to assess the toxicity and bioavailability of PAHs, however, evidence has shown that their performance could be affected by different carbon sources.

For biological monitoring, technologies such as those described for chemical analysis are not yet available. A series of methods could be performed on site using portable equipment and organizing on site laboratory facilities.

Highly Sensitive Methods for Tracking Sources of PAH Contamination

DNA adducts are commonly recognized as a biomarker of exposure to PAHs and are frequently applied in biomonitoring studies [3]. The presence and formation of DNA adducts have been extensively studied in many organisms [3, 5, 97]. Their detection in aquatic organisms is commonly used as exposure indicators since the 90s [98]. Moreover, the presence of DNA adducts represents an evaluation of the mutagenic/carcinogenic risk to the aquatic organisms themselves. In North America the technique has been widely used, with more than 30 marine and freshwater species being studied, and guidelines for implementation are published in an ICES TIMES technical document [99]. Across the OSPAR maritime area, the assay has been used in several biological effects monitoring programs using a range of indicator species including blue mussels (*Mytilus* sp.), perch (*Perca fluviatilis*), dab (*Limanda limanda*), European flounder (*Platichthys flesus*), eelpout (*Zoarces viviparous*), and Atlantic cod (*Gadus morhua*). Studies from both North America and Europe have clearly demonstrated that, when using non-migratory fish, the levels of DNA adducts strongly correlates with the concentration of PAHs found in the sediment [65].

At present, DNA adducts are analyzed by the ^{32}P-postlabelling assay [100]. This technique is highly sensitive for quantification, however, the lack of specific structural information regarding the identity of the carcinogen-DNA adducts makes this approach limited in the identification of the source of the contamination. The development of new mass spectrometry (MS) based methods for identifying DNA adducts in complex matrices has been suggested as a necessary mission in toxicology [101 - 105]. This approach has been very challenging and it is only with the latest developments in mass spectrometry that the methods have started to reach the necessary sensitivity required in order to conduct this type of analysis. The detection limits for MS analysis of DNA adducts are now so good that the sensitivity of the method is approaching the sensitivity seen for the ^{32}P-postlabeling method. With the additional structural information gained from MS, this analysis will be able to provide the necessary information for tracking PAH contamination sources. This will represent an important step for biomonitoring studies.

Proteomics is a relatively recently introduction for environmental monitoring and ecotoxicology [106 - 108], however, in human health monitoring proteomics has proved to be a useful tool [109, 110]. The low detection limits, which are commonly found in proteomic analysis, make this a suitable method to detect irregularities at an early stage. PAH protein adducts with organic electrophiles have been studied for more than 30 years in human health research and monitoring. Due to improvement in instrument sensitivity in MS, the use of protein adducts as biological markers of exposure is constantly increasing. As part of ongoing research on the effect of PAHs on Atlantic cod (*Gadus morhua*), aimed at increasing the understanding of the implications caused by PAHs exposure on fish [3, 108], the studies of PAH-protein adducts in fish biological fluid/tissue is considered suitable as a multibiomarker of exposure to PAH for biomonitoring activities. These new sensitive biological markers in the form of expressed proteins affected by PAH exposure (*i.e.* PAH-protein adducts) are therefore suggested, with the major advantage of this approach being the ability to determine the source of PAH contamination in field samples. The detected protein adducts will in fact represent the fingerprint of the contamination source.

CONCLUDING REMARKS

It is clearly necessary to evaluate the environmental impact of petrogenic PAHs in the aquatic environment through monitoring activity at a global level, especially due to their toxic properties and their carcinogenic potential. In addition, accidental events (*e.g.* oil spills, boat accidents, accidental discharges) are constantly exposing the environment to an extraordinary risk, and monitoring tools can help in discovering and evaluating these accidents. Measurements of PAH concentrations in water, sediment and biota and of biological effects related to PAH contamination can give an estimation of their presence, bioavailability and reveal the real environmental risk. Monitoring activities, including both chemical and biological approaches, of petrogenic PAHs in the aquatic environment were described in this chapter aiming to show the great research achievements and the current challenges. In recent years, the increased knowledge and technological improvement are showing the potential of new monitoring strategies. Soon it will be possible to track sources of PAH contaminations, helping reduce their presence in the aquatic environment. Moreover, real-time monitoring will support interventions for accidents, assisting in reducing the environmental impact.

Finally, highly sensitive methods for tracking the source of PAH contamination in the form of both PAH-DNA adducts and PAH-protein adducts are very promising tools that will require more research and development prior to becoming applicable in biomonitoring studies.

CONFLICT OF INTEREST

The author confirms that the author has no conflict of interest to declare for this publication.

ACKNOWLEDGEMENTS

Declared none.

REFERENCES

[1]   Altenburg, R.; Ait-Aissa, S.; Antczak, P.; Backhaus, T.; Barceló, D.; Seiler, T.B.; Brion, F.; Bush, W.; Chipman, K.; de Alda, M.L.; de Aragão Umbuzeiro, G.; Escher, B.I.; Falciani, F.; Faust, M.; Focks, A.; Hilscherova, K.; Hollender, J.; Hollert, H.; Jäger, F.; Jahnke, A.; Kortenkamp, A.; Krauss, M.; LAmkine, G. F.; Munthe, J.; Neumann, S.; Schymanski, E. L.; Scrimshaw, M.; Segner, H.; Slobodnik, J.; Smedes, F.; Kughathas, S.; Teodorovic, I.; Tindall, A. J.; Tollefsen, K. E.; Walz, K. H.; Williams, T. D.; Van den Brink, P. J.; van Gils, J.; Vrana, B.; Zhang, X.; Brack, W. Future water quality monitoring - Adapting tools to deal with mixtures of pollutants in water resource management. *Sci. Total Environ.,* **2015**, *512-513*, 540-551.
[PMID: 25644849]

[2]   Crone, T.J.; Tolstoy, M. Magnitude of the 2010 Gulf of Mexico oil leak. *Science,* **2010**, *330*(6004), 634.
[http://dx.doi.org/10.1126/science.1195840] [PMID: 20929734]

[3]   Pampanin, D.M.; Sydnes, M.O. Polycyclic aromatic hydrocarbons a constituent of petroleum: Presence and influence in the aquatic environment. In: *Hydrocarbons*; Vladimir, K.; Kolesnikov, A., Eds.; InTech: Rijeka, **2013**; pp. 83-118.

[4]   Abdel-Shafy, H.I.; Mansour, S.M. A review on polycyclic aromatic hydrocarbons: source, environmental impact, effect on human health and remediation. *Egypt. J. Petrol.,* **2016**, *25*, 107-123.
[http://dx.doi.org/10.1016/j.ejpe.2015.03.011]

[5]   van der Oost, R.; Beyer, J.; Vermeulen, N.P. Fish bioaccumulation and biomarkers in environmental risk assessment: a review. *Environ. Toxicol. Pharmacol.,* **2003**, *13*(2), 57-149.
[http://dx.doi.org/10.1016/S1382-6689(02)00126-6] [PMID: 21782649]

[6]   Committee on pyrene and selected analogues. Polycyclic aromatic hydrocarbons from natural and stationary anthropogenic sources and their atmospheric concentrations. In: *Polycyclic Aromatic Hydrocarbons: Evaluation of Sources and Effects*; National In Academies Press (US): Washington, D.C., **1983**. http://www.ncbi.nlm.nih.gov/books/NBK217758/

[7]   Olivella, M.A.; Ribalta, T.G.; de Febrer, A.R.; Mollet, J.M.; de Las Heras, F.X. Distribution of polycyclic aromatic hydrocarbons in riverine waters after Mediterranean forest fires. *Sci. Total Environ.,* **2006**, *355*(1-3), 156-166.
[http://dx.doi.org/10.1016/j.scitotenv.2005.02.033] [PMID: 15885751]

[8]   Haritash, A.K.; Kaushik, C.P. Biodegradation aspects of polycyclic aromatic hydrocarbons (PAHs): a review. *J. Hazard. Mater.,* **2009**, *169*(1-3), 1-15.
[http://dx.doi.org/10.1016/j.jhazmat.2009.03.137] [PMID: 19442441]

[9]   Nilsen, E.B.; Rosenbauer, R.J.; Fuller, C.C.; Jaffe, B.J. Sedimentary organic biomarkers suggest detrimental effects of PAHs on estuarine microbial biomass during the 20th century in San Francisco Bay, CA, USA. *Chemosphere,* **2015**, *119*, 961-970.
[http://dx.doi.org/10.1016/j.chemosphere.2014.08.053] [PMID: 25303655]

[10] Shahi, A.; Aydin, S.; Ince, B.; Ince, O. Evaluation of microbial population and functional genes during the bioremediation of petroleum-contaminated soil as an effective monitoring approach. *Ecotoxicol. Environ. Saf.,* **2016**, *125*, 153-160.
[http://dx.doi.org/10.1016/j.ecoenv.2015.11.029] [PMID: 26685788]

[11] Baussant, T.; Krolicka, A.; Boccadoro, C.; Mæland, M.; Preston, C.; Birch, J.; Scholin, C. Detection of oil leaks by quantifying hydrocarbonoclastic bacteria in cold marine environments using the environmental sample processor. *Proceedings of the Thirty-seventh AMOP Technical Seminar on Environmental Contamination and Response,* Alberta, Canada, **2014**, pp. 791-807.

[12] Vila, J.; Tauler, M.; Grifoll, M. Bacterial PAH degradation in marine and terrestrial habitats. *Curr. Opin. Biotechnol.,* **2015**, *33*, 95-102.
[http://dx.doi.org/10.1016/j.copbio.2015.01.006] [PMID: 25658648]

[13] Ite, A.E.; Ibok, U.J.; Ite, M.U.; Petters, S.W. Petroleum exploration and production: past and present environmental issues in the Nigeria's Niger Delta. *Am. J. Environ. Prot.,* **2013**, *1*, 78-90.
[http://dx.doi.org/10.12691/env-1-4-2]

[14] Bakke, T.; Klungsøyr, J.; Sanni, S. Environmental impacts of produced water and drilling waste discharges from the Norwegian offshore petroleum industry. *Mar. Environ. Res.,* **2013**, *92*, 154-169.
[http://dx.doi.org/10.1016/j.marenvres.2013.09.012] [PMID: 24119441]

[15] Hylland, K.; Beyer, J.; Berntssen, M.; Klungsøyr, J.; Lang, T.; Balk, L. May organic pollutants affect fish populations in the North Sea? *J. Toxicol. Environ. Health A,* **2006**, *69*(1-2), 125-138.
[http://dx.doi.org/10.1080/15287390500259392] [PMID: 16291566]

[16] Hylland, K.; Tollefsen, K.E.; Ruus, A.; Jonsson, G.; Sundt, R.C.; Sanni, S.; Røe Utvik, T.I.; Johnsen, S.; Nilssen, I.; Pinturier, L.; Balk, L.; Barsiene, J.; Marigòmez, I.; Feist, S.W.; Børseth, J.F. Water column monitoring near oil installations in the North Sea 20012004. *Mar. Pollut. Bull.,* **2008**, *56*(3), 414-429.
[http://dx.doi.org/10.1016/j.marpolbul.2007.11.004] [PMID: 18158163]

[17] Brooks, S.J.; Harman, C.; Grung, M.; Farmen, E.; Ruus, A.; Vingen, S.; Godal, B.F.; Barsiene, J.; Andreikenaite, L.; Skarpheðinsdottir, H.; Liewenborg, B.; Sundt, R.C. Water column monitoring of the biological effects of produced water from the Ekofisk offshore oil installation from 2006 to 2009. *J. Toxicol. Environ. Health A,* **2011**, *74*(7-9), 582-604.
[http://dx.doi.org/10.1080/15287394.2011.550566] [PMID: 21391100]

[18] Sundt, R.C.; Ruus, A.; Jonsson, H.; Skarphéðinsdóttir, H.; Meier, S.; Grung, M.; Beyer, J.; Pampanin, D.M. Biomarker responses in Atlantic cod (*Gadus morhua*) exposed to produced water from a North Sea oil field: laboratory and field assessments. *Mar. Pollut. Bull.,* **2012**, *64*(1), 144-152.
[http://dx.doi.org/10.1016/j.marpolbul.2011.10.005] [PMID: 22070981]

[19] Sundt, R.C.; Pampanin, D.M.; Grung, M.; Baršienė, J.; Ruus, A. PAH body burden and biomarker responses in mussels (*Mytilus edulis*) exposed to produced water from a North Sea oil field: laboratory and field assessments. *Mar. Pollut. Bull.,* **2011**, *62*(7), 1498-1505.
[http://dx.doi.org/10.1016/j.marpolbul.2011.04.009] [PMID: 21558042]

[20] Utvik, T.I. Chemical charaterisation of produced water from four offshore oil production platforms in the North Sea. *Chemosphere,* **1999**, *39*, 2593-2606.
[http://dx.doi.org/10.1016/S0045-6535(99)00171-X]

[21] Røe, T.I.; Johnsen, S. *Discharges of produced water to the North Sea; Effects in the water column. Produced water 2. Environmental Issues and Mitigation Technologies. S. Johnsen*; Plenum Press: New York, **1996**, pp. 13-25.

[22] Bakke, T.; Hameedi, J.; Kimstach, V.; Macdonald, R.; Melnikov, S.; Robertson, A.; Shearer, R.; Thomas, D. Petroleum hydrocarbons. *AMAP report,* **1998**, 661-701.

[23] Neff, J.M.; Durell, G.S. Bioaccumulation of petroleum hydrocarbons in arctic amphipods in the oil development area of the Alaskan Beaufort Sea. *Integr. Environ. Assess. Manag.,* **2012**, *8*(2), 301-319. [http://dx.doi.org/10.1002/ieam.1247] [PMID: 22006590]

[24] Nigeria, F.M. *Nigerian Conservation Foundation. Niger delta natural resource damage assessment and restoration project. Phase 1 – Scoping report. WWF UK*; CEESP – IUCN, **2006**, pp. 1-13.

[25] Neff, J.M.; Page, D.S.; Boehm, P.D. Exposure of sea otters and harlequin ducks in Prince William Sound, Alaska, USA, to shoreline oil residues 20 years after the Exxon Valdez oil spill. *Environ. Toxicol. Chem.,* **2011**, *30*(3), 659-672. [http://dx.doi.org/10.1002/etc.415] [PMID: 21298711]

[26] Sammarco, P.W.; Kolian, S.R.; Warby, R.A.; Bouldin, J.L.; Subra, W.A.; Porter, S.A. Distribution and concentrations of petroleum hydrocarbons associated with the BP/Deepwater Horizon Oil Spill, Gulf of Mexico. *Mar. Pollut. Bull.,* **2013**, *73*(1), 129-143. [http://dx.doi.org/10.1016/j.marpolbul.2013.05.029] [PMID: 23831318]

[27] Beyer, J.; Trannum, H.C.; Bakke, T.; Hodson, P.V.; Collier, T.K. Environmental effects of the Deepwater Horizon oil spill: A review. *Mar. Pollut. Bull.,* **2016**, *110*(1), 28-51. [http://dx.doi.org/10.1016/j.marpolbul.2016.06.027] [PMID: 27301686]

[28] Montero, N.; Belzunce-Segarra, M.J.; Menchaca, I.; Garmendia, J.M.; Franco, J.; Nieto, O.; Etxebarria, N. Integrative sediment assessment at Atlantic Spanish harbours by means of chemical and ecotoxicological tools. *Environ. Monit. Assess.,* **2013**, *185*(2), 1305-1318. [http://dx.doi.org/10.1007/s10661-012-2633-x] [PMID: 22544172]

[29] Merhaby, D.; Net, S.; Halwani, J.; Ouddane, B. Organic pollution in surficial sediments of Tripoli harbour, Lebanon. *Mar. Pollut. Bull.,* **2015**, *93*(1-2), 284-293. [http://dx.doi.org/10.1016/j.marpolbul.2015.01.004] [PMID: 25619918]

[30] Mali, M.; Dell'Anna, M.M.; Mastrorilli, P.; Damiani, L.; Ungaro, N.; Gredilla, A.; de Vallejuelo, S.F. Identification of hot spots within harbor sediments through a new cumulative hazard index. Case study: Port of Bari, Italy. *Ecol. Indic.,* **2016**, *60*, 548-556. [http://dx.doi.org/10.1016/j.ecolind.2015.07.024]

[31] de los Ríos, A.; Pérez, L.; Echavarri-Erasun, B.; Serrano, T.; Barbero, M.C.; Ortiz-Zarragoitia, M.; Orbea, A.; Juanes, J.A.; Cajaraville, M.P. Measuring biological responses at different levels of organisation to assess the effects of diffuse contamination derived from harbour and industrial activities in estuarine areas. *Mar. Pollut. Bull.,* **2016**, *103*(1-2), 301-312. [http://dx.doi.org/10.1016/j.marpolbul.2015.11.056] [PMID: 26707886]

[32] Chapman, P.M.; Anderson, J. A decision-making framework for sediment contamination. *Integr. Environ. Assess. Manag.,* **2005**, *1*(3), 163-173. [http://dx.doi.org/10.1897/2005-013R.1] [PMID: 16639882]

[33]   Bebianno, M.J.; Pereira, C.G.; Rey, F.; Cravo, A.; Duarte, D.; DErrico, G.; Regoli, F. Integrated approach to assess ecosystem health in harbor areas. *Sci. Total Environ.,* **2015**, *514*, 92-107.
[http://dx.doi.org/10.1016/j.scitotenv.2015.01.050] [PMID: 25659308]

[34]   Zakaria, M.P.; Takada, H.; Tsutsumi, S.; Ohno, K.; Yamada, J.; Kouno, E.; Kumata, H. Distribution of polycyclic aromatic hydrocarbons (PAHs) in rivers and estuaries in Malaysia: a widespread input of petrogenic PAHs. *Environ. Sci. Technol.,* **2002**, *36*(9), 1907-1918.
[http://dx.doi.org/10.1021/es011278+] [PMID: 12026970]

[35]   Keith, L.H. The cource of U.S. EPA's sixteen PAH priority pollutants. *Pol. Arom. Hydroc.,* **2015**, *35*, 147-160.
[http://dx.doi.org/10.1080/10406638.2014.892886]

[36]   Rahmanpour, S.; Farzaneh Ghorghani, N.; Lotfi Ashtiyani, S.M. Polycyclic aromatic hydrocarbon (PAH) in four fish species from different trophic levels in the Persian Gulf. *Environ. Monit. Assess.,* **2014**, *186*(11), 7047-7053.
[http://dx.doi.org/10.1007/s10661-014-3909-0] [PMID: 25004856]

[37]   Dosunmu, M.I.; Oyo-Ita, I.O.; Oyo-Ita, O.E. Risk assessment of human exposure to polycyclic aromatic hydrocarbons *via* shrimp (*Macrobrachium felicinum*) consumption along the Imo River catchments, SE Nigeria. *Environ. Geochem. Health,* **2016**, *38*(6), 1333-1345.
[http://dx.doi.org/10.1007/s10653-016-9799-z] [PMID: 26792660]

[38]   Fent, K. Ecotoxicological problems associated with contaminated sites. *Toxicol. Lett.,* **2003**, *140-141*, 353-365.
[http://dx.doi.org/10.1016/S0378-4274(03)00032-8] [PMID: 12676484]

[39]   Harris, K.L.; Banks, L.D.; Mantey, J.A.; Huderson, A.C.; Ramesh, A. Bioaccessibility of polycyclic aromatic hydrocarbons: relevance to toxicity and carcinogenesis. *Expert Opin. Drug Metab. Toxicol.,* **2013**, *9*(11), 1465-1480.
[http://dx.doi.org/10.1517/17425255.2013.823157] [PMID: 23898780]

[40]   Meador, J.P.; Stein, J.E.; Reichert, W.L.; Varanasi, U. Bioaccumulation of polycyclic aromatic hydrocarbons by marine organisms. *Rev. Environ. Contam. Toxicol.,* **1995**, *143*, 79-165.
[http://dx.doi.org/10.1007/978-1-4612-2542-3_4] [PMID: 7501868]

[41]   Boitsov, S.; Petrova, V.; Jensen, H.K.; Kursheva, A.; Litvinenko, I.; Klungsøyr, J. Sources of polycyclic aromatic hydrocarbons in marine sediments from southern and northern areas of the Norwegian continental shelf. *Mar. Environ. Res.,* **2013**, *87-88*, 73-84.
[http://dx.doi.org/10.1016/j.marenvres.2013.03.006] [PMID: 23623160]

[42]   Nagpal, N.K. *Ambient Water Quality Criteria For Polycyclic Aromatic Hydrocarbons (PAHs)*; Ministry of Environment, Lands and Parks Province of British Columbia, **1993**.

[43]   Chen, G.; White, P.A. The mutagenic hazards of aquatic sediments: a review. *Mutat. Res.,* **2004**, *567*(2-3), 151-225.
[http://dx.doi.org/10.1016/j.mrrev.2004.08.005] [PMID: 15572285]

[44]   Li, Y.; Duan, X. Polycyclic aromatic hydrocarbons in sediments of China Sea. *Environ. Sci. Pollut. Res. Int.,* **2015**, *22*(20), 15432-15442.
[http://dx.doi.org/10.1007/s11356-015-5333-6] [PMID: 26341340]

[45] Yunker, M.B.; Macdonald, R.W.; Vingarzan, R. PAHs in the Fraser River basin: a critical appraisal of PAH ratios as indicators of PAH source and composition. *Org. Geochem.,* **2002**, *33*, 489-515. [http://dx.doi.org/10.1016/S0146-6380(02)00002-5]

[46] Warner, W.; Ruppert, H.; Licha, T. Application of PAH concentration profiles in lake sediments as indicators for smelting activity. *Sci. Total Environ.,* **2016**, *563-564*, 587-592. [http://dx.doi.org/10.1016/j.scitotenv.2016.04.103] [PMID: 27176930]

[47] Andelman, J.B.; Suess, M.J. Polynuclear aromatic hydrocarbons in the water environment. *Bull. Org. mond. Sante. Bull. Wld. Hlth. Org.,* **1970**, *43*, 479-508.

[48] Wurl, O.; Obbard, J.P. A review of pollutants in the sea-surface microlayer (SML): a unique habitat for marine organisms. *Mar. Pollut. Bull.,* **2004**, *48*(11-12), 1016-1030. [http://dx.doi.org/10.1016/j.marpolbul.2004.03.016] [PMID: 15172807]

[49] Wang, Z.; Li, K.; Fingas, M.; Sigouin, L.; Ménard, L. Characterization and source identification of hydrocarbons in water samples using multiple analytical techniques. *J. Chromatogr. A,* **2002**, *971*(1-2), 173-184. [http://dx.doi.org/10.1016/S0021-9673(02)01003-8] [PMID: 12350112]

[50] Huang, S.; He, S.; Xu, H.; Wu, P.; Jiang, R.; Zhu, F.; Luan, T.; Ouyang, G. Monitoring of persistent organic pollutants in seawater of the Pearl River Estuary with rapid on-site active SPME sampling technique. *Environ. Pollut.,* **2015**, *200*, 149-158. [http://dx.doi.org/10.1016/j.envpol.2015.02.016] [PMID: 25732847]

[51] Kim, M.; Yim, U.H.; Hong, S.H.; Jung, J.H.; Choi, H.W.; An, J.; Won, J.; Shim, W.J. Hebei Spirit oil spill monitored on site by fluorometric detection of residual oil in coastal waters off Taean, Korea. *Mar. Pollut. Bull.,* **2010**, *60*(3), 383-389. [http://dx.doi.org/10.1016/j.marpolbul.2009.10.015] [PMID: 19942234]

[52] Lambert, P. A literature review of portable fluorescence-based oil-in-water monitors. *J. Hazard. Mater.,* **2003**, *102*(1), 39-55. [http://dx.doi.org/10.1016/S0304-3894(03)00201-2] [PMID: 12963282]

[53] Ternon, E.; Tolosa, I. Comprehensive analytical methodology to determine hydrocarbons in marine waters using extraction disks coupled to glass fiber filters and compound-specific isotope analyses. *J. Chromatogr. A,* **2015**, *1404*, 10-20. [http://dx.doi.org/10.1016/j.chroma.2015.05.029] [PMID: 26054558]

[54] Sun, R.X.; Lin, Q.; Ke, C.L.; Du, F.Y.; Gu, Y.G.; Cao, K.; Luo, X.J.; Mai, B.X. Polycyclic aromatic hydrocarbons in surface sediments and marine organisms from the Daya Bay, South China. *Mar. Pollut. Bull.,* **2016**, *103*(1-2), 325-332. [http://dx.doi.org/10.1016/j.marpolbul.2016.01.009] [PMID: 26778499]

[55] Baumard, P.; Budzinski, H.; Garrigues, P.; Narbonne, J.F.; Burgeot, T.; Michel, X.; Bellocq, J. Polycyclic aromatic hydrocarbon (PAH) burden of mussels (*Mytilus* sp.) in different marine environments in relation with sediment PAH contamination, and bioavailability. *Mar. Environ. Res.,* **1999**, *47*, 415-439. [http://dx.doi.org/10.1016/S0141-1136(98)00128-7]

[56] Luna-Acosta, A.; Budzinski, H.; Le Menach, K.; Thomas-Guyon, H.; Bustamante, P. Persistent

organic pollutants in a marine bivalve on the Marennes-Oléron Bay and the Gironde Estuary (French Atlantic Coast) - part 1: bioaccumulation. *Sci. Total Environ.,* **2015**, *514*, 500-510.
[http://dx.doi.org/10.1016/j.scitotenv.2014.08.071] [PMID: 25440063]

[57]   Yang, X.; Yu, L.; Chen, Z.; Xu, M. Bioavailability of polycyclic aromatic hydrocarbons and their potential application in eco-risk assessment and source apportionment in urban river sediment. *Sci. Rep.,* **2016**, *6*, 23134.
[http://dx.doi.org/10.1038/srep23134] [PMID: 26976450]

[58]   Srogi, K. Monitoring of environmental exposure to polycyclic aromatic hydrocarbons: a review. *Environ. Chem. Lett.,* **2007**, *5*, 169-195.
[http://dx.doi.org/10.1007/s10311-007-0095-0]

[59]   OSPAR. Report of the Fourth ICES/OSPAR Workshop on Integrated Monitoring of Contaminants and their Effects in Coastal and Open Sea Areas (WKIMON IV) ICES CM 2008/ ACOM, **2008**, *49*, p. 82.

[60]   UNEP/RAMOGE. *Manual on the biomarkers recommended for the MED POL biomonitoring programme*; UNEP: Athens, **1999**, p. 92.

[61]   Martinez-Gomez, C.; Campillo, J. A.; Leon, V.; Fernandez, B.; Benedicto, J. Biomonitoring strategy to assess the effects of chemical pollution along the Iberian Mediterranean Coast: Present state and future development. *ICES CM/2007/I:11,* **2007**, p. 16.

[62]   Report of the ICES/OSPAR workshop on integrated monitoring of contaminants and their effects in coastal and open-sea areas. *ICES Document CM 2007/ACME,* **2007**, *01*, p. 209.

[63]   Martinez-Gomez, C.; Vethaak, A.D.; Hylland, K.; Burgeot, T.; Kohler, A.; Lyons, B.P.; Thain, J.; Gubbins, M.J.; Davies, M. A guide to toxicity assessment and monitoring effects at lower levels of biological organization following marine oil spills in European waters. In: *International Council for the Exploration of the Sea*; Published by Oxford Journals, **2010**; p. 14.

[64]   Goksøyr, A.; Husøy, A.M.; Larsen, H.E.; Klungsøyr, J.; Wilhelmsen, S.; Maage, A.; Brevik, E.M.; Andersson, T.; Celander, M.; Pesonen, M. Environmental contaminants and biochemical responses in flatfish from the Hvaler Archipelago in Norway. *Arch. Environ. Contam. Toxicol.,* **1991**, *21*(4), 486-496.
[http://dx.doi.org/10.1007/BF01183869] [PMID: 1759844]

[65]   Davies, I.M.; Vethaak, A.D. Integrated marine environmental monitoring of chemicals and their effects. *ICES Cooperative Research Report No. 315,* **2012**, p. 277.

[66]   Pampanin, D.M.; Kemppainen, E.K.; Skogland, K.; Jørgensen, K.B.; Sydnes, M.O. Investigation of fixed wavelength fluorescence results for biliary metabolites of polycyclic aromatic hydrocarbons formed in Atlantic cod (*Gadus morhua*). *Chemosphere,* **2016**, *144*, 1372-1376.
[http://dx.doi.org/10.1016/j.chemosphere.2015.10.013] [PMID: 26492423]

[67]   Peakall, D.B. A practical basis for using physiological indicators in environmental assessment. *Ambio,* **1992**, *21*, 437-438.

[68]   Hylland, K. Polycyclic aromatic hydrocarbon (PAH) ecotoxicology in marine ecosystems. *J. Toxicol. Environ. Health A,* **2006**, *69*(1-2), 109-123.
[http://dx.doi.org/10.1080/15287390500259327] [PMID: 16291565]

[69]   Sundt, R.C.; Brooks, S.; Grøsvik, B.E.; Pampanin, D.M.; Farmen, E.; Harman, C.; Meier, S. *Water column monitoring of offshore produced water discharges*, **2010**.

[70] Collier, T.; Connor, S.D.; Eberhart, B.T.; Anulacion, B.F.; Goksøyr, A.; Varanasi, U. Using cytochrome P450 to monitor the aquatic environment: Initial results from regional and national surveys. *Mar. Environ. Res., 1992, 34,* 195-199.
[http://dx.doi.org/10.1016/0141-1136(92)90107-W]

[71] Elliot, M.; Hemmingwa, K.L.; Krueger, D.; Thiel, R.; Hylland, K.; Arukwe, A.; Forlin, L.; Sayer, M. From the individual to the population and community responses to pollution. In: *Effects of pollution on fish. Molecular effects and population responses*; Elliot, M.; Hemmingwa, K.L., Eds.; Blackwell Publishing, 2003; pp. 221-255.
[http://dx.doi.org/10.1002/9780470999691.ch6]

[72] Gauthier, P.T.; Norwood, W.P.; Prepas, E.E.; Pyle, G.G. Metal-PAH mixtures in the aquatic environment: a review of co-toxic mechanisms leading to more-than-additive outcomes. *Aquat. Toxicol., 2014, 154,* 253-269.
[http://dx.doi.org/10.1016/j.aquatox.2014.05.026] [PMID: 24929353]

[73] Torres, R.J.; Cesar, A.; Pastor, V.A.; Pereira, C.D.; Choueri, R.B.; Cortez, F.S.; Morais, R.D.; Abessa, D.M.; do Nascimento, M.R.; Morais, C.R.; Fadini, P.S.; Casillas, T.A.; Mozeto, A.A. A critical comparison of different approaches to sediment-quality assessments in the Santos Estuarine System in Brazil. *Arch. Environ. Contam. Toxicol., 2015, 68*(1), 132-147.
[http://dx.doi.org/10.1007/s00244-014-0099-2] [PMID: 25398222]

[74] Calmano, W. Risk assessment of aquatic sediments – recommendations and proposals for an integrated approach. In: *SedNet Workshop: Chemical analysis and risk assessment of emerging contaminants in sediments and dredged material*; Barcelona, Spain, 2002; pp. November 28-30;102-106.

[75] Riley, A.K.; Chernick, M.; Brown, D.R.; Hinton, D.E.; Di Giulio, R.T. Hepatic responses of juvenile fundulus heteroclitus from pollution-adapted and nonadapted populations exposed to Elizabeth river sediment extract. *Toxicol. Pathol., 2016, 44*(5), 738-748.
[http://dx.doi.org/10.1177/0192623316636717] [PMID: 26992886]

[76] Luna-Acosta, A.; Bustamante, P.; Budzinski, H.; Huet, V.; Thomas-Guyon, H. Persistent organic pollutants in a marine bivalve on the Marennes-Oléron Bay and the Gironde Estuary (French Atlantic Coast) - part 2: potential biological effects. *Sci. Total Environ., 2015, 514,* 511-522.
[http://dx.doi.org/10.1016/j.scitotenv.2014.10.050] [PMID: 25666833]

[77] Ohe, T.; Watanabe, T.; Wakabayashi, K. Mutagens in surface waters: a review. *Mutat. Res., 2004, 567*(2-3), 109-149.
[http://dx.doi.org/10.1016/j.mrrev.2004.08.003] [PMID: 15572284]

[78] Iversen, P.E.; Lind, M.J.; Ersvik, M.; Rønning, I.; Skaare, B.B.; Green, A.M.; Bakke, T.; Lichtenthaler, R.; Klungsøyr, J.; Hylland, K. *Guidelines for environmental monitoring of petroleum activities on the Norwegian continental shelf*; Norwegian Environmental Agency, 2015, p. 65.

[79] Di Giulio, R.T.; Clark, B.W. The Elizabeth River story: a case study in evolutionary toxicology. *J. Toxicol. Environ. Health B Crit. Rev., 2015, 18*(6), 259-298.
[http://dx.doi.org/10.1080/15320383.2015.1074841] [PMID: 26505693]

[80] Hanson, N.; Larsson, A. Experiences from a biomarker study on farmed rainbow trout (*Oncorhynchus mykiss*) used for environmental monitoring in a Swedish river. *Environ. Toxicol. Chem., 2009, 28*(7), 1536-1545.
[http://dx.doi.org/10.1897/08-501.1] [PMID: 19249877]

[81]   Barrixk, A.; Chatel, A.; Marion, J.M.; Perrein-Ettajani, H.; Bruneau, M.; Mouneyrac, C. A novel methodology for the determination of biomarker baseline levels in the marine polychaete *Hediste diversicolor. Mar. Pollut. Bull.,* **2016**.
[http://dx.doi.org/10.1016/j.marpolbul.2016.04.056]

[82]   Gray, J.S. Perceived and real risks: produced water from oil extraction. *Mar. Pollut. Bull.,* **2002**, *44*(11), 1171-1172.
[http://dx.doi.org/10.1016/S0025-326X(02)00357-0] [PMID: 12523515]

[83]   Zhu, S.; King, S.C.; Haasch, M.L. Biomarker induction in tropical fish species on the Northwest Shelf of Australia by produced formation water. *Mar. Environ. Res.,* **2008**, *65*(4), 315-324.
[http://dx.doi.org/10.1016/j.marenvres.2007.11.007] [PMID: 18187187]

[84]   Canestro, D.; Raimondi, P.T.; Reed, D.C.; Schrnitt, R.J.; Holbrook, S.J. A study of methods and techniques for detecting ecological impacts. In: *American Academy of Underwater Sciences*; AAUS, **1996**.

[85]   Hylland, K.; Lang, T.; Vethaak, D. *Biological Effects of Contaminants in Marine Pelagic Ecosystems*; SETAC Press, **2006**, p. 475.

[86]   Carls, M.G.; Short, J.W.; Payne, J. Accumulation of polycyclic aromatic hydrocarbons by Neocalanus copepods in Port Valdez, Alaska. *Mar. Pollut. Bull.,* **2006**, *52*(11), 1480-1489.
[http://dx.doi.org/10.1016/j.marpolbul.2006.05.008] [PMID: 16814326]

[87]   Nilssen, J.; Bakke, T. Water column monitoring of offshore oils and gas activities on the Norwegian Continental shelf: past, present and future. In: *Produced Water*; Lee, K.; Neff, J., Eds., **2011**; pp. 431-439.
[http://dx.doi.org/10.1007/978-1-4614-0046-2_23]

[88]   Rodriguez-Mozaz, S.; Lopez de Alda, M.J.; Barceló, D. Biosensors as useful tools for environmental analysis and monitoring. *Anal. Bioanal. Chem.,* **2006**, *386*(4), 1025-1041.
[http://dx.doi.org/10.1007/s00216-006-0574-3] [PMID: 16807703]

[89]   Spier, C.R.; Vadas, G.G.; Kaattari, S.L.; Unger, M.A. Near real-time, on-site, quantitative analysis of PAHs in the aqueous environment using an antibody-based biosensor. *Environ. Toxicol. Chem.,* **2011**, *30*(7), 1557-1563.
[http://dx.doi.org/10.1002/etc.546] [PMID: 21547938]

[90]   Li, K.; Woodward, L.A.; Karu, A.E.; Li, Q.X. Immunochemical detection of polycyclic aromatic hydrocarbons and 1-hydroxypyrene in water and sediment samples. *Anal. Chim. Acta,* **2000**, *419*, 1-8.
[http://dx.doi.org/10.1016/S0003-2670(00)00989-2]

[91]   Knopp, D.; Seifert, M.; Vaananen, V.; Niessner, R. Determination of polycyclic aromatic hydrocarbons in contaminated water and soil samples by immunological and chromatographic methods. *Environ. Sci. Technol.,* **2000**, *34*, 2035-2041.
[http://dx.doi.org/10.1021/es991215f]

[92]   Moore, E.J.; Kreuzer, M.P.; Pravda, M.; Guilbault, G.G. Development of a rapid single-drop analysis biosensor for screening of phenanthrene in water samples. *Electroanalysis,* **2004**, *16*, 1653-1659.
[http://dx.doi.org/10.1002/elan.200303033]

[93]   Liu, M.; Li, Q.X.; Rechnitza, G.A. Flow injection immunosensing of polycyclic aromatic hydrocarbons with a quartz crystal microbalance. *Anal. Chim. Acta,* **1999**, *387*, 29-38.
[http://dx.doi.org/10.1016/S0003-2670(99)00030-6]

[94]   Dostálek, J.; Pribyl, J.; Homola, J.; Skládal, P. Multichannel SPR biosensor for detection of endocrine-disrupting compounds. *Anal. Bioanal. Chem.,* **2007**, *389*(6), 1841-1847.
[http://dx.doi.org/10.1007/s00216-007-1536-0] [PMID: 17906855]

[95]   Gobi, K.V.; Miura, N. Highly sensitive and interference-free simultaneous detection of two polycyclic aromatic hydrocarbons at parts-per-trillion levels using a surface plasmon resonance immunosensor. *Sens. Actuators B Chem.,* **2004**, *103*, 265-271.
[http://dx.doi.org/10.1016/j.snb.2004.04.076]

[96]   Goryacheva, I.Y.; Eremin, S.A.; Shutaleva, E.A.; Suchanek, M.; Niessner, R.; Knopp, D. Development of a fluorescence polarization immunoassay for polycyclic aromatic hydrocarbons. *Anal. Lett.,* **2007**, *40*, 1445-1460.
[http://dx.doi.org/10.1080/00032710701297034]

[97]   Pampanin, D.M.; Le Goff, J.; Grøsvik, B.E. DNA adducts in haddock and cod exposed to Produced Water (WCM 2011 – laboratory exposure). In: *IRIS report,* **2013**; p. 111.

[98]   Kurelec, B.; Gupta, R.C. Biomonitoring of aquatic systems. In: *Postlabelling Methods for Detection of DNA adducts*; Phillips, D.H.; Castegnaro, M.; Bartsch, H., Eds.; International Agency for Research on Cancer, IARC: Lyon, **1993**; pp. 365-372.

[99]   Reichert, W.L.; French, B.L.; Stein, J.E. Biological effects of contaminants: Measurements of DNA adducts in fish by ^{32}P-postlabelling. *ICES Techniques in Marine Environmental Sciences,* **1999**, *25*, 45.

[100] Phillips, D.H. On the origins and development of the (32)P-postlabelling assay for carcinogen-DNA adducts. *Cancer Lett.,* **2013**, *334*(1), 5-9.
[http://dx.doi.org/10.1016/j.canlet.2012.11.027] [PMID: 23178450]

[101] Koc, H.; Swenberg, J.A. Applications of mass spectrometry for quantitation of DNA adducts. *J. Chromatogr. B Analyt. Technol. Biomed. Life Sci.,* **2002**, *778*(1-2), 323-343.
[http://dx.doi.org/10.1016/S1570-0232(02)00135-6] [PMID: 12376138]

[102] Singh, R.; Farmer, P.B. Liquid chromatography-electrospray ionization-mass spectrometry: the future of DNA adduct detection. *Carcinogenesis,* **2006**, *27*(2), 178-196.
[http://dx.doi.org/10.1093/carcin/bgi260] [PMID: 16272169]

[103] Koivisto, P.; Peltonen, K. Analytical methods in DNA and protein adduct analysis. *Anal. Bioanal. Chem.,* **2010**, *398*(6), 2563-2572.
[http://dx.doi.org/10.1007/s00216-010-4217-3] [PMID: 20922519]

[104] Tretyakova, N.; Goggin, M.; Sangaraju, D.; Janis, G. Quantitation of DNA adducts by stable isotope dilution mass spectrometry. *Chem. Res. Toxicol.,* **2012**, *25*(10), 2007-2035.
[http://dx.doi.org/10.1021/tx3002548] [PMID: 22827593]

[105] Klaene, J.J.; Sharma, V.K.; Glick, J.; Vouros, P. The analysis of DNA adducts: the transition from (32)P-postlabeling to mass spectrometry. *Cancer Lett.,* **2013**, *334*(1), 10-19.
[http://dx.doi.org/10.1016/j.canlet.2012.08.007] [PMID: 22960573]

[106] Bohne-Kjersem, A.; Bache, N.; Meier, S.; Nyhammer, G.; Roepstorff, P.; Saele, O.; Goksøyr, A.; Grøsvik, B.E. Biomarker candidate discovery in Atlantic cod (*Gadus morhua*) continuously exposed to North Sea produced water from egg to fry. *Aquat. Toxicol.,* **2010**, *96*(4), 280-289.
[http://dx.doi.org/10.1016/j.aquatox.2009.11.005] [PMID: 20031237]

[107] Karim, M.; Puiseux-Dao, S.; Edery, M. Toxins and stress in fish: proteomic analyses and response network. *Toxicon,* **2011**, *57*(7-8), 959-969.
[http://dx.doi.org/10.1016/j.toxicon.2011.03.018] [PMID: 21457724]

[108] Pampanin, D.M.; Larssen, E.; Oysæd, K.B.; Sundt, R.C.; Sydnes, M.O. Study of the bile proteome of Atlantic cod (*Gadus morhua*): Multi-biological markers of exposure to polycyclic aromatic hydrocarbons. *Mar. Environ. Res.,* **2014**, *101*, 161-168.
[http://dx.doi.org/10.1016/j.marenvres.2014.10.002] [PMID: 25440786]

[109] Törnqvist, M.; Fred, C.; Haglund, J.; Helleberg, H.; Paulsson, B.; Rydberg, P. Protein adducts: quantitative and qualitative aspects of their formation, analysis and applications. *J. Chromatogr. B Analyt. Technol. Biomed. Life Sci.,* **2002**, *778*(1-2), 279-308.
[http://dx.doi.org/10.1016/S1570-0232(02)00172-1] [PMID: 12376136]

[110] Rappaport, S.M.; Li, H.; Grigoryan, H.; Funk, W.E.; Williams, E.R. Adductomics: characterizing exposures to reactive electrophiles. *Toxicol. Lett.,* **2012**, *213*(1), 83-90.
[http://dx.doi.org/10.1016/j.toxlet.2011.04.002] [PMID: 21501670]

CHAPTER 3

Oil Spill Fingerprinting – Identification of Crude Oil Source of Contamination

Magne O. Sydnes[*]

Faculty of Science and Technology, Department of Mathematics and Natural Science, University of Stavanger, Stavanger, Norway

Abstract: Source identification of oil with its thousands of compounds has always been a great challenge for analytical chemists. With the significant improvements in analytical instrumentation and work-up techniques this challenge has started to become possible to overcome. Lately, the development of new mass spectrometry techniques and the increased sensitivity of the instruments have made it possible to analyze a big fraction of compounds found in oil in one single analysis. The analytical field conducting oil fingerprinting is now commonly referred to as petroleomics. The main focus of this chapter is directed towards how analyses are used in environmental investigations at present and how the technology can be used in the future for oil spill forensics.

Keywords: Analytical chemistry, Crude oil, Mass spectrometry (MS), Oil fingerprinting, Petroleomics, Polycyclic aromatic hydrocarbons (PAHs).

INTRODUCTION

Crude oil enters the environment through both natural processes and accidental spills often caused by human error [1]. Natural oil seeps are found unevenly distributed around the world and crude oil has been leaking from its reservoirs for as long as oil has existed [2]. Initially it was through oil seeps on land where oil naturally had gathered on the surface that originally lead to the discovery of oil.

---

[*] **Corresponding author Magne O. Sydnes**: Faculty of Science and Technology, Department of Mathematics and Natural Science, University of Stavanger, Stavanger, Norway; Tel: +47 51831761; E-mail: magne.o.sydnes@uis.no

Daniela M. Pampanin and Magne O. Sydnes (Eds.)
All rights reserved-© 2017 Bentham Science Publishers

The ecosystem around oil seeps, both on land and in the ocean, is adjusted to the conditions caused by slow release of hydrocarbons. In fact, organisms living in such areas, predominantly bacteria, have methods for detoxifying the various compounds found in crude oil. For many of these bacteria the oil becomes an energy source [2]. Bacteria degrades the lighter fractions much faster than the heavier compounds leaving behind the latter [3].

Oil spills caused by human activities generally have larger consequences for the environment then natural oil seeps, due to the large volume of oil entering a location simultaneously. Such spills are caused by blow-outs, leakage during development and production of an oil field, accidents with oil tankers during loading and transportation, and leakage from fuel tanks. For example, two large incidents that recently occurred is the *Exxon Valdez* spill in Alaska in 1989 and the *Deep Water Horizon* blow-out in the Gulf of Mexico in 2010 [4]. Even though it is more than 25 years ago since the *Exxon Valdez* accident took place in Prince Williams Sound, Alaska, there is still research published describing the long term effects of this spill, particularly on fish populations [4].

The two mentioned incidents were very large oil spills, which rarely take place. However, smaller oil spills are taking place quite frequently. Data from the Republic of Korea covering a 19 year period (1990-2008) showed that there were 354 oil spills larger than 50,000 L taking place over that period of time in waters of the coast of the country [5]. This averages out to more than 18 oil spills per year in that area. So oil spills are more common than one might think. Small oil spills are taking place around on the seven seas several times per day (*e.g.* small scale spills take place on almost daily bases in the oceans of the Republic of Korea [5]). Many of these smaller events are not reported and when they are discovered it is difficult to point out the source of the oil contamination. Nevertheless, small oil spills in a sensitive environment, or at a sensitive developmental stage for an organism for example, can have large impact for that species at the site of the spill.

The development in analytical instrumentation over the last 10-15 years has now brought chemists in a position where it is possible to do source identification of crude oil [6]. By such means, it is possible to identify the oil's fingerprint and use that in order to find out where the oil originated from.

Crude Oil

Crude oil is a complex mixture of alkenes (paraffin's), cycloalkanes (naphthenes), aromatics, and asphaltics [7]. The distribution between the different chemical groups varies from oil to oil, and the distribution between the various compounds categorized in the same group also varies significantly between oils from different sources. The variations is so large that petroleum from different origins has its own unique combination and concentration of compounds, *i.e.* a unique fingerprint, which is possible to identify due to modern analytical techniques. Table **1** summarizes the average distribution between the major compound groups found in crude oil and highlights the large composition range that is found in crude oil.

Table 1. Composition by weight of compound groups found in crude oil [7].

Hydrocarbon	Average	Range
Alkanes (paraffins)	30%	15-60%
Cycloalkanes (naphthenes)	49%	30-60%
Aromatics	15%	3-30%
Asphaltics	6%	remainder

The aromatic fraction of crude oil can vary from only 3% to 30% (Table **1**). Benzene, toluene, and xylene (*o*-xylene, *m*-xylene, and *p*-xylene) make up the majority of the quantity of aromatic compounds found in crude oil. Polycyclic aromatic hydrocarbons (PAHs) only accounts for a minute amount of the aromatic fraction [1, 8]. However, PAHs are the compounds of most concern due to their persistence in the environment and their toxicological properties (*i.e.* carcinogenicity) [9 - 11]. Similarly to the large variation of aromatic compounds in crude oil, there is also a large difference in the concentration of the various PAHs found in oil [1]. The concentration of the different PAHs strongly influence the toxicity of an oil since the toxicity of the various PAHs is greatly different, or more precisely the toxicity of the metabolites formed *in vivo* [1, 9, 11]. In general, the higher the concentration of PAHs with more than 3 rings is in the oil the more toxic the PAH fraction of the oil will be after metabolization [11].

Advancement in Analytical Chemistry

The full characterization of compounds found in petroleum has been a long standing challenge for analytical chemists with thousands of compounds making up crude oil [12]. Such analysis has been named petroleomics by Alan Marshall [13], and petroleomics was referred to in 2004 as the next grand challenge for chemical analysis by Marshall and Rodgers [14].

A range of different mass spectrometry (MS) techniques have been used for petroleomics, *i.e.* oil fingerprinting. This type of analysis has also greatly benefitted from the constant development and improvements of mass spectrometers. Gas chromatography mass spectrometry (GC-MS) generally gives useful structural information for compounds with mass lower than 400 Da [15 - 21], which accounts for a fair amount of the compounds found in oil. There is still vast amounts of compounds in petroleum that will not be identified by this method due to too high boiling point. The GC columns in fact does not tolerate temperatures higher than 400 °C due to the issue of decomposition of the column material [22 - 25], and by such means ruling out the analysis of many of the components found in crude oil. In addition the GC column is not very suitable for the analysis of polar compounds that are formed by weathering processes taking place in an oil spill [26]. Larger PAHs, with 4 rings or more in addition to the sulfur-containing and partly the nitrogen-containing PAHs, are not readily analyzed by GC-MS [27]. An improved GC-MS method, namely GC vacuum ultraviolet photoionization MS (GC/VUV-MS), was recently used for the analysis of crude oil from the Gulf of Mexico. This method offers the opportunity to get more structural information due to the fact that it overcomes some of the limitations of isomer resolution in GC of complex mixtures [28].

GC-MS analysis has some clear limitations [29]. However, a lot of useful structural information can be obtained by this technique. In order to broaden the analytical window for petroleomics new MS techniques are now introduced for this type of analysis. Ion mobility mass spectrometry [30, 31], direct analysis in real time-mass spectrometry (DART-MS) in combination with other MS techniques [32], and furrier transform ion cyclotron resonance mass spectrometry (FT-ICR-MS) [33, 34] are now more commonly used. By far the most commonly used technique for petroleomics among the three is FT-ICR MS [24, 29, 34 - 43], which originally was invented by Comisarow and Marshall [44] and further developed for the use in petroleomics by Marshall [33, 34, 45, 46]. Among the instruments currently available on the market, FT-ICR MS offers the highest accuracy, highest mass resolving power, and the highest broadband mass resolution [47]. For detailed description of the instrumental developments facilitating petroleomics, the readers are referred to a few review articles on the topic [6, 33, 40, 47, 48].

The original interest towards petroleomics came from the oil refinery industry who due to increasing oil prices in combination with the global supply of crude oil, is becoming heavier, more acidic in addition to having higher sulfur content were interested in more detailed structural information about the feed stock – crude oil [47]. In order to have a more fine-tuned refinery process, better knowledge regarding the crude product is an advantage, since oils with different compositions require different strategies in the refining process [6, 13, 47]. Thus far, petroleomics has predominantly been used in order to provide the petroleum industry with a better understanding of the composition of the crude oil. However, the knowledge developed for the refinery industry can equally well be used for environmental studies and identification of the source of contamination.

Environmental use of Petroleomics

When an oil spill has taken place, detailed information regarding the type of oil that is spilled is important, especially in order to determine measures to use for the clean-up [29]. Light fuel oils are more easily dealt with than heavier fuel oils and crude oil. The same strategies used to unravel structural information for the refinery industry can also be used in order to determine the type of oil in an oil spill. Although, the challenge here is even larger, since once oil is released to the environment a range of processes, which collectively is called weathering, starts taking place [5]. These processes, which in the marine environment includes spreading, evaporation, emulsification, dissolution, microbial, and photochemical degradation, changes the characteristics of the oil significantly [8, 39, 49 - 52]. By such means the weathering of oil also changes its physical properties and consequently the clean-up method needs to be changed in order to accommodate for the new physical properties of the oil.

An additional complicating issue, relevant when analyzing samples from areas with natural oil seeps and/or a lot of human activities, is to be able to distinguish the spilled oil from the background hydrocarbons already present in the ocean prior to the oil spill [5, 53]. This issue does not influence the clean-up process, but it is highly relevant when it comes to determining the extent of the damages from a particular oil spill linked to a particular accident [5, 53, 54]. Occasionally oil spills from different accidents enters the water in the same geographical region, but due to differences in the local current at the particular spill site are spread in different directions [54, 55].

In order to distinguish between crude oil, and a processed oil the analytical chemist can look for the presence of 2-methylanthracene, a PAH that is formed under refining of crude oil, and not commonly present in crude oil [56]. Another strategy is to conduct a fingerprint analysis of diamondoids (Fig. **1**) [57]. A refined oil will in fact have an overweight of lighter diamondoids compared to crude oil, which will have a higher concentration of heavier diamondoids.

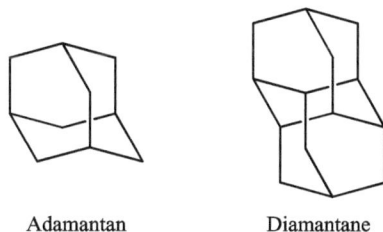

Adamantan   Diamantane

Fig. (1). Structure of two light diamondoids.

In order to evaluate how fresh an oil spill is, it is possible to compare the ratio between lighter and heavier fractions. A freshly spilled oil have most of the lighter components still present (including lighter PAHs). However, over time these will be lost due to evaporation, dispersion, and microbial degradation [8, 58]. In an oil spill simulation it was found that 22-30% of the oil had evaporated and 0.7-10% of the oil was dispersed after four days [8]. Furthermore, changes in the benzo[*a*]anthracene/chrysene (Fig. **2**) ratio are used as indicator of photodegradation [58 - 60]. The weathering of an oil will depend on a range of factors at the spill sight, *e.g.* wind conditions (waves), season (temperature and sunlight), and current. All these factors will naturally be different at each spill site and therefore each oil spill will be unique [29].

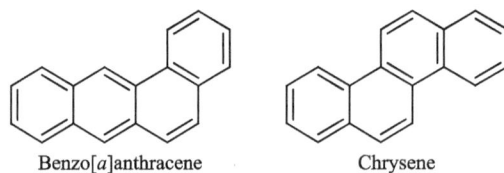

Benzo[*a*]anthracene   Chrysene

Fig. (2). Structure of benzo[*a*]anthracene and chrysene.

Quick and accurate information regarding the type and weathering condition of an oil is crucial in order to minimize the environmental impact of a spill by initiating the correct clean-up strategies immediately. Accurate knowledge regarding the type of oil is also helpful when determining how severe the effect of a spill will be. The type of oil and its molecular make-up is also important for evaluating its toxicity towards the ecosystem. The latter information can be crucial for example for the aquaculture industry who has farmed fish in pens in affected oceans. These pens can in fact be moved if required, in order to keep the fish away from an oil spill.

Oil tankers often carry oil from more than one source [5], and ships generally have fuel oil from several sources in their tanks [61]. Petroleomics was used in order to determine which fuel tank an oil spill in San Francisco Bay originated from when two fuel tanks on M/V *Cosco Busan* were ruptured in November 2007 [58, 61]. In the initial report by Lemkau *et al.*, GC-FID and GC-MS were used in order to determine that the fuel predominantly leaked from fuel tank number four. In a later study this finding was confirmed by utilizing principal component analysis of electrospray ionization fourier transform ion cyclotron resonance MS [61]. The two mentioned studies in addition to a third study, by the same researcher group resulted in significant data regarding weathering of the heavy oil from M/V *Cosco Busan* [58, 61, 62].

In a spill taking place five miles off Taean on the west coast of the Republic of Korea, in December 2007, three tanks transporting crude oil on M/V *Hebei Spirit* were ruptured. Three different types of crude oil, all originating from the Middle

East, entered the coastal waters and polluted about 400 km of coastline [5]. Oil fingerprinting was used in order to distinguish between coastal areas affected by oil from the M/V *Hebei Spirit* and other sources. Similarly a few years earlier petroleomics was used in order to determine that oil from the tanker *Erika* only was to blame for the oil washing up on the north part of the Atlantic Coast of France and not for the oil arriving on beaches further south [54].

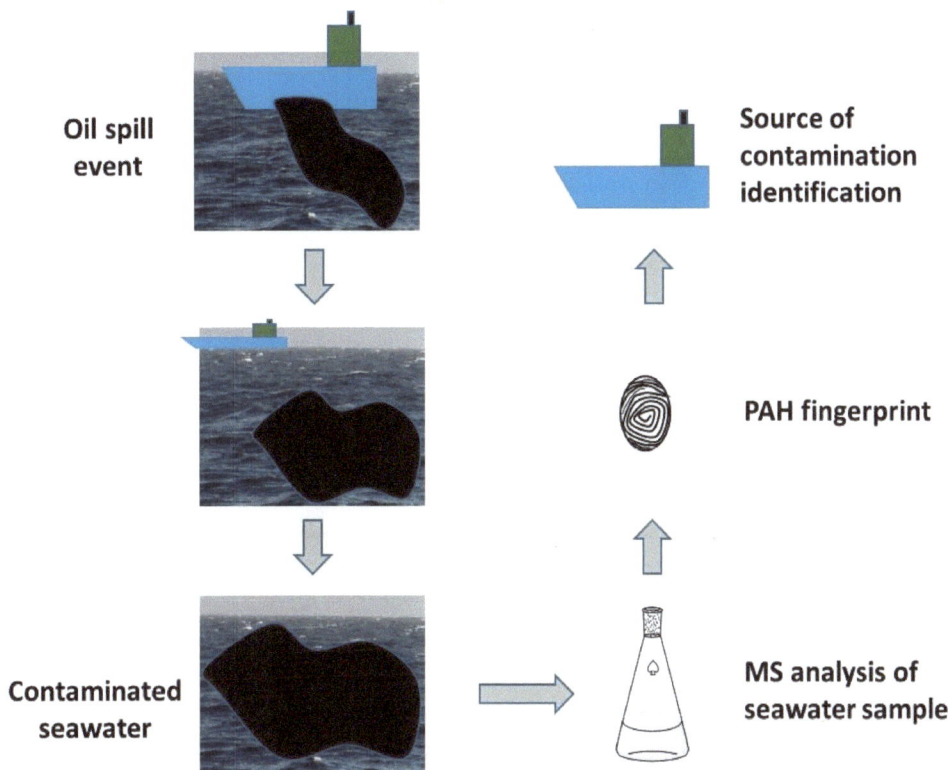

Oil spill event

Source of contamination identification

PAH fingerprint

Contaminated seawater

MS analysis of seawater sample

Fig. (3). Schematic outline of the use of petroleomics for determining the source of unreported oil spills.

Future Directions

In the previous examples, the ships where the oil originated from were known making it possible to obtain reference oil from the tanks where the oil had leaked from. Thus making it possible to verify the results from the field samples. The potential for petroleomics is however even greater. Petroleomics can in the future be used in the investigation of environmental crime, *e.g.* investigation of unreported oil spills. There is a underreporting of oil spills, especially small and medium sized events, since admitting to an oil spill usually results in a claim to cover the cost for the clean-up and eventually damages. With the advancement in petroleomics it should in the future be possible to identify the origin of an oil spill, which at the first glance does not have an obvious source (Fig. **3**).

To have and effective strategy it would be required to have a database of oil fingerprints available for searching. The more comprehensive such a database is the better. This would require that oil fingerprinting data for oils should be routinely collected. This does not necessarily require a lot of extra work, since a lot of this data is already available from crude oil analysis around the world.

The PAHs that are found in the oil are also part of the oil fingerprint. The combination of various PAHs in crude oil is also specific to the type of oil. In the future it is a target to be able to utilize that information in order to determine based on PAH adducts found in biological samples (*e.g.* fish bile) what oil the studied organism has been exposed to.

CONCLUDING REMARKS

The development of MS has made it possible to elucidate the structure of a vast amount of the compounds found in crude oil and refined oil products. The analysis, which often are referred to as petroleomics, has made it possible to establish the source of an oil spill with good confidence. So far this has been done when knowledge regarding the spilled oil was available. With continued steady progress in petroleomics, it should be possible to track down the source of an oil spill on the open sea without previous knowledge regarding the source of contamination.

CONFLICT OF INTEREST

The author confirms that the author has no conflict of interest to declare for this publication.

ACKNOWLEDGEMENTS

The University of Stavanger and the research program Bioactive is acknowledged for providing funding for the ongoing research in the group.

REFERENCES

[1]   Pampanin, D.M.; Sydnes, M.O. Polycyclic aromatic Hydrocarbons a constituent of Petroleum: Presence and Influence in the Aquatic Environment. In: *Hydrocarbons*; Kutcherov, V.; Kolesnikov, A., Eds.; InTech: Rijeka, **2013**; pp. 83-118.

[2]   Head, I.M.; Jones, D.M.; Röling, W.F. Marine microorganisms make a meal of oil. *Nat. Rev. Microbiol.,* **2006**, *4*(3), 173-182.
[http://dx.doi.org/10.1038/nrmicro1348] [PMID: 16489346]

[3]   Pond, K.L.; Huang, Y.; Wang, Y.; Kulpa, C.F. Hydrogen isotopic composition of individual *n*-alkanes as an intrinsic tracer for bioremediation and source identification of petroleum contamination. *Environ. Sci. Technol.,* **2002**, *36*(4), 724-728.
[http://dx.doi.org/10.1021/es011140r] [PMID: 11878389]

[4]   Kemsley, J. Oil-Spill Lessons. *Chem. Eng. News,* **2015**, *93*(28), 8-12.
[http://dx.doi.org/10.1021/cen-09328-cover]

[5]   Yim, U.H.; Kim, M.; Ha, S.Y.; Kim, S.; Shim, W.J. Oil spill environmental forensics: the *Hebei Spirit* oil spill case. *Environ. Sci. Technol.,* **2012**, *46*(12), 6431-6437.
[http://dx.doi.org/10.1021/es3004156] [PMID: 22582823]

[6]   Rodgers, R.P.; Schaub, T.M.; Marshall, A.G. Petroleomics: MS Returns to Its Roots. *Anal. Chem.,* **2005**, *77*, 20A-27A.
[http://dx.doi.org/10.1021/ac053302y]

[7]   Hyne, N.J. The Nature of Gas and Oil. In: *Non-technical Guide to Petroleum Geology, Exploration, Drilling & Production,* 3rd edition; PennWell Corporation: Tulsa, Oklahoma, **2012**.

[8]   Yim, U.H.; Ha, S.Y.; An, J.G.; Won, J.H.; Han, G.M.; Hong, S.H.; Kim, M.; Jung, J-H.; Shim, W.J. Fingerprint and weathering characteristics of stranded oils after the Hebei Spirit oil spill. *J. Hazard. Mater.,* **2011**, *197*, 60-69.
[http://dx.doi.org/10.1016/j.jhazmat.2011.09.055] [PMID: 21996619]

[9]   Jacob, J. The significance of polycyclic aromatic hydrocarbons as environmental carcinogens. 35 Years research on PAH – a retrospective. *Polycycl. Aromat. Compd.,* **2008**, *28*, 242-272.
[http://dx.doi.org/10.1080/10406630802373772]

[10]   White, K.L. An overview of immunotoxicology and carcinogenic polycyclic aromatic hydrocarbons. *J. Environ. Sci.,* **2012**, *2*, 163-202.

[11]   Jacob, J. The significance of polycyclic aromatic hydrocarbons as environmental carcinogens. *Pure Appl. Chem.,* **1996**, *68*, 301-308.
[http://dx.doi.org/10.1351/pac199668020301]

[12]   Pym, J.G.; Ray, J.E.; Smith, G.W.; Whitehead, S.E. Petroleum triterpene fingerprinting of crude oils. *Anal. Chem.,* **1975**, *47*, 1617-1622.
[http://dx.doi.org/10.1021/ac60359a052]

[13]   Henry, C.M. Fine look at crude oil. *Chem. Eng. News,* **2003**, *81*(13), 39.
[http://dx.doi.org/10.1021/cen-v081n044.p039]

[14]   Marshall, A.G.; Rodgers, R.P. Petroleomics: the next grand challenge for chemical analysis. *Acc.*

Chem. Res., **2004**, *37*(1), 53-59.
[http://dx.doi.org/10.1021/ar020177t] [PMID: 14730994]

[15] Mahé, L.; Dutriez, T.; Courtiade, M.; Thiébaut, D.; Dulot, H.; Bertoncini, F. Global approach for the selection of high temperature comprehensive two-dimensional gas chromatography experimental conditions and quantitative analysis in regards to sulfur-containing compounds in heavy petroleum cuts. *J. Chromatogr. A,* **2011**, *1218*(3), 534-544.
[http://dx.doi.org/10.1016/j.chroma.2010.11.065] [PMID: 21168139]

[16] Dutta, S.; Bhattacharya, S.; Raju, S.V. Biomarker signatures from Neoproterozoic-Early Cambrian oil, western India. *Org. Geochem.,* **2013**, *56*, 68-80.
[http://dx.doi.org/10.1016/j.orggeochem.2012.12.007]

[17] Ho, S.J.; Wang, C.Y.; Luo, Y.M. GC-MS analysis of two types of mixed oils, a comparison of composition and weathering patterns. *Mar. Pollut. Bull.,* **2015**, *96*(1-2), 271-278.
[http://dx.doi.org/10.1016/j.marpolbul.2015.05.013] [PMID: 25982414]

[18] Sun, P.; Bao, M.; Li, F.; Cao, L.; Wang, X.; Zhou, Q.; Li, G.; Tang, H. Sensitivity and identification indexes for fuel oils and crude oils based on the hydrocarbon components and diagnostic ratios using principal component analysis (PCA) biplots. *Energy Fuels,* **2015**, *29*, 3032-3040.
[http://dx.doi.org/10.1021/acs.energyfuels.5b00443]

[19] Sun, P.; Gao, Z.; Cao, L.; Wang, X.; Zhou, Q.; Zhao, Y.; Li, G. Application of a step-by-step fingerprinting identification method on a spilled oil accident in the bohai sea area. *J. Ocean Univ. China,* **2011**, *10*, 35-41.
[http://dx.doi.org/10.1007/s11802-011-1716-6]

[20] Khan, N.A.; Engle, M.; Dungan, B.; Holguin, F.O.; Xu, P.; Carroll, K.C. Volatile-organic molecular characterization of shale-oil produced water from the Permian Basin. *Chemosphere,* **2016**, *148*, 126-136.
[http://dx.doi.org/10.1016/j.chemosphere.2015.12.116] [PMID: 26802271]

[21] Byer, J.D.; Siek, K.; Jobst, K. Distinguishing the C_3 *vs* SH_4 mass split by comprehensive two-dimensional gas chromatography-high resolution time-of-flight mass spectrometry. *Anal. Chem.,* **2016**, *88*(12), 6101-6104.
[http://dx.doi.org/10.1021/acs.analchem.6b01137] [PMID: 27269256]

[22] Shen, J. Minimization of interferences from weathering effects and use of biomarkers in identification of spilled crude oils by gas chromatography/mass spectrometry. *Anal. Chem.,* **1984**, *56*, 214-217.
[http://dx.doi.org/10.1021/ac00266a021]

[23] Krahn, M.M.; Stein, J.E. Assessing exposure of marine biota and habitats to petroleum compounds. *Anal. Chem.,* **1998**, *70*, 186A-192A.
[http://dx.doi.org/10.1021/ac981748r]

[24] Fernandez-Lima, F.A.; Becker, C.; McKenna, A.M.; Rodgers, R.P.; Marshall, A.G.; Russell, D.H. Petroleum crude oil characterization by IMS-MS and FTICR MS. *Anal. Chem.,* **2009**, *81*(24), 9941-9947.
[http://dx.doi.org/10.1021/ac901594f] [PMID: 19904990]

[25] Fernández-Varela, R.; Andrade, J.M.; Muniategui, S.; Prada, D. Selecting a reduced suite of diagnostic ratios calculated between petroleum biomarkers and polycyclic aromatic hydrocarbons to characterize

a set of crude oils. *J. Chromatogr. A,* **2010**, *1217*(52), 8279-8289.
[http://dx.doi.org/10.1016/j.chroma.2010.10.043] [PMID: 21081235]

[26]   Aeppli, C.; Carmichael, C.A.; Nelson, R.K.; Lemkau, K.L.; Graham, W.M.; Redmond, M.C.; Valentine, D.L.; Reddy, C.M. Oil weathering after the deepwater horizon disaster led to the formation of oxygenated residues. *Environ. Sci. Technol.,* **2012**, *46*(16), 8799-8807.
[http://dx.doi.org/10.1021/es3015138] [PMID: 22809266]

[27]   Hagazi, A.H.; Andersson, J.T. Limitations to GC.MS determination of sulfur-containing polycyclic aromatic compounds in geochemical, petroleum, and environmental investigations. *Energy Fuels,* **2007**, *21*, 3375-3384.
[http://dx.doi.org/10.1021/ef700362v]

[28]   Worton, D.R.; Zhang, H.; Isaacman-VanWertz, G.; Chan, A.W.; Wilson, K.R.; Goldstein, A.H. Comprehensive chemical characterization of hydrocarbons in NIST standard reference material 2779 gulf of Mexico crude oil. *Environ. Sci. Technol.,* **2015**, *49*(22), 13130-13138.
[http://dx.doi.org/10.1021/acs.est.5b03472] [PMID: 26460682]

[29]   McKenna, A.M.; Nelson, R.K.; Reddy, C.M.; Savory, J.J.; Kaiser, N.K.; Fitzsimmons, J.E.; Marshall, A.G.; Rodgers, R.P. Expansion of the analytical window for oil spill characterization by ultrahigh resolution mass spectrometry: beyond gas chromatography. *Environ. Sci. Technol.,* **2013**, *47*(13), 7530-7539.
[http://dx.doi.org/10.1021/es305284t] [PMID: 23692145]

[30]   Fasciotti, M.; Lalli, P.M.; Klitzke, C.F.; Corilo, Y.E.; Pudenzi, M.A.; Pereira, R.C.; Bastos, W. Daroda, R. J.; Eberlin, M. N. Petroleomics by Traveling Wave Ion Mobility-Mass Spectrometry Using CO_2 as a Drift Gas. *Energy Fuels,* **2013**, *27*, 7277-7286.
[http://dx.doi.org/10.1021/ef401630b]

[31]   Santos, J.M.; Galverna, R. de S.; Pudenzi, M.A.; Schmidt, E.M.; Sanders, N.L.; Kurulugarma, R.T.; Mordehai, A.; Stafford, G.C.; Wisniewski, A., Jr; Eberlin, M.N. Petroleomics by ion mobility mass spectrometry: resolution and charactherization of contaminants and additives in crude oils and petrofuels. *Anal. Methods,* **2015**, *7*, 4450-4463.
[http://dx.doi.org/10.1039/C5AY00265F]

[32]   Romão, W.; Tose, L.V.; Vaz, B.G.; Sama, S.G.; Lobinski, R.; Giusti, P.; Carrier, H.; Bouyssiere, B. Petroleomics by direct analysis in real time-mass spectrometry. *J. Am. Soc. Mass Spectrom.,* **2016**, *27*(1), 182-185.
[http://dx.doi.org/10.1007/s13361-015-1266-z] [PMID: 26432579]

[33]   Marshall, A.G.; Hendrickson, C.L.; Jackson, G.S. Fourier transform ion cyclotron resonance mass spectrometry: a primer. *Mass Spectrom. Rev.,* **1998**, *17*(1), 1-35.
[http://dx.doi.org/10.1002/(SICI)1098-2787(1998)17:1<1::AID-MAS1>3.0.CO;2-K]   [PMID: 9768511]

[34]   Nikolaev, E.N.; Kostyukevich, Y.I.; Vladimirov, G.N. Fourier transform ion cyclotron resonance (FT ICR) mass spectrometry: Theory and simulations. *Mass Spectrom. Rev.,* **2016**, *35*(2), 219-258.
[http://dx.doi.org/10.1002/mas.21422] [PMID: 24515872]

[35]   Marshall, A.G. Milestones in Fourier transform ion cyclotron resonance mass spectrometry technique development. *Int. J. Mass Spectrom.,* **2000**, *200*, 331-356.
[http://dx.doi.org/10.1016/S1387-3806(00)00324-9]

[36] Hughey, C.A.; Rodgers, R.P.; Marshall, A.G. Resolution of 11,000 compositionally distinct components in a single electrospray ionization Fourier transform ion cyclotron resonance mass spectrum of crude oil. *Anal. Chem.,* **2002**, *74*(16), 4145-4149.
[http://dx.doi.org/10.1021/ac020146b] [PMID: 12199586]

[37] Hegazi, A.H.; Fathalla, E.M.; Panda, S.K.; Schrader, W.; Andersson, J.T. High-molecular weight sulfur-containing aromatics refractory to weathering as determined by Fourier transform ion cyclotron resonance mass spectrometry. *Chemosphere,* **2012**, *89*(3), 205-212.
[http://dx.doi.org/10.1016/j.chemosphere.2012.04.016] [PMID: 22560701]

[38] Podgorski, D.C.; Corilo, Y.E.; Nyadong, L.; Lobodin, V.V.; Bythell, B.J.; Robbins, W.K.; McKenna, A.M.; Marshall, A.G.; Rodgers, R.P. Heavy petroleum composition. 5. compositional and structural continuum of petroleum revealed. *Energy Fuels,* **2013**, *27*, 1268-1276.
[http://dx.doi.org/10.1021/ef301737f]

[39] Hegazi, A.H.; Fathalla, E.M.; Andersson, J.T. Weathering trend characterization of medium-molecular weight polycyclic aromatic disulfur heterocycles by Fourier transform ion cyclotron resonance mass spectrometry. *Chemosphere,* **2014**, *111*, 266-271.
[http://dx.doi.org/10.1016/j.chemosphere.2014.04.029] [PMID: 24997927]

[40] Cho, Y.; Ahmed, A.; Islam, A.; Kim, S. Developments in FT-ICR MS instrumentation, ionization techniques, and data interpretation methods for petroleomics. *Mass Spectrom. Rev.,* **2015**, *34*(2), 248-263.
[http://dx.doi.org/10.1002/mas.21438] [PMID: 24942384]

[41] Nagornov, K.O.; Kozhinov, A.N.; Tsybin, O.Y.; Tsybin, Y.O. Ion trap with narrow aperture detection electrodes for Fourier transform ion cyclotron resonance mass spectrometry. *J. Am. Soc. Mass Spectrom.,* **2015**, *26*(5), 741-751.
[http://dx.doi.org/10.1007/s13361-015-1089-y] [PMID: 25773900]

[42] Pudenzi, M.A.; Eberlin, M.N. Assessing relative electrospray ionization, atmospheric pressure photoionization, atmospheric pressure chemical ionization, and atmospheric pressure photo- and chemical ionization efficiencies in mass spectrometry petroleomic analysis *via* pools and pairs of selected polar compound standards. *Energy Fuels,* **2016**, *30*(9), 7125-7133.
[http://dx.doi.org/10.1021/acs.energyfuels.6b01403]

[43] Kim, E.; Cho, E.; Moon, C.; Ha, J.; Cho, E.; Ha, J-H.; Kim, S. Correlation among petroleomics data obtained with high-resolution mass spectrometry and elemental and NMR analyses of maltene fractions of atmospheric pressure residues. *Energy Fuels,* **2016**, *30*(9), 6958-6967.
[http://dx.doi.org/10.1021/acs.energyfuels.6b01047]

[44] Comisarow, M.B.; Marshall, A.G. Fourier transform ion cyclotron resonance spectroscopy. *Chem. Phys. Lett.,* **1974**, *25*, 282-283.
[http://dx.doi.org/10.1016/0009-2614(74)89137-2]

[45] Marshall, A.G. Fourier transform ion cyclotron resonance mass spectrometry. *Acc. Chem. Res.,* **1985**, *18*, 316-322.
[http://dx.doi.org/10.1021/ar00118a006]

[46] Schaub, T.M.; Hendrickson, C.L.; Quinn, J.P.; Rodgers, R.P.; Marshall, A.G. Instrumentation and

method for ultrahigh resolution field desorption ionization fourier transform ion cyclotron resonance mass spectrometry of nonpolar species. *Anal. Chem.,* **2005**, *77*(5), 1317-1324.
[http://dx.doi.org/10.1021/ac048766v] [PMID: 15732913]

[47]   Marshall, A.G.; Rodgers, R.P. Petroleomics: chemistry of the underworld. *Proc. Natl. Acad. Sci. USA,* **2008**, *105*(47), 18090-18095.
[http://dx.doi.org/10.1073/pnas.0805069105] [PMID: 18836082]

[48]   Xian, F.; Hendrickson, C.L.; Marshall, A.G. High resolution mass spectrometry. *Anal. Chem.,* **2012**, *84*(2), 708-719.
[http://dx.doi.org/10.1021/ac203191t] [PMID: 22263633]

[49]   Douglas, G.S.; Bence, A.E.; Prince, R.C.; McMillen, S.J.; Butler, E.L. Environmental stability of selected petroleum hydrocarbon source and weathering ratios. *Environ. Sci. Technol.,* **1996**, *30*, 2332-2339.
[http://dx.doi.org/10.1021/es950751e]

[50]   Wang, Z.; Fingas, M.; Blenkinsopp, S.; Sergy, G.; Landriault, M.; Sigouin, L.; Foght, J.; Semple, K.; Westlake, D.W. Comparison of oil composition changes due to biodegradation and physical weathering in different oils. *J. Chromatogr. A,* **1998**, *809*(1-2), 89-107.
[http://dx.doi.org/10.1016/S0021-9673(98)00166-6] [PMID: 9677713]

[51]   Rodgers, R.P.; Blumer, E.N.; Freitas, M.A.; Marshall, A.G. Jet fuel chemical composition, weathering, and identification as a contaminant at a remediation site, determined by fourier transform ion cyclotron resonance mass spectrometry. *Anal. Chem.,* **1999**, *71*, 5171-5176.
[http://dx.doi.org/10.1021/ac9904821]

[52]   Seidel, M.; Kleindienst, S.; Dittmar, T.; Joye, S.B.; Medeiros, P.M. Biodegradation of crude oil and dispersants in deep seawater from the Gulf of Mexico: Insights from ultra-high resolution mass spectrometry. *Deep Sea Res. Part II Top. Stud. Oceanogr.,* **2016**, *129*, 108-118.
[http://dx.doi.org/10.1016/j.dsr2.2015.05.012]

[53]   Bence, A.E.; Kvenvolden, K.A.; Kennicutt, M.C., II Organic geochemistry applied to environmental assessments of Prince William Sound, Alaska, after the *Exxon Valdez* oil spill – a review. *Org. Geochem.,* **1996**, *24*, 7-42.
[http://dx.doi.org/10.1016/0146-6380(96)00010-1]

[54]   Mazeas, L.; Budzinski, H. Molecular and stable carbon isotopic source identification of oil residues and oiled bird feathers sampled along the Atlantic Coast of France after the Erika oil spill. *Environ. Sci. Technol.,* **2002**, *36*(2), 130-137.
[http://dx.doi.org/10.1021/es010726a] [PMID: 11827045]

[55]   Wang, Z.; Fingas, M.; Landriault, M.; Sigouin, L.; Castle, B.; Hostetter, D.; Zhang, D.; Spencer, B. Identification and linkage of tarballs from the coasts of vancouver Island and Northern California using GC/MS and isotopic techniques. *J. High Resolut. Chromatogr.,* **1998**, *21*, 383-395.
[http://dx.doi.org/10.1002/(SICI)1521-4168(19980701)21:7<383::AID-JHRC383>3.0.CO;2-3]

[56]   Uhler, A.D.; Stout, S.A.; Douglas, G.S. Chemical heterogeneity in modern marine residual fuel oils. In: *Oil Spill Environmental Forensics: Fingerprinting and Source Identification*; Wang, Z.; Stout, S.A., Eds.; Academic Press: Boston, **2007**; pp. 327-348.
[http://dx.doi.org/10.1016/B978-012369523-9.50014-8]

[57]   Wang, Z.; Yang, C.; Hollebone, B.; Fingas, M. Forensic fingerprinting of diamondoids for correlation and differentiation of spilled oil and petroleum products. *Environ. Sci. Technol.,* **2006**, *40*(18), 5636-5646.
[http://dx.doi.org/10.1021/es060675n] [PMID: 17007120]

[58]   Lemkau, K.L.; Peacock, E.E.; Nelson, R.K.; Ventura, G.T.; Kovecses, J.L.; Reddy, C.M. The M/V *Cosco Busan* spill: source identification and short-term fate. *Mar. Pollut. Bull.,* **2010**, *60*(11), 2123-2129.
[http://dx.doi.org/10.1016/j.marpolbul.2010.09.001] [PMID: 20888014]

[59]   Radović, J.R.; Aeppli, C.; Nelson, R.K.; Jimenez, N.; Reddy, C.M.; Bayona, J.M.; Albaigés, J. Assessment of photochemical processes in marine oil spill fingerprinting. *Mar. Pollut. Bull.,* **2014**, *79*(1-2), 268-277.
[http://dx.doi.org/10.1016/j.marpolbul.2013.11.029] [PMID: 24355571]

[60]   King, S.M.; Leaf, P.A.; Olson, A.C.; Ray, P.Z.; Tarr, M.A. Photolytic and photocatalytic degradation of surface oil from the Deepwater Horizon spill. *Chemosphere,* **2014**, *95*, 415-422.
[http://dx.doi.org/10.1016/j.chemosphere.2013.09.060] [PMID: 24139429]

[61]   Corilo, Y.E.; Podgorski, D.C.; McKenna, A.M.; Lemkau, K.L.; Reddy, C.M.; Marshall, A.G.; Rodgers, R.P. Oil spill source identification by principal component analysis of electrospray ionization Fourier transform ion cyclotron resonance mass spectra. *Anal. Chem.,* **2013**, *85*(19), 9064-9069.
[http://dx.doi.org/10.1021/ac401604u] [PMID: 24033143]

[62]   Lemkau, K.L.; McKenna, A.M.; Podgorski, D.C.; Rodgers, R.P.; Reddy, C.M. Molecular evidence of heavy-oil weathering following the M/V *Cosco Busan* spill: insights from Fourier transform ion cyclotron resonance mass spectrometry. *Environ. Sci. Technol.,* **2014**, *48*(7), 3760-3767.
[http://dx.doi.org/10.1021/es403787u] [PMID: 24559181]

CHAPTER 4

Carcinogenicity of Petrogenic PAHs

Jérémie Le Goff[*]

ADn'tox SAS, Centre de Lutte contre le Cancer François Baclesse, Caen, France

Abstract: The carcinogenicity associated with PAH contamination in the aquatic environment has been a topic of prime importance since neoplasia was described in fish living in multi-sources contaminated areas, 50 years ago. Since then, a whole array of studies were conducted in order to better characterize the fate of PAHs, their bioavailability for biota and their effects in term of neoplastic lesions. Genotoxicity assessment is at the heart of the matter, as benzo[*a*]pyrene, the leader of PAHs, is described as ubiquitous genotoxic and carcinogenic compound, acting in particular by the occurrence of reactive diol metabolite that forms DNA adducts. The causal relationship between exposure to PAHs and occurrence of neoplasia in organisms involves to explore the biological plausibility of the association. Biomarkers of genotoxicity/mutagenicity are a central part of tools that measure the biological plausibility. DNA adducts and DNA strand breaks are lesions that can be advantageously used in sentinel organisms.

Keywords: DNA, DNA adducts, Genotoxicity, Neoplasia, Polycyclic aromatic hydrocarbons (PAHs).

INTRODUCTION

Many polycyclic aromatic hydrocarbons (PAHs) are known to be toxic for humans and other organisms, with well described acute and chronic effects [1]. In general, it is admitted that one-, two- or three-ring aromatic compounds have predominately acute effects on living organisms, while higher molecular weight PAHs are mainly responsible for chronical health impacts, among them carcinogenicity in relation (or not) with genotoxic/mutagenic potential [2]. The

[*] **Corresponding author Jérémie Le Goff**: ADn'tox, Centre François Baclesse, 14076 Caen cedex 5, France; Tel/Fax: +33 231 455218; E-mail: j.legoff@adntox.com

Daniela M. Pampanin and Magne O. Sydnes (Eds.)
All rights reserved-© 2017 Bentham Science Publishers

exploration of a structure-activity relationship suggests that carcinogenicity in PAH family is promoted by a structural arrangement in a so-called fjord (or bay) region that requires molecules of at least four aromatic rings [1]. This distinction is far from being absolute, especially because interactions between PAHs, and PAHs and other pollutants, are of wide variety and are contributing to highly variable qualitative and quantitative global toxic effects [3]. Dioxin- like responses in interaction with the aryl hydrocarbon receptor (AhR) are also observed in different biological models exposed to PAH mixtures or isolated compounds. Results are still inconsistent and mechanisms are still relatively unknown [4, 5]. Many PAHs and derivatives, notably alkylated compounds that are predominant among PAHs in crude oils, are supposed to be endocrine disrupters and, at least partly, exert carcinogenic effects through non genotoxic mechanisms including tumor promotion [6]. Retene (1-methyl-7-isopropyl phenanthrene), an alkyl PAH used in chemical fingerprinting of crude oils, has for example been responsible for the embryotoxicity and other dioxin like adverse effects in marine medaka, probably by a toxic mechanism different from TCDD and its parent phenanthrene [7].

The determination of the mutagenic potential associated with single chemical or mixtures can be considered as a crucial endpoint in highlighting the causal relation between exposure and cancer development (even though numerous carcinogens are not mutagens). In this way, some work realized on *in vitro* and *in vivo* models of toxicology reflect the mutagenic/carcinogenic potential of certain petrogenic PAHs. Seven complex mixtures of petroleum derived PAHs and numerous isolated PAHs (among them phenanthrene, anthracene, methylanthrancène, dimethylbenzo[*c*]phenanthrene) have proven to be mutagen by the *in vitro* reverse mutation bacterial test (Ames test), in presence of S9 fraction (used to mimic the mammalian metabolic conditions) [8]. In a similar assay, the mutagenic potential of benz[*a*]anthracene, benzo[*a*]pyrene (BaP) and 25 methylated metabolites was directly associated to the chemical formula and more precisely to the positioning of methyl groups around the aromatic structure [9]. An activation of the cellular response to DNA damage including the phosphorylation of the P53 protein is observed when rat liver stem-like cells are exposed to methylated BaPs [10]. DNA adducts, as marker of genotoxicity, have been observed too. Non genotoxic mechanisms of carcinogenicity are potentially associated to 1-MeBaP and 3-MeBaP as agonists for AhR. Interestingly, mutagenic potential of PAHs in mixture can be different than the sum of the effect caused by each agent alone. By measuring mutations of the hprt gene in Hepa-1c1c7 cells (a mouse hepatoma cell line), Huang *et al.* showed that fluoranthene is able to reduce significantly the mutagenic impact of BaP when cells are co-exposed [11]. This effect is not dependent of the AhR nor the CYP1A1. This *in vitro* study highlights once again the complexity of global effects of pollutant mixtures.

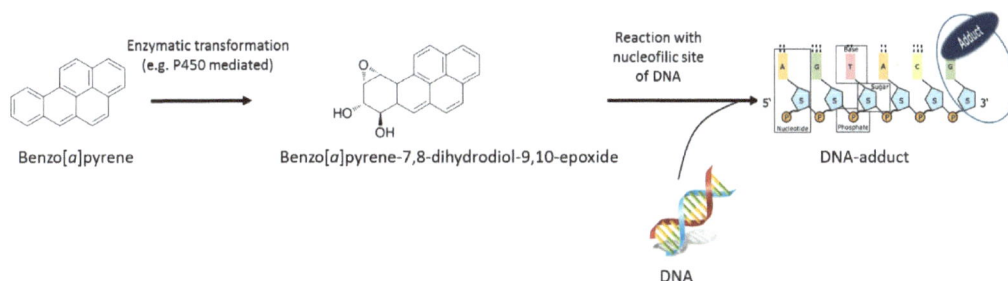

Fig. (1). Example of a PAH metabolisation that generates an electrophilic form that can directly react with a nucleophilic site of DNA.

On *in vivo* experimental models, some well characterized PAHs proved to be carcinogens, mutagens and teratogens. Some of them are responsible for respiratory and cardiovascular diseases too. On human, epidemiologic studies tend to corroborate these experimental data. BaP is considered the leader of carcinogenic PAHs. It is also the most studied to date, being used as a PAH model compound, notably as regards the metabolism pathways. In numerous toxicologic model organisms, PAHs are not direct carcinogens, and necessitate the enzymatic biotransformation by cytochrome P450 monooxygenases in order to form *in situ* highly reactive metabolites. Phase I enzymes, such as cytochrome P450s, catalyse the monooxygenation of PAHs leading to phenols and epoxides. As ever mentioned, the best documented example is BaP, which metabolisation generates the well-known benzo[*a*]pyrene-r-7,t-8-dihydrodiol-t-9,10-epoxide (BPDE), an electrophilic diol epoxide that can directly react with nucleophilic sites of DNA, RNA, lipids and proteins (Fig. **1**). These interactions with DNA and the formation of adducts are typically presented as a mechanism of initiation in the complex process of the chemical carcinogenesis [12].

It is to note now that concerning chronic toxicity of PAHs including carcinogenicity, lots of data are associated to parent compounds. Alkylated (methylated) PAHs and hetero PAHs are less studied at the time, even though differences in structure seem to be associated to differences in bioactivation ways and therefore toxicity [13, 14].

In view of the accumulation of experimental and epidemiologic scientific data, IARC classified numerous PAHs according to their carcinogenicity for human, recently in class 1 (*i.e.* carcinogenic compound) for BaP (in 2012), class 2A (probably carcinogenic to humans) for dibenz[*a,h*]anthracene and dibenzo [*a,l*] pyrene, or class 2B (possibly carcinogenic to humans) for benz[*a*]anthracene, benzo[*b*]fluoranthene, benzo[*j*]fluoranthene, benzo[*k*]fluoranthene, benzo[*c*] ph-enanthrene, chrysene, dibenzo [*a,h*]pyrene, dibenzo[*a,i*]pyrene, indeno [1,2,3-*cd*] pyrene. More frequent cancers in humans are localized in lung and bladder, as indicated by a literature review focused on occupational exposure to PAH, with a dose dependent relationship between exposure measurement and cancer risk factor [15].

As part of environmental concerns and considering their toxic potential and human exposure rates, 16 PAHs are a part of the "priority pollutants" listing of the US Environmental Protecting Agency (US EPA) and European Commission (Regulation EC No 166/2006) [16].

Crude oil is a complex mixture of organic and inorganic compounds, among them hydrocarbons: aliphatics, alicyclics and PAHs. The carcinogenicity of the product has been evaluated on different organisms, especially in the context of environmental impact studies [17]. This complex mixture is mutagenic *in vitro* in bacteria. *In vivo*, crude oil induces chromosomal aberrations in mice and is carcinogenic to mice *via* dermal route. Numerous petrogenic PAHs has been studied in an isolated manner (see later).

The problematic of cancer occurrence in aquatic vertebrate organisms begins with studies in fish. Epizootics of neoplasia in wild fish is described since the end of

1800's, beginning of 1900's. Numerous studies has been conducted, especially in regions with large pollutant contaminations, among them PAHs, as the great Lakes (freshwater system) or Puget sound (salt water system) [18]. Nevertheless, there has always been difficulties to relate exposure to a pollutant or pollutants family with long-term effects (chronic toxicity) on the organisms living in the aquatic environment.

Numerous tumors in fish are known to be related first to infection by viruses. The main viruses that affect fish in association with tumor development are from Herpesviridae, Papillomaviridae, and Retrovirida [19]. Seasonality of the infections and diseases are commonly reported, probably in association with cyclic variation of certain environmental factors as the water temperature. Different techniques have been developed to reveal the viruses and their implication in neoplasic diseases, from transmission electron microscopy for detection of virus like particles to molecular approaches by measuring the reverse transcriptase. To illustrate, it has been observed that epidermal papillomas in White sucker may reach high prevalence (above 50%) in some polluted locations, suggesting first pollutant related effects. However, this hypothesis was not confirmed in laboratory studies, showing how difficult the etiological diagnosis is.

Apart from fish, carcinogenicity studies on PAHs are less abundant on other aquatic organisms, especially on mammals. These scientific works are difficult to undertake given the need for abundant data for a statistical point of view, hardly compatible with field studies in these animals. Various types of cancer affect many species living in marine environments (see [20] for a review). And one more time, many etiological factors are discussed: virus infection, genetic and immune system suppression chemical carcinogens (*e.g.* PCB, DDT, PAHs). Specifically, a study of beluga in the estuary of the Saint Lawrence was conducted to suspect the exposure to PAHs as a major cause of the observed cancers [21]. Neoplastic lesions are also described in some invertebrates including bivalves. The study of these sentinel organisms is presenting many advantages in the context of health monitoring in the marine environment (easy to collect, relative low mobility limiting multi-site interactions of pollutant effects, high exposure to sediment contaminants) and some disadvantages too (significance of the result, extrapolation to other organisms including humans).

Isolated PAHs are never encountered in the aquatic environment. Environmental exposures are always associated to complex mixtures of PAHs and derivatives such as alkylated PAHs or the more polar oxygenated PAHs, for which specific toxicity and exact mode of action are still unclear at the moment [22], and nitro-PAHs. According to their origin, the chemical structure of PAHs and their ratio may vary considerably. This variability leads to variability in the toxicity of complex mixtures of PAHs, which poses problem in the way of the fine risk assessment [3]. In particular, the carcinogenicity of PAHs can vary to a very large extend, despite their relative structural similarities. In order to assess risks arising from exposure to PAH mixtures, two approaches are commonly used: measurement of BaP as a reference marker for PAH exposure and component-based approach associated to the determination of individual toxic equivalency factors (TEF) in reference to BaP (TEF = 1).

As mentioned, the biological impact of more polar PAH derivatives (alkyl-PAH, nitro-PAH and oxy-PAH) is probably facilitated by their higher water solubility (*i.e.* higher bioavailability for living organisms) compared to parent compounds and generally present in higher concentrations in crude oil and in the environment. Bioavailability is an essential parameter for better assessment of pollutant toxicity and ecological risks [23]. New methods are needed for a better evaluation of the bioavailable fraction of pollutants in sediments and aqueous media. Recent works on these derivatives call into question the relevance of the EPA priority PAH list that is limited to unsubstituted compounds [24].

Thanks to a better understanding of the toxic mechanisms of action related to PAHs, diverse biomarkers have been developed and are applicable to humans and other organisms [25]. Many of these markers are used in the marine world. They are now contributing to ameliorate the understanding of the relation between exposure to pollutants like PAHs and tumor development in aquatic organisms, as for example fish [26]. They also provide an important added value in the monitoring of bioremediation [27].

CARCINOGENICITY OF PETROGENIC PAHS IN THE MARINE ENVIRONMENT: FOCUS ON FISH AS SENTINEL ORGANISMS

1. Neoplasia in Fish Associated to Polluted Areas

In the assessment of ecotoxicological problems that are associated to contaminated sites, one of the most worrying matters are the possible long-term chronic effects of pollutants on aquatic ecosystems [28]. Carcinogenicity is one of preoccupant chronic toxicity for organisms.

Historically, epizootics of neoplasia in wild fish is described since the end of the 19th century. These diseases were first linked to environmental contaminants in the '60s. In 1964, Dawe *et al.* observed hepatic neoplasms in 3 among 12 white suckers and in one brown bullhead sampled in the Deep Creek Lake, Maryland USA [29]. The implication of some carcinogenic hydrocarbons associated to motorboat exhausts was suspected, among other etiologic factors, from chemical origins (rotenone, pesticide, DDT) or not (parasitic infestation by unidentified protozoan parasite for example). Exposure of fish to carcinogenic hydrocarbons through contaminated sediments at the bottom of the lack was also evocated, even though at that time, no measurement in sediment has been realized.

In the '70s, a study focused on the potential environmental carcinogens as part of the etiology of papillomas, epidermal and/or oral tumors caused by papillomaviruses in white suckers of the Great Lakes [30]. It was notified that the frequency of papilloma in this species increases in a large extend in the vicinity of urbanized areas, which led to suspect anthropogenic pollutants in the disease development.

Over the same decade, tumors were observed in 4.38% of 2121 fish from the Fox River watershed (Wisconsin, USA), an interconnected series of lakes, on the banks of which cities, villages and farm lands have developed over the last 100 years [31]. By comparison, the incidence of tumors was only 1.03% of the 4639 fish caught in the Lake of the Woods (Ontario, Canada), the chosen control site. In detail, interspecies variation are observed, with 12% of the brown bullhead affected by predominant hepatic tumors. For all species together, tumors of liver represent 9% of observed neoplasm in the control site to reach 14% at the Fox

River watershed. For both sites, the environmental characteristics concerning oxygen content of water, temperature and nutritional status were similar. The main inter-site variation was associated to chemical contaminants (in absence of virological examination). Moreover, seasonal effect was noted, with higher prevalence of tumors from May to October, in relation with the seasonal bacterial fluctuation. The authors concluded that if pollutions (including PAHs, and among them benzanthracene (oil by-product), naphthalene, crude oil, gasoline) were not proved to be *per se* the cause of neoplasia, the presence of these pollutants probably enhances the prevalence of such fish diseases.

Thereafter, studies on fish cancers in fresh water and marine environment and the etiologies have multiplied all over the world. These studies have shown that most of neoplasms in fish associated to chemical pollutants are mainly of epithelial origin (with liver and skin tumors first of all). Numerous studies have demonstrated elevated tumor prevalence in a variety of species across North America [32], including brown bullhead and white sucker populations from a wide range of urbanized areas in bays and tributaries of the Great Lakes in both Canadian and United States waters [33]. The studied species in the region has progressively grown to include fish species like bowfin, freshwater drum, sauger, mummichog, English sole [32]. A summary of the studies revealed that epizootics of neoplasm with more than 6% prevalence in fish mainly affect skin and liver, and are most often suspected to be associated to PAHs. Most often species that exhibit tumors are benthic feeders.

Field studies are quickly confronted with the difficulty of linking exposure to a pollutant (or a class of pollutants) with long-term effects (chronic toxicity) on the organisms living in the marine environment, in particular because of the variety of pollutants sources in an unclosed environment that is open to a variety of disturbing factors (*e.g.* various abiotic stressors, other pollutants). The mobility of organisms, making it even more difficult to assess accurate exposures to pollutants, this is also an element that must be taken into account in such an approach.

In the exploration for in field relationship between the detection of pollutants in the environment and an observed tumor development in aquatic organisms like fish, many confounding factors have been described: other causal agents of cancer (especially some oncogenic viruses), inter-individual and inter-species variations (from genetic and/or environmental origins), age and sex of targeted organisms, bias in the effective level of exposure (*e.g.* what is the bioavailability of the pollutant to the studied organisms? what is the fate of pollutant in the studied environment?).

To illustrate, no less than 30 proliferative diseases are associated to viral contamination in fish [34]. Concerning cancer, several family of viruses have been involved in tumor development in fish. Such is the case, for instance, of the Herpesviridae, the Papillomaviridae or the Retroviridae. Thus, numerous papilloma and sarcoma observed in fish were associated to retrovirus, as in the case of freshwater fish walleye Atlantic salmon (see [34] for a review).

2. The Relationship between PAHs and Observed Tumor Development in Fish

By the early '80s, PAHs were more precisely suspected to cause cancer in aquatic organisms because of the improvement of knowledge related to the toxic effects of isolated compounds, including the leader BaP. Carcinogenic effects are consistently demonstrated in animal models, especially in rodents, simultaneously with the description of fundamental elements of chemical carcinogenesis as the production of pro-carcinogenic metabolites and associated DNA damages [35 - 37].

These findings are supported by the development of laboratory studies on fish carcinogenesis models. Rainbow trout proved to be a very good model of carcinogenesis. BaP was a hepatic carcinogen in this latter fish model [38]. In laboratory, tests of carcinogenicity have been conducted for example on Medaka and Guppy [39]. It was demonstrated that both BaP and DMBA (7,12-dimethylbenzo[*a*]anthracene) are carcinogenic for both fish species by inducing hepatocellular neoplasm, predominantly for DMBA in Medaka, at low concentrations (comparable to environmental levels) and short but repetitive exposures. The comparison of metabolites profiles in rodents and fish is performed on certain PAHs including BaP, which show some similarities in the metabolic pathways of these environmental toxicants, in particular the formation of mutagenic metabolites [40].

In vivo, methylated BaP metabolites showed the capacity of inducing skin tumors in the traditional two-stage skin carcinogenesis model on mice (initiation is accomplished by the application of a sub-carcinogenic dose of a carcinogen, promoted by the repeated application of a tumor promoting agent such as the phorbol ester, 12-O-tetradecanoylphorbol-13-acetate (TPA) [41].

Another key point in the attempt to demonstrate the carcinogenicity of PAHs is the laboratory experiment. Many laboratory studies confirm the relationship between exposure to PAHs and the development of cancer in aquatic environment, both from isolated pollutants (*e.g.* BaP) or contaminated sediments. These works are conducted on the basis of experimental results in animal models of carcinogenicity, applied to specific characteristics of aquatic species and environments.

It was shown for example that the continuous dietary exposure of rainbow trout to 1000 ppm BaP caused 15% of incidence of hepatocarcinomas after 12 months of exposure and reached 25% of incidence after 18 months, while no tumors was observed in the control group [38]. More generally, rainbow trout has revealed interesting insight for studying chemical carcinogenesis. A major focus of research effort has been to propose also the English sole as a laboratory model for investigating the biochemical mechanisms of chemical carcinogenesis in fish, with an emphasis on understanding the processes involved in the activation and detoxication of carcinogenic PAHs [42]. Numerous other fish models have been proposed thereafter.

In large scale mesocosms, the exposure over 3 years of flounder to contaminated sediments that contained PAHs, originated from a dredge spoil of Rotterdam harbor (the Netherlands), induce liver neoplasia and other preneoplastic lesions as observed by histopathological analysis. At the time, this was the first study that demonstrated liver neoplasia in fish under simulated field conditions, at contaminant levels near the ones found in the environment [43].

More recently, other laboratory studies show new results on PAH carcinogenicity in the aquatic environment, trying to assess more precisely the modes of toxic action, taking into account the mixture effects and the origin of contamination. In order to explore the effects of PAH mixtures in environmental conditions, zebra fish were chronically exposed through the diet to different PAH fractions from different origins: pyrolitic, petrogenic light oil and petrogenic heavy oil [44]. Preneoplastic and neoplastic disorders of the bile duct epithelium and germ cells were the most frequently observed abnormalities with the three PAH fractions. The frequency of these neoplasms was associated to PAH concentration and exposure duration. Neoplasms were more frequent with petrogenic heavy oil. Surprisingly, no genotoxicity was detected with use of both haemocytes micronucleus and erythrocytes comet assays, leading the author to conclude the carcinogenicity of PAHs without genotoxic mechanisms in the model. To explain such results, PAHs might chronically increase the duct cell proliferation rate leading to indirect neoplastic development (illustrating an epigenetic mechanism of carcinogenesis).

In marine area, the continuation of field, mesocosm and laboratory studies has improved the understanding of the relationship between exposure to PAHs and the occurrence of tumors in numerous organisms including fish species.

A large field study in Puget Sound (pacific coast of USA) was conducted during the '80s with the objective to determine urban associated disease in fish [45, 46]. The main results in the flatfish English sole caught at 70 stations of the area between 1979 and 1984 revealed a correlation between the concentration of PAHs in sediment and the prevalence of neoplasm. Interestingly, at 11 stations, the bile levels of BaP metabolites were positively correlated to the prevalence of hepatic lesions such as neoplasms, which seems to underlie the major role of these particular PAHs in carcinogenicity. The observed link between sediment and tissue PAH concentrations and hepatic diseases emerged in favor of a cause to effect relationship. However, the coexistence of other pollutants, identified or not, and the probable cocktail effects of PAHs in mixtures that results in specific interactions surely complicates the task of establishing the aforesaid causal relationship.

Marine remediation and bioremediation using for example living cells or biological systems are promising responses to environment degradation by pollutants like PAHs [47]. An important point in the search for a causal relationship between exposure to PAHs and the incidence of certain cancers is the analysis of remediation and its environmental health effects. Thus, after a rapid decline of polluting activities in the vicinity of the Black River in Lorain County (Ohio, USA) between 1980 and 1987, the level of PAHs in both sediments and brown bullhead catfish tissue decreased rather significantly [48]. For total PAHs, the reduction was of a factor of 250 in sediment and of a factor of 15 in whole-body of 3 years old catfish. At the same time, liver cancer frequency declined to about one quarter the 1982 value.

At the beginning of the '90s, the highly polluted Eagle Harbor located in Puget Sound was cleaned up by capping of PAH contaminated sediments [27]. A second cap was added in 2002. The bioremediation was associated with the drastic reduction of PAH exposure for English sole captured in the area, as revealed by the use of the exposure biomarkers fluorescent aromatic compounds in bile and hepatic DNA adducts. Hepatic lesions, including neoplasm, were strongly limited too. The prevalence for any toxicopathic lesion in sole liver was reduced from 83% in 1983-1986 to less than 10% in the 2000s.

These two examples are in line with a probable cause-effect relationship between PAH exposure and cancer incidence in sentinel fish.

The essential concepts of bioavailability of PAHs, their metabolic processes in exposed organisms, the interspecies difference in susceptibility to toxic effects and the numerous confusing factors were more precisely described, especially in relation with the development of biomarkers of genotoxicity such as mutagenic PAHs and metabolites in fish bile [49], DNA adducts [42, 50] or cytogenetic damages like micronuclei [51]. These new approaches are designed to evaluate the biological plausibility in the relation between exposure to PAHs and cancer.

One last important point in the assessment of the relationship between PAH exposure and cancer development in aquatic organisms is the study of the biological plausibility of the suspected association, especially through the development of biomarkers. This topic is taken up in the next paragraph.

CARCINOGENICITY OF PAHS IN AQUATIC ORGANISMS OTHER THAN FISH

Apart from fish species, the capacity of PAHs to induce neoplasia have been evaluated in several other aquatic organisms, vertebrates and invertebrates.

In bivalve mollusks, two predominant types of neoplasms are generally described: disseminated neoplasia (characterized by abnormal circulating cells, also called hemic neoplasia) and gonadal neoplasia (involving the proliferation of undifferentiated germ cells). The origin of these pathologies is still unclear, and generally attributed to viruses, biotoxins, stressors or environmental contaminants, and/or their combinations [52].

Typically, invertebrates including bivalves are considered less capable than fish to metabolise, bioactivate and eliminate hydrophobic PAHs. This is due at least in part to microsomal enzyme systems, and cytochromes P450 proteins, that seem to exhibit a lower activity in invertebrates, resulting in different balances between metabolic pathways. PAHs and other pollutants that necessitate enzymatic transformations in order to increase significantly their elimination by increasing hydrophylicity tend to bioaccumulate in these organisms. The impact in terms of preneoplastic and/or neoplastic lesions is still partially unknown [53].

More recently, the occurrence of neoplasia in the Baltic clam (*Macoma balthica*) from the gulf of Gdansk (Poland) has been highlighted by histology and cytogenetic responses [54]. The prevalence of neoplastic disorders was around 30% of individuals between 1995 and 2002 with an influence of the sampling site in the Golf. The greatest proportions of affected clams were observed in a polluted site and during warm months, leading to the assumption that pollutants, among them cyclic hydrocarbons, could have at least an indirect involvement in neoplasia. The suspected seasonal effect was supposed to be attributed to the cyclic variation in metabolic activity of bivalves and the consequence of more elevated water temperature in summer that seems to stimulate neoplasia and mortality processes. The disease have a negative effect on bivalve population in the Golf by decreasing the reproduction capacities, inducing potential genetic alterations and causing elevated mortality (up to 80%). Finally, this epizootic disease is supposed to constitute a real hazard for all benthic populations in the region.

The causal relationship between exposure to pollutants such as PAHs and the incidence of cancer diseases in bivalves is particularly difficult to establish. As previously stated, the hemic neoplasia is a blood cell disorder neoplasia resembling leukemia that affects bivalve mollusks. The pathology is endemic to common mussel *Mytilus edulis* in Puget Sound, Washington, USA, with observed prevalence ranged from 0 to 30%, depending on the sampling site [55]. The involvement of exposure to pollutants has been evocated to explain such inter-site variations, and no relationship between body burden of the main environmental contaminants (including PAHs) and prevalence of neoplasia was found. To go further in this etiologic hypothesis, mussels were fed microencapsulated PAHs or PCBs in both short-term (30 days) and long-term (180 days) exposure protocols. While an increase of hemic neoplasia is observed in exposed mussel groups, the difference of prevalence with control group was not significant.

In other cases, the relationship between pollutants and neoplasms in bivalves seems more obvious. For example, neoplastic disorders in eastern oyster (*Crassostrea virginica*) were notified in association with laboratory and field exposure to chemically contaminated sediment from the Black Rock Harbor (BRH), Bridgeport, Connecticut, USA [56]. In a laboratory experiment, neoplasms were observed after 30 and 60 days of continuous exposure of suspended sediment (20 mg/L). Overall, 13.6% of BRH exposed oyster showed neoplasm, with higher occurrence of tumor in renal excretory epithelium, followed by gill, gonad and gastrointestinal tissue. Chemical analyses of BRH sediment revealed the presence of numerous PAHs among various classes of pollutants, some of which were already characterized as carcinogenic in experimental studies (*e.g.* benzo[*a*]anthracene, BaP, benzofluoranthene). Interestingly in the same study, winter flounders fed contaminated blue mussels exhibit in their tissues the same pollutants than analyzed in bivalves. Neoplasms from renal and pancreatic origins and hepatic preneoplastic lesions were also found in fish. These last results illustrate a trophic transfer of sediment-bound carcinogens and their toxic effects through the marine food chain. Complex mechanisms of carcinogenicity related to exposure to pollutants like PAHs are gradually explored in marine invertebrates. Such work may improve the suspected causal relationship between exposure to certain pollutants and disease by investigating its "biological plausibility" in these invertebrates. As an example, tissue-specific modulation of the expression of the p53 tumor suppressor gene and the proto-oncogene ras in response to BaP has been reported in blue mussel (*Mytilus edulis*) [57]. DNA strand breaks and tissue damage were observed too. According to the authors, the biological responses and their interpretation in this model organism could be applied advantageously to gain an understanding of carcinogenicity in more evolved species including humans.

At the other end of the trophic chain, the issue of PAHs carcinogenicity to marine mammals is a very specific topic. More generally, the studies that are carried out on marine mammal neoplasia are exceptionally complicated to achieve in the wild, particularly given the difficulty of sampling. These organisms are complexes, have most often a long life span and surely constitute good sentinels for the marine environment health, especially in regard to the exposure to pollutants [58]. Despite the difficulties, research for neoplasia in marine mammals have been realized on pinnipeds (sea lions, seals, walrus), cetaceans (beluga whales, dolphins, manatees), sea otter, or polar bears. The tumors observed in these wild species are generally similar to those seen in more common domestic species (see [20], for a review). Published data from over 40 years of researcher of varying quality, as mentioned in a recent review compiling 426 publications, with a focus on the prevalence of urogenital carcinoma in California sea lions. It was concluded that disease trends in marine mammals is particularly uneasy to evaluate correctly, first because of the absence of centralized data reporting organization [59]. In the same period, reported data from The Marine Mammal Center (TMMC, California, USA) indicate an increasing rate of cases of this particular cancer. To date, little is known about the etiologies of neoplasia in marine mammals. Chemical carcinogens, that are susceptible to be present in the environment of these animals, are only one among a variety of possible causes. Genetic immune system suppression or oncogenic virus infections are for example other etiologic factors.

The intervention of PAHs in mammal cancer development is regularly evocated among the overall suspected environmental carcinogens. This hypothesis is supported in part by the results that are emerging from molecular epidemiology approach based on the analyses of specific biomarkers of genotoxicity (see later in the chapter), and also through the fact that PAHs in environment can reach concentrations near those associated to carcinogenesis in mammals in laboratory experiments.

Among 129 beluga carcasses collected in the estuary of Saint Lawrence river (Canada) in the period 1983-1999, cancer has been designated as the primary cause of death for 18% of cetaceans, the second source of mortality after respiratory and gastrointestinal infections, and the first cause in adults [21]. 30% of neoplasms were adenocarcinomas observed in the intestine close to the stomach. It is to note that cancers of the proximate intestine are generally rare in mammals except in certain bovine an ovine livestock, very probably in relation with the chronic exposure to carcinogenic herbicides, especially the 2,4-dichlorophenoxyacetic acid (2,4-D). In this particular case, as for belugas, the combination of exposure of animals to carcinogenic compounds and infection with certain viruses seems to widely promote cancer development. Many of the collected belugas had gastric papillomatosis, a disease caused by papillomaviruses. Throughout their lifetime, belugas are in direct contact with sediments they dig for their food, looking for invertebrates that have the potential to bioaccumulate pollutants like PAHs. Saint Lawrence estuary is indeed contaminated with numerous pollutants, particularly PAHs released mainly from the aluminum industry. Thus, it was hypothesize that high prevalence of intestine cancers in beluga would be due to PAHs exposure, given that intestinal adenocarcinomas were similarly observed after mouse exposure to coal tar (a PAH rich product) in the animal 2 years assay of carcinogenicity. Interestingly, it was proposed that high levels of the CYP1A and CYP2B enzymes observed in beluga, and probably associated to PCB exposure [60], enhance the carcinogenic potential of PAHs by increasing their enzymatic bioactivation and the generation of reactive metabolites. If that was the case, this example perfectly illustrates the synergistic action of pollutants that are most often present in the environment as complex mixtures. To finish, human populations that live in the area of the Saint Lawrence estuary have elevated number of incidences of cancers associated with PAH exposure compared to the rest of the population in Quebec. This is clearly another argument in favor of incriminating environmental pollutants in cancer incidence in that region of Quebec, illustrating the close connection between human and ecosystem health at the same time.

MODE OF TOXIC ACTION ASSOCIATED TO PAHS: TOWARDS THE "BIOLOGICAL PLAUSIBILITY" OF THE ASSOCIATION BETWEEN EXPOSURE AND NEOPLASIA

Carcinogenesis involves complex biological mechanisms. At the cellular level, carcinogenesis is a multistep process that allows a cell to escape the control of its environment [61 - 63].

Despite the accumulation of scientific data, these mechanisms remain largely unknown. To date, at least five models of carcinogenesis have been described, such as for example the well described «mutational» model, that focus on a lot of chemical carcinogens that can modify the DNA structure, or the more recent «non genotoxic» model, which attempts to explain the mode of action of some chemicals that are modulators of cancer development, without acting by direct DNA alterations (epigenetic carcinogens) [64].

It has been said that the causal relationship between exposure to a substance (a fortiori to a chemical mixture) and the occurrence of cancer is very difficult to definitively confirm. Researchers have to check a beam of nine main arguments (called the Bradford Hill criteria, 1965), which requires the acquisition of matching data on the biological plausibility of the association, in the context of a risk assessment for humans but also for the environment [65]. The approach is combining different scientific disciplines (from epidemiology to molecular biology) and evolves continuously to integrate the emerging research tools [66].

The biological plausibility goes through the opening of the "black box", the overall events that occur between the exposure to a carcinogen and the final effect observed in the form of tumor. This is the area of the relatively recent molecular epidemiology science (see [67], for a review on human biomonitoring). This discipline is now widely used to study the chemicals and environmental carcinogenesis [12, 68]. Its application in sentinel organisms is based on the fact that biological changes associated with the process of chemical carcinogenesis can be used as indicators of exposure or risk to carcinogens. So, when they are measured in the context of potential toxic exposure, theses biological changes are called biomarkers. Their interest for the marine pollution monitoring is well recognized [69]. Numerous biomarkers have been applied in numerous marine sentinel species. For example, mutations in critical genes involved in cell cycle regulation or in the management of lesions of the cell are considered important events in the process of carcinogenesis. Many recognized carcinogens are also mutagenic and genotoxic, *i.e.* able to alter DNA commonly very rapidly following exposure.

The presence of mutagens in surface waters has also become a topic of major concern [70]. Among 128 scientific studies published between 1990 and 2004, the breakdown of genotoxicity/mutagenicity assays, that has been applied to surface water biomonitoring, reveals two major technical approaches: the bacterial assays (including the Ames test, that concerns 37% of overall studies) and aquatic organism assays (like micronucleus, DNA adducts or comet assay). Heavy metals, PAHs, heterocyclic amines or pesticides are the toxins that are most frequently cited as mutagenic contaminants.

1. *In vitro* Genotoxic/Mutagenic Potential of PAHs in Relation to the Aquatic Environment

Toward the assessment of the PAH carcinogenicity in aquatic environments, either intrinsically or when they are adsorbed in the sediments, many mutagenesis tests were conducted, most often on the base on traditional tools such as the bacterial reverse mutation assay on different *Salmonella typhimurium* strains (Ames test [71],). The first studies on isolated compounds date from the early '80s. They permit to demonstrate the importance of the structural bay region in the mutagenicity of certain PAHs, and the role of biotransformation enzymes [72, 73]. In another study, the genotoxicity/mutagenicity of 61 unsubstituted and nitrated polycyclic aromatic hydrocarbons has been evaluated using the *Escherichia coli* PQ37 SOS-Chromotest [74]. benzo[*ghi*]fluoranthene, benzo[*j*] fluoranthene, benzo[*c*]phenanthrene, benzo[*a*]pyrene, chrysene, dibenzo[*a,1*] pyrene, fluoranthene and triphenylene exhibit high genotoxic potentials in this bacterial system.

In the light of environmental issues associated to PAHs derived from petroleum products (petrogenic PAH), mutagenicity tests were also more specifically performed on some major compounds in these particular mixtures.

An optimized Salmonella mutagenicity assay (Ames test) was applied to seven petroleum-derived mixtures and 29 individual PAHs [75]. Mutagenicity was observed in a vast majority of PAHs (23 compounds among 29) and in the overall 7 mixtures.

The methyl substituted PAHs, some derived PAH compounds especially abundant in petrogenic PAH mixtures, revealed a potential of mutagenicity in relation to the chemical structure and the methylation patterns [13]. The mutagenicity of 6-methylbenzo[*a*]pyrene was for example twice that of the parent BaP, while 7-, 8-, 9- and 10-methylbenzo[*a*]pyrene were less mutagenic. These differences are associated to the metabolism of the compounds and the formation of reactive metabolites, biological processes that can differ according to the chemical structure of the initial compound. Later, metabolic activation pathways of methyl PAH have been proposed, including the formation of aralkyl DNA adducts [14]. In a study conducted on benzo[*a*]anthracene, BaP and 25 methylated derivates, it was notified that the observed mutagenicity was not in accordance with the *in vivo* carcinogenic potential of the compounds, probably for methodological reasons [9].

Gradually, the genotoxicity/mutagenicity of PAHs in complex mixtures and the synergistic effect is also evaluated. By the use of the Ames assay, Hermann demonstrated that some non-mutagenic 2 and 3 rings hydrocarbons enhanced the mutagenicity of BaP, while most mutagenic PAH strongly reduce it [76]. The later effect is dependent on the dose of PAHs, as an activating effect of BaP mutagenicity is observed for low concentrations of PAHs. The observed effect is supposed to be associated to PAH interactions at the enzymatic biotransformation level. More recently, mixtures containing up to 16 PAHs were tested in the SOS chromotest [77]. Interestingly, the reconstituted mixtures were consistent with PAHs measured in mussel tissues from the Saint Lawrence estuary. At the lowest doses, the observed genotoxicity of the PAH mixtures was in accordance with additivity of individual effects. It may be reasonable to think that the mutagenic risk posed by a mixture of PAHs can be estimated as the sum of the mutagenic effects of each component.

It is worth noting that the mutagenicity by the Ames test of environmental and complex mixtures of PAHs has been largely examined in relation to air pollution, particularly in the context of the risk assessment for human health. Numerous air samples were tested for mutagenicity, as for example in Taichung, Taiwan [78]. It has been showed that the observed mutagenicity in air samples on the salmonella TA98 strain was correlated with the amount of 10 measured PAH, and more especially with benzo[*g,h,i*]perylene, an hydrocarbon derived almost exclusively from both diesel and gasoline motor exhausts. This result leads to criminalize automobile pollution as the dominant factor in the mutagenicity of air samples. However, it can be noted that PAHs are only contributors among other pollutants to the mutagenicity of the urban air samples [79]. In order to evaluate the PAH mixture effects, an interesting approach has consisted to measure the mutagenicity of organic extracts of diesel exhaust particles following the artificial addition of individual PAHs. Once again, it was notified that the four PAHs used in the assay acted additively in the mutagenic response of the Ames test [80].

The interaction between PAHs in term of mutagenicity has been also evaluated in models of eukaryotic cells. In such models, it was shown for example that fluoranthene, a minor genotoxic PAH compared to BaP, is capable of enhancing p53 expression and decreases the mutagenesis induced by BaP in mouse hepatoma cell lines [11]. This interaction between both PAHs seems to be compound and cell specific. The mechanism involved is still unclear. This topic appears essential given the existence of PAHs in the environment as complex mixtures.

In order to better understand the behavior of pollutants in aquatic systems, the mutagenicity/genotoxicity of PAHs has been tested on aquatic specific *in vitro* models, like the micronucleus assay on the fish liver cell line RTL-W1. As an example, this assay reveals genotoxicity by inducing micronuclei of six heterocyclic PAHs for the first time in a vertebrate system [81]. Dibenzofurane was positive in that assay while it was tested negative in previous mammalian models. Authors concluded that genetic toxicity of heterocyclic PAHs is probably underestimate and most of them should be reconsidered as being of priority interest. It is to note that RTL-W1 cells were recently used to evaluate the genotoxicity potential and DNA repair capacity impairment of coal tar from pavement, a mixture rich in PAHs, on the basis of modified comet assays, with positive results too [82].

These results of PAH *in vitro* mutagenicity are consistent with the genotoxicity/mutagenicity of the sediments contaminated with these pollutants (see [83], for a review). The same *in vitro* assays were applied on sediments, in particular the Ames test. In 2004, 41.1% of mutagenicity data relative to PAH contaminated sediments was obtained with the Ames test. Overall, a positive relation is generally observed between the PAH concentrations in sediment and mutagenic response on the TA98 and TA100 strains, in association with frameshift mutation and base-pair substitution mode of toxic action. The SOS chromotest was for example used to evaluate the genotoxicity of sediments in the Seine estuary, Normandie, France [84]. A genotoxicity gradient in the organic extracts of sediments was observed with higher genotoxicity at the upper part of the Seine estuary. The genotoxic profile has enabled the incriminate PAHs among the various classes of pollutants detected in sediments. Finally, attention is drawn on the potential risks for living organisms that are exposed to such genotoxic pollutants.

In addition, specific fish or mammal cell lines were developed in the aim to better assess sediment genotoxicity/mutagenicity. Thus, the continuous cell line FG (derived from the gill tissues of flounder) was used in a variant of the comet assay to test the genotoxicity of three sediment extracts from Qingdao coastal areas (China). The response rates were in accordance with measured PAH concentrations [85].

If the mutagenicity assays on sediments permit to assess an overall toxic effect in an environmental matrix, they can also identify more precisely the compounds that are more toxicologically significant. Thus, the mutagenicity of sediments from different estuaries of the United Kingdom was evaluated by the mutatox test (dark mutant strain of *Photobacterium phosphoreum*) after the fractionation of samples [86]. Interestingly, no mutagenicity was detected in the pore water, which means the absence of hydrosoluble mutagens. The "Toxicity Identification Evaluation" (TIE) characterization on residual particulates that present mutagenicity leads to identify some mutagens among other alkyl substituted naphthalenes, two to five-ring PAHs, oxygenated-PAHs. Given the negative results in the pore water, the authors concluded that mutagens detected were probably weakly bioavailable, even though the more polar compounds identified in sediment were more likely to enter in the food chain than nonpolar ones.

The genotoxicity/mutagenicity on sediments has been applied in relation with the more specific oil pollutions. Two years after the oil spill of Hebei Spirit tanker near Taean, South Korea, 10 among 22 sediment samples revealed a genotoxic potential in the chicken DT40 mutant cells bioassay [87]. The strongest responses occur with mutant cells deficient in XPA, RAD54, and REV3 pathways, with the same pattern as that obtained in direct assay on crude oil spill. These similarities suggests the direct involvement of the oil spill in the genotoxicity observed two years later in the sediments, confirming the extended life of pollutants such PAHs in the matrix.

2. Carcinogenic Effects of PAHs in the Marine Environment: Use of Biomarkers of Genotoxicity/Mutagenicity in Sentinel Organisms

In 1987, the Committee of Biological markers of the National Research Council (NRC, Washington, USA) introduced the concept of biomarkers in connection with the little understood pathway which leads from toxicant exposure of human and nonhuman populations to damage to health like diseases, through the use of molecular technologies [88]. Biomarkers are defined as "the measurements of body fluids, cells, or tissues that indicate in biochemical or cellular terms the presence of contaminants or the magnitude of the host response" [89].

In order to explore the carcinogenic mechanisms that are associated with PAHs in the aquatic media, and thus to evaluate the biological plausibility between exposure and biological effects, biomarkers responses to PAHs can be directly appreciate in sentinel organisms, species that are living in the contaminated sites. In this way, it is obvious that the use of mutagenicity as a biomarker of PAH exposure in sentinel organisms promotes greater relevance for environmental risk assessment than the use of bacterial model in the Ames assay. However, some technical and methodological problems emerge and data can appear more difficult to interpret correctly.

Sarkar *et al.* have described more precisely the significance and the fields of application of biomarkers in marine pollution monitoring [69]. Biomarkers, classically divided into biomarkers of the internal dose (pollutant has penetrated into the organism), biomarkers of the biological effective dose (pollutant has interacted with the molecular target in the organism), biomarkers of early biological effects and altered structure (pollutant has begun to exert its toxicity) and biomarkers of susceptibility (pollutant effect dependent on the host) are considered as "early warning" indicators of a degraded environment resulting from contamination by numerous pollutants, in particular the highly persistent pollutants such as polychlorinated biphenyls (PCB), polychlorinated dibenzo-dioxins (PCDD) or PAHs. The cytochrome P4501A induction or measurement of DNA integrity were for example two widely used biomarkers in this field.

It has for example been shown that DNA adduct levels measured in marine fish (especially in English sole) collected in PAH contaminated areas are associated with increased hepatic lesions including tumors [90]. These lesions were considered as predictive early indicators of different degenerative and preneoplastic liver diseases in fish. Interestingly, interspecies variation of this biomarker was related to interspecies difference of the disease prevalence.

Biomarkers are also valuable tools for the understanding of the bioavailability and the fate (in term of bioaccumulation, biotransformation, toxicokinetic and toxicodynamic processes) of PAHs for sentinel organisms in different exposure conditions. Thus 1-hydroxypyrene metabolite in bile has been used as indicator of biotransformation, for example in flatfish species turbot (*Scophthalmus maximus*) [91]. In that study, biotransformation activity was correlated with DNA damage in fish erythrocytes (revealed by the comet assay), illustrating that biotransformation is probably necessary to the genotoxic potential of PAHs. Other appropriate examples of the use of biomarker are described in next paragraph.

2.1. CYP Enzyme as Biomarker of Exposure to PAHs

The genotoxic/mutagenic potential of PAHs, as the main representative BaP, goes through a metabolic bioactivation and the involvement of metabolizing enzymes, particularly cytochrome P450 1A (CYP1A) monooxygenase enzymes (phase I of biotransformation). Mainly described in mammalian as humans, these enzymatic systems exist in marine vertebrates as fish [69], as well as invertebrates like bivalves [92]. The main function of the microsomal CYP enzymes is the oxidation of substrates like PAHs (and some other persistent pollutants, *e.g.* dioxins, polychlorinated biphenyls, furans) in order to facilitate their elimination by increasing their hydrophilicity and facilitating the conjugation to endogenous molecules (phase II of biotransformation). In some cases, the multistep metabolic biotransformation can lead to the apparition of electrophilic reactive metabolites such as epoxides, leading finally to a net increase of the toxicity. This is especially the case for many PAHs, like BaP which forms the mutagenic well known benzopyrene diolepoxide. These enzymes are also potentially inducible by certain pollutants, especially PAHs, the induction depending to organism too. The mechanism involves an increase in protein synthesis by interaction at gene transcription level, implicating the AhR, a ligand dependent transcription factor. The measurement of the activity of these enzymes (more often by the 7-ethoxy-resorufin-O-deethylase (EROD) O-dealkylation activity of the CYP1A) is generally considered as biomarker of exposure to PAHs [93]. In a laboratory study, the intraperitoneal injection of BaP to polar cod has made it possible to monitor the variation in time of mRNA expression of CYP 1A1 gene in liver, the encoded protein level and EROD activity for a better understanding of underlying mechanisms [94]. It is to note that the measurement tends to be lacking of specificity to PAHs pollutants and is probably influenced by different environmental parameters as the water temperature [95]. As an example, Oliva *et al.* failed to associate EROD activity in gills of Senegal sole (*Solea senegalensis*) to PAH concentrations in heavy metal and PAH polluted estuary [96]. Alternatively, a positive correlation of EROD activity was found with heavy metals. The absence of correlation between this activity and PAH exposure has also been described in dab, a frequently used sentinel species [97].

Regardless, since the end of the '80s, cytochrome P450 activity in response to

PAH exposure has been widely used in marine biomonitoring, in very different biological models and sentinel species, in very different conditions of exposure (including laboratory controlled exposures) and in very different geographic areas. Much valuable information can be provided by this easy-to-use biomarker [98]. This is especially the case for the improvement of environmental health impact associated to large-scale oil-spills. Ten years after the grounding of the American oil tanker *Exxon Valdez*, elevated CYP1A protein (immunohistochemical dosage) and EROD activity in liver of near shore fish species are still measured, indicating the on-going exposure of biota to pollutants including PAHs originated from contaminated sediments [99].

Preneoplastic lesions and hepatic neoplasms related to pollutants exposure among them PAH were explored in the two benthic fish English sole and Starry flounder [100]. In order to elucidate the higher prevalence of such neoplasms in the English sole, the activity of enzymes implicated in biotransformation of PAHs were measured in both species and compared. Overall, hepatic activity of the aryl hydrocarbon hydroxylase (AHH, encoded by the CYP1A1 gene) was up to 2 fold higher in English sole compared to flounder. The opposite trend was observed for epoxide hydrolase and glutathione S transferase. These data appeared in accordance with higher metabolic activation and lower detoxication of PAHs by the English sole. That can explain at least in part the species susceptibility of English sole to PAH associated neoplasm.

Considering that the measurement of PAHs in tissues is probably a poor reflection of fish exposure and generally of very low levels, Wessel *et al.* have compared the EROD activity in the liver of juvenile sole exposed to a mixture of three PAHs with the rate of hydroxylated PAH metabolites in bile and genotoxicity in erythrocytes [101]. A positive correlation was observed between metabolite levels and genotoxicity but no effect on EROD activity was found. Interestingly, authors concluded that interaction and competitive phenomenon between PAHs in their biotransformation by CYP enzymes would possibly explained the latter result, suggesting that measurement of bile metabolites is finally a better biomarker of PAH exposure.

2.2. DNA Adducts as Biomarkers of "Biological Effective Dose" to Genotoxic PAHs in the Marine Environment

DNA adducts are resulting from the covalent liaison between an electrophilic chemical (generally a xenobiotic metabolite) and DNA. As such, DNA adducts are biomarkers of the biological effective dose (BED), reflecting the active fraction of genotoxicant(s) that actually interact with the exposed organism. DNA adducts are considered as a crucial biomarker of exposure in fish, especially for there early emergence after a genotoxic exposure, which may play a key role in establishing a mode of action for cancer [102]. Thus, in 1987, Dunn *et al.* measured significant DNA adduct levels in livers of wild Brown bullheads sampled from sites in the Buffalo and Detroit Rivers, in association with exposure of fish to high concentrations of PAHs [103]. Since these early works, a large range of fish species was studied, in a large panel of applications (laboratory and field studies).

In the marine environment, numerous published works are focused on a few sentinel fish species as the flounder, haddock or Atlantic cod. Most of them indicate that adducts are detected in the liver when fish are exposed to environmental genotoxicants, especially PAHs [104]. Data are available in relation with controlled laboratory exposures and environmental field studies [105 - 107]. In a laboratory study, Varanasi *et al.* showed as early as 1989 a dose response relationship between exposure of English sole to BaP (2-100 mg BaP/kg body wt., intraperitoneal injection) and BPDE DNA adduct formation in liver [108]. In 2000, Aas *et al.* has analysed the evolution through time of cytochrome P4501A expression, PAH metabolites in bile and DNA adducts in Atlantic cod after laboratory exposure to dispersed crude oil at three relevant concentrations (0.06, 0.25 and 1 ppm) [109]. Relation has been revealed between PAH concentration in liver, PAH metabolite in fish (biotransformation process) and DNA adduct levels (mainly at 1 ppm). It has been reported an exposure dependent response for the three biomarkers. In field, numerous studies were focused in European seas including the North Sea, taking into account the offshore oil and gas exploitation and the question of its environmental impacts. To illustrate, one of them was conducted on haddock (*Melanogrammus aeglefinus*) and Atlantic cod (*Gadus morhua*) caught in two areas of the North Sea with extensive oil production: Tampen and Sleipner [110]. From 2001 to 2004 fish campaigns, Balk *et al.* revealed significant higher levels of hepatic adducts in haddock from the Tampen area compared to a control site located in southwest Norway (Egersund bank). Qualitatively, the so called typical radioactive diagonal zone (DRZ) observable by using the ^{32}P postlabelling method (Fig. **2**), that is supposed to be associated to exposures to PAH mixtures, was described at the Tampen site. Complementary, other biomarkers as EROD activity in liver or bile metabolites of PAHs supported the suspicion of PAHs in the observed damage.

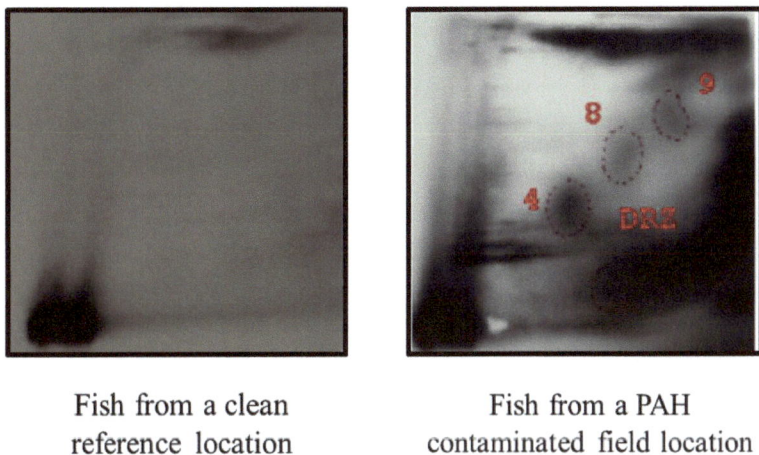

Fish from a clean
reference location

Fish from a PAH
contaminated field location

Fig. (2). DNA adducts analysed in fish liver using the ^{32}P-postlabelling method. The figure shows TLC (thin layer chromatography) of samples, both single spots (*i.e.* DNA adducts 4, 8, 9) and the typical PAH radioactive diagonal zone (DRZ) are highlighted.

As expected, these biomarkers have been associated to neoplasia in fish. A comparative field study indicates that higher DNA adduct levels were measured in the liver of flounders caught in polluted upper Seine estuary (France), associated with high frequencies of preneoplastic lesions [111].

In other cases, the relation of DNA adducts caused by PAHs with tumors in fish remains difficult to elucidate. No correlation was found between the concentration of DNA adducts in the liver of brown bullhead and liver or skin tumor prevalence and exposure to PAHs in Chesapeake Bay (USA) [112]. Tumor occurrence was more especially correlated with covariates like fish length or sex. More generally, high levels of DNA adducts are not systematically predictive of tumor development in fish [113]. Such observations might be explained by the fact that DNA damage that initiate the process conducting to the emergence of cancerous cells is probably due to particular adducts that are not assessed by the global DNA adducts measurement. Interestingly, it has also been noted that neoplasia cells generally exhibit lower CYP1A induction and lower DNA adduct levels compared to normal cells in response to PAH exposure, indicating a possible "resistant" phenotype to PAH bioactivation [114].

In the way of a better environmental risk assessment, the determination of reliable threshold values for biomarkers is becoming a crucial issue. For some years now, the topic is under discussion concerning DNA adducts, as mentioned in the report of the study group on integrated monitoring of contaminants and biological effects dated 14-18 march 2011 [115]. The proposed BAC (Background Assessment Concentrations) and EAC (Environmental Assessment Criteria) values for DNA adducts in some fish species are of specific interest in risk assessment evaluations.

DNA adducts were also measured in marine mammals, with the objective to correlate the biomarker with exposure to certain pollutants and/or with tumor prevalence (see [20] for a review). In the context of the awareness-raising campaign for protection of the sensitive Artic environment, DNA adducts were measured with significant elevated levels in Beluga [116], while this area is often considered as free from significant pollutants. Previously, BaP DNA adducts were detected in 10 out of 11 Saint Lawrence beluga whales found stranded between 1983 and 1990, but not in four others found in the artic [117]. These results were reconciled with the occurrence of cancer in such population exposed to numerous pollutants as PAHs.

By a more technical approach, attention is drawn to the condition of sampling, and more particularly to the postmortem thermal history, as mammals collected for DNA adducts analysis are generally found dead [118].

There are other biomarkers of genotoxicity currently used in the aquatic environment, in association with contaminated areas by PAHs: micronuclei, sister chromatid exchange, chromosome aberration and comet assay.

2.3. Other Biomarkers of Genotoxicity Associated to PAHs in the Marine Environment

In complement to DNA adducts, numerous other biomarkers of genotoxicity has been used in relation to the marine areas contamination by PAHs, revealing complementary modes of action for these pollutants. The well-known comet assay (or Single Cell Gel Electrophoresis (SCGE)), based on strand breaks detection in its standard form, has been successfully applied on sentinel organisms like invertebrates (on mussel hemocytes or gill cells) or fish, and considered as the most sensitive method for DNA damage when compared to micronuclei and sister chromatid exchanges [119].

This biomarker is generally in direct relation to exposure, and completes the kinetic and bioavailability approaches dedicated to pollutant fate in the environment. In a laboratory study, a four days exposure of turbot (*Scophthalmus maximus*) to complex mixtures of PAHs was sufficient to cause a significant increase of DNA strand break in erythrocytes [91]. These results were in line with the relative fast biotransformation of PAHs as indicated by the early bile production of some metabolites.

In another example, the laboratory exposure of clams to artificial sediment spiked with two environmental doses of phenanthrene (a non-carcinogenic PAH for human) or benzo[*b*]fluoranthene (a carcinogenic PAH) shows genotoxic impacts by the alkaline comet assay on gill cells, even at the lower concentrations [120]. These results raise the questions regarding the real bioavailability for filter-feeders of unpolar contaminants that are present in sediment, and the real genotoxic hazard for environmental of human carcinogenic and non-carcinogenic pollutants.

The response to pollutants in sentinel species can be difficult to interpret correctly, and sometimes unexpected. In a combined laboratory and field study, Large *et al.* showed no difference in the amount of DNA strand breaks in gill and digestive gland cells of blue mussels collected at two distinct locations, while the level of PAHs (including BaP) in mussel was seven fold higher in one location [121]. In laboratory studies, the continuous exposure of mussels to high concentrations of BaP in food (*via* pre-exposed *Isochrysis galbana*) led to a temporary increase of the DNA damage that returned to levels similar to the controls after 14 days. Such results clearly highlight the adaptive capabilities of mussels to resist pollutant genotoxic effects.

The comet assay is also used in marine fish in order to better understand pollutant interactions and the global genotoxic effect of mixtures. Thus, considered separately, benzo[*b*]fluoranthene and phenanthrene induced similar strand breaks but without clear dose-response patterns in sea bass exposed to spiked sediments for 28 days [122]. When both PAHs were mixed together, a supra-additive effect was observed. Such results reveal the complexity found when interpreting the effect of mixture of pollutants on the simple basis of individual toxicological data.

In zebrafish chronically exposed to different PAH mixtures (pyrogenic and petrogenic oils) *via* spiked diet at different concentrations, preneoplastic lesions and neoplasms were observed in relation with exposure duration and concentrations, especially with petrogenic heavy oils [44]. However, negative results resulting from comet and micronucleus assays on erythrocytes led the investigators to consider the carcinogenicity of PAHs without genotoxic effects in this particular model.

The comet assay was also used very recently in order to enhance knowledge about the genotoxicity of poorly studied oxygenated PAHs (oxy-PAHs) like acenaphthenequinone and 7,12-benzo[*a*]anthracenquinone. When Medaka were exposed to these two pollutants, significant DNA damage were observed from the first two days; even at the lower tested concentration [123]. Damage levels were similar to those caused by the corresponding non-oxygenated PAH acenaphthene and benzo[*a*]anthracene and by BaP. These results seem to confirm genotoxic hazard associated to oxy-PAHs and the need to conduct further investigations.

Mechanistically, the comet assay as marker of genotoxicity/clastogenicity permits the detection of DNA alterations in response to the exposure to PAHs that are different from DNA adduct formation, involving different ways of genotoxic action. In bivalve model blue mussel, the controlled exposure to BaP at a sublethal concentration in water (56 µg/L) for 12 days results in DNA strands breaks in hemocytes, histological abnormalities and increased expression of well-known tumor regulating genes p53 and ras, with tissue specific differences [57]. Interestingly, these two genes contain highly conserved sequences amongst living organisms (including humans) and are recognized as implicated in chemical carcinogenesis in laboratory animals and humans.

In particular, p53 is a tumor suppressor gene highly studied in the field of chemical carcinogenesis, especially in mammalian models. Data related to lower vertebrates like fish are less abundant and supposed to be comparable to mammalians. Functions of the p53 encoded protein in cells are numerous and globally associated to maintaining the genome stability by facilitating DNA damage repair or inducing cell death by apoptosis when damage is too severe (see [124] for a review). The p53 induction is indicating cell stress (*e.g.* DNA damage, hypoxia, oxidative stress) and has been used as genotoxic assay predicting the carcinogenic potential of a substance [125].

Information of the p53 gene regulation and its protein activity under chemical stress is lacking in lower vertebrates like fish, and is inconsistent. In the common whitefish (*Coregonus lavaretus*), BaP is able to induce P53 expression, especially in the liver cells, as revealed by the measurement of RNA messengers by real time PCR method [126].

In fish cell model PHLC-1, p53 is induced by none of the five tested chemotherapeutics that had shown induction in mammalian cells [127].

A diminished p53 transcription rate has been described in the head-kidney highly proliferating cells of Rainbow trout after 24 hours exposure to BaP or cyclopenta[*c*]phenanthrene [128]. It was hypothesized that some PAHs would influence cell physiology by down regulating expression of such key genes, perhaps *via* the modification of methylation rate in promotor sequence.

Moreover, and as mentioned, the p53 induction/repression is not specifically resulting from exposure to genotoxic substances. Environmental conditions can act as cell stressors able to influence p53 expression as well (Fig. **3**). One example is the water temperature. Cheng *et al.* showed recently that high water temperature of 34 °C induces ROS production and apoptosis in red blood cell of pufferfish in association with overexpression of different genes implicated in apoptosis: P53, Bax, caspase 9 and caspase 3 [129].

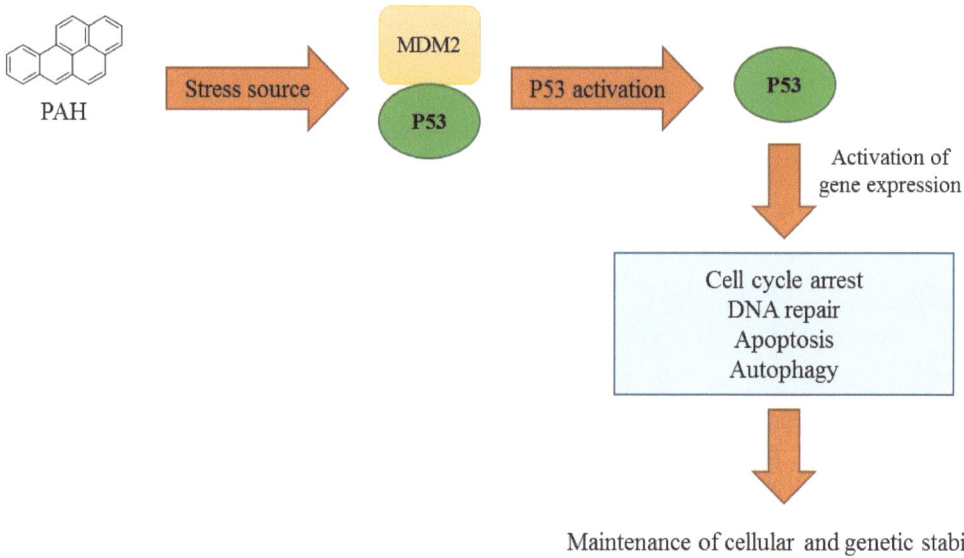

Fig. (3). P53 pathway. Under normal conditions, P53 protein is inactivated by negative regulators, *e.g.* MDM2. Following a stress input (*e.g.* PAH exposure, oxidative stress, hypoxia), P53 protein gets activated and induces various mechanisms to protect the organism.

Due to its key role in cell response to different stressors including genotoxicity, modifications in the sequence of the p53 gene by mutation process that alter the structure and/or functions of protein encoded have been associated with cancer development at each different stages of initiation, promotion, aggressiveness, and metastasis [130]. The gene is mutated in about half of the human cancers. Carcinogenic PAHs, like BaP, are able to alter the p53 sequence by the way of DNA adduct formation in the gene sequence and induction of mutations in described hot-spot [131]. In fish, the sequence of p53 gene is thought to vary greatly between animals in the same species, indicating genetic polymorphism. However, a specific mutation (T:A to A:T transversion at codon 147) that can alter the function of p53 protein was detected in hyperplastic lesion of a flounder and is thought to be partly implicated in neoplasia development in fish exposed to pollutants including PAHs [132]. By comparing different fish species (3 salmonids, puffer fish and the barbell) five highly conserved regions of P53 gene in mammals, it was argued that the DNA sequencing of four regions would be used as good biomarker in the aquatic environment by providing fingerprint of genotoxic pollutants [133].

Recently, the emphasis is on the use of a battery of biomarkers in order to evaluate genotoxicity of pollutants such PAHs on new relevant living models. As an example, the shanny *Lipophrys pholis* was tested to monitor oil spills in European marine ecosystems, by using the induction in erythrocytic nuclear abnormalities (ENA), a biomarker of genotoxicity, in complement to EROD activity and fluorescent aromatic compounds in bile [134]. The objective is to propose "cost effective, rapid and easy to use" new tools for using in the future European Water Policy legislation.

CONCLUDING REMARKS

Greater knowledge and understanding of the carcinogenicity of petrogenic PAHs in the aquatic environment and underlying biological mechanisms is a very important issue, considering the very large sources of pollution all around the world, the chemical complexity of this pollutant family, and the special fate in environment of these generally low water-soluble compounds. It is now recognized that certain PAHs are carcinogenic for humans and 16 PAHs are recognized by US EPA as priority-pollutants with regard to their toxicity, potential human exposure, frequency of occurrence in contaminated sites. But the list is probably not definitive, and questions are being posed also regarding the implication of more polar (and thus more bioavailable) PAH derivatives containing nitro or oxygen groups.

In order to better understand the real carcinogenic potential of PAHs, considered separately or more importantly in combinations, as always in the environment, the bioavailability for threatened marine organisms and biological mechanisms implicated in cancer cell development need to be well documented.

Since the '60s, numerous sentinel organisms were studied in relation with PAH exposure, from invertebrates like filter feeders to higher vertebrates as whales species. Studies are conducted in laboratory, mesocosms and field. Neoplastic lesions are regularly described.

The causal relationship between exposure to PAHs and neoplastic lesions can be addressed by the way of biomarker use. DNA adducts are alterations that are often described and an interesting marker of the effective biological dose associated to genotoxic compounds. The DNA strand breaks, another DNA damage, are also measured in numerous organisms, particularly by using the well-known comet assay.

Some biomarkers more specifically associated to early biological adverse effects need to be more widely applied to marine sentinels. The P53 gene, called "guardian of the genome" and mutated in around half of human cancers, has been little studied to date in environmental topics.

The better understanding of carcinogenic potential of PAHs and more globally their chronic effects will contribute to protect more effectively the aquatic environment and its resources. Validated biomarkers will be very helpful in this complex task, reporting on the biological plausibility of the causal relationship between exposure to PAHs and health effects as neoplasia. However, generated data are often difficult to exploit. Emerging biological and bioinformatics tools associated to omics would be the next field of investigation of carcinogenicity of PAHs in the aquatic environment. Bivalve omics for the monitoring of toxic compounds is ever a promising approach [135].

CONFLICT OF INTEREST

The author confirms that the author has no conflict of interest to declare for this publication.

ACKNOWLEDGEMENTS

Declared none.

REFERENCES

[1]   Boström, C-E.; Gerde, P.; Hanberg, A.; Jernström, B.; Johansson, C.; Kyrklund, T.; Rannug, A.; Törnqvist, M.; Victorin, K.; Westerholm, R. Cancer risk assessment, indicators, and guidelines for polycyclic aromatic hydrocarbons in the ambient air. *Environ. Health Perspect.,* **2002,** *110* Suppl. 3, 451-488.
[http://dx.doi.org/10.1289/ehp.02110s3451] [PMID: 12060843]

[2]   Kim, K-H.; Jahan, S.A.; Kabir, E.; Brown, R.J. A review of airborne polycyclic aromatic hydrocarbons (PAHs) and their human health effects. *Environ. Int.,* **2013,** *60,* 71-80.

[http://dx.doi.org/10.1016/j.envint.2013.07.019] [PMID: 24013021]

[3]   Jarvis, I.W.; Dreij, K.; Mattsson, Å.; Jernström, B.; Stenius, U. Interactions between polycyclic aromatic hydrocarbons in complex mixtures and implications for cancer risk assessment. *Toxicology*, **2014**, *321*, 27-39.
[http://dx.doi.org/10.1016/j.tox.2014.03.012] [PMID: 24713297]

[4]   Villeneuve, D.L.; Khim, J.S.; Kannan, K.; Giesy, J.P. Relative potencies of individual polycyclic aromatic hydrocarbons to induce dioxinlike and estrogenic responses in three cell lines. *Environ. Toxicol.*, **2002**, *17*(2), 128-137.
[http://dx.doi.org/10.1002/tox.10041] [PMID: 11979591]

[5]   Otte, J.C.; Keiter, S.; Faßbender, C.; Higley, E.B.; Rocha, P.S.; Brinkmann, M.; Wahrendorf, D-S.; Manz, W.; Wetzel, M.A.; Braunbeck, T.; Giesy, J.P.; Hecker, M.; Hollert, H. Contribution of priority PAHs and POPs to Ah receptor-mediated activities in sediment samples from the River Elbe Estuary, Germany. *PLoS One*, **2013**, *8*(10), e75596.
[http://dx.doi.org/10.1371/journal.pone.0075596] [PMID: 24146763]

[6]   Kummer, V.; Masková, J.; Zralý, Z.; Neča, J.; Simecková, P.; Vondrácek, J.; Machala, M. Estrogenic activity of environmental polycyclic aromatic hydrocarbons in uterus of immature Wistar rats. *Toxicol. Lett.*, **2008**, *180*(3), 212-221.
[http://dx.doi.org/10.1016/j.toxlet.2008.06.862] [PMID: 18634860]

[7]   Mu, J.; Wang, J.; Jin, F.; Wang, X.; Hong, H. Comparative embryotoxicity of phenanthrene and alkyl-phenanthrene to marine medaka (*Oryzias melastigma*). *Mar. Pollut. Bull.*, **2014**, *85*(2), 505-515.
[http://dx.doi.org/10.1016/j.marpolbul.2014.01.040] [PMID: 24559736]

[8]   Carver, J.H.; Machado, M.L.; MacGregor, J.A. Application of modified Salmonella/microsome prescreen to petroleum-derived complex mixtures and polynuclear aromatic hydrocarbons (PAH). *Mutat. Res.*, **1986**, *174*(4), 247-253.
[http://dx.doi.org/10.1016/0165-7992(86)90042-4] [PMID: 3526138]

[9]   Utesch, D.; Glatt, H.; Oesch, F. Rat hepatocyte-mediated bacterial mutagenicity in relation to the carcinogenic potency of benz(a)anthracene, benzo(a)pyrene, and twenty-five methylated derivatives. *Cancer Res.*, **1987**, *47*(6), 1509-1515.
[PMID: 3545447]

[10]   Trilecová, L.; Krčková, S.; Marvanová, S.; Pěnčíková, K.; Krčmář, P.; Neča, J.; Hulinková, P.; Pálková, L.; Ciganek, M.; Milcová, A.; Topinka, J.; Vondráček, J.; Machala, M. Toxic effects of methylated benzo[*a*]pyrenes in rat liver stem-like cells. *Chem. Res. Toxicol.*, **2011**, *24*(6), 866-876.
[http://dx.doi.org/10.1021/tx200049x] [PMID: 21604763]

[11]   Huang, M-C.; Chen, F-Y.; Chou, M-T.; Su, J-G. Fluoranthene enhances p53 expression and decreases mutagenesis induced by benzo[*a*]pyrene. *Toxicol. Lett.*, **2012**, *208*(3), 214-224.
[http://dx.doi.org/10.1016/j.toxlet.2011.11.011] [PMID: 22120587]

[12]   Wogan, G.N.; Hecht, S.S.; Felton, J.S.; Conney, A.H.; Loeb, L.A. Environmental and chemical carcinogenesis. *Semin. Cancer Biol.*, **2004**, *14*(6), 473-486.
[http://dx.doi.org/10.1016/j.semcancer.2004.06.010] [PMID: 15489140]

[13]   Santella, R.; Kinoshita, T.; Jeffrey, A.M. Mutagenicity of some methylated benzo[*a*]pyrene derivatives. *Mutat. Res.*, **1982**, *104*(4-5), 209-213.
[http://dx.doi.org/10.1016/0165-7992(82)90146-4] [PMID: 6287248]

[14] Xue, W.; Warshawsky, D. Metabolic activation of polycyclic and heterocyclic aromatic hydrocarbons and DNA damage: a review. *Toxicol. Appl. Pharmacol.,* **2005**, *206*(1), 73-93.
[http://dx.doi.org/10.1016/j.taap.2004.11.006] [PMID: 15963346]

[15] Mastrangelo, G.; Fadda, E.; Marzia, V. Polycyclic aromatic hydrocarbons and cancer in man. *Environ. Health Perspect.,* **1996**, *104*(11), 1166-1170.
[http://dx.doi.org/10.1289/ehp.961041166] [PMID: 8959405]

[16] Wenzl, T.; Simon, R.; Anklam, E.; Kleiner, J. Analytical methods for polycyclic aromatic hydrocarbons (PAHs) in food and the environment needed for new food legislation in the European Union. *TrAC. Trends Analyt. Chem.,* **2006**, *25*, 716-725.
[http://dx.doi.org/10.1016/j.trac.2006.05.010]

[17] Word, Microsoft *Microsoft Word - SN256_Crude Oil_March_2011.doc - category_crude oil_march_2011.pdf.,* **2011**.

[18] Malins, D.C.; Krahn, M.M.; Brown, D.W.; Rhodes, L.D.; Myers, M.S.; McCain, B.B.; Chan, S.L. Toxic chemicals in marine sediment and biota from Mukilteo, Washington: relationships with hepatic neoplasms and other hepatic lesions in English sole (*Parophrys vetulus*). *J. Natl. Cancer Inst.,* **1985**, *74*(2), 487-494.
[PMID: 3856055]

[19] Baumann, P.C. The use of tumors in wild populations of fish to assess ecosystem. *J. Aquat. Ecosyst. Health,* **1992**, *1*, 21-31.
[http://dx.doi.org/10.1007/BF00044045]

[20] Newman, S.J.; Smith, S.A. Marine mammal neoplasia: a review. *Vet. Pathol.,* **2006**, *43*(6), 865-880.
[http://dx.doi.org/10.1354/vp.43-6-865] [PMID: 17099143]

[21] Martineau, D.; Lemberger, K.; Dallaire, A.; Labelle, P.; Lipscomb, T.P.; Michel, P.; Mikaelian, I. Cancer in wildlife, a case study: beluga from the St. Lawrence estuary, Québec, Canada. *Environ. Health Perspect.,* **2002**, *110*(3), 285-292.
[http://dx.doi.org/10.1289/ehp.02110285] [PMID: 11882480]

[22] Lundstedt, S.; White, P.A.; Lemieux, C.L.; Lynes, K.D.; Lambert, I.B.; Oberg, L.; Haglund, P.; Tysklind, M. Sources, fate, and toxic hazards of oxygenated polycyclic aromatic hydrocarbons (PAHs) at PAH-contaminated sites. *Ambio,* **2007**, *36*(6), 475-485.
[http://dx.doi.org/10.1579/0044-7447(2007)36[475:SFATHO]2.0.CO;2] [PMID: 17985702]

[23] Yang, X.; Yu, L.; Chen, Z.; Xu, M. Bioavailability of polycyclic aromatic hydrocarbons and their potential application in eco-risk assessment and source apportionment in urban river sediment. *Sci. Rep.,* **2016**, *6*, 23134.
[http://dx.doi.org/10.1038/srep23134] [PMID: 26976450]

[24] Andersson, J.T.; Achten, C. Time to say goodbye to the 16 EPA PAHs? toward an up-to-date use of PACs for environmental purposes. *Polycycl. Aromat. Compd.,* **2015**, *35*(2-4), 330-354.
[http://dx.doi.org/10.1080/10406638.2014.991042] [PMID: 26823645]

[25] Franco, S.S.; Nardocci, A.C.; Günther, W.M. PAH biomarkers for human health risk assessment: a review of the state-of-the-art. *Cad. Saude Publica,* **2008**, *24* Suppl. 4, s569-s580.
[http://dx.doi.org/10.1590/S0102-311X2008001600009] [PMID: 18797723]

[26] Pikney, A.E.; Harshbarger, J.C.; Rutter, M.A. Tumors in brown bullheads in the Chesapeake Bay watershed: analysis of survey data from 1992 through 2006. *J. Aquat. Anim. Health,* **2009**, *21*(2), 71-81.
[http://dx.doi.org/10.1577/H08-037.1] [PMID: 19873828]

[27] Myers, M.S.; Anulacion, B.F.; French, B.L.; Reichert, W.L.; Laetz, C.A.; Buzitis, J.; Olson, O.P.; Sol, S.; Collier, T.K. Improved flatfish health following remediation of a PAH-contaminated site in Eagle Harbor, Washington. *Aquat. Toxicol.,* **2008**, *88*(4), 277-288.
[http://dx.doi.org/10.1016/j.aquatox.2008.05.005] [PMID: 18571248]

[28] Fent, K. Ecotoxicological problems associated with contaminated sites. *Toxicol. Lett.,* **2003**, *140-141*, 353-365.
[http://dx.doi.org/10.1016/S0378-4274(03)00032-8] [PMID: 12676484]

[29] Dawe, C.J.; Stanton, M.F.; Schwartz, F.J. Hepatic neoplasms in native bottom-feeding fish of Deep Creek Lake, Maryland. *Cancer Res.,* **1964**, *24*, 1194-1201.
[PMID: 14216151]

[30] Sonstegard, R.A. Environmental carcinogenesis studies in fishes of the Great Lakes of North America. *Ann. N. Y. Acad. Sci.,* **1978**, *298*, 261-269.
[http://dx.doi.org/10.1111/j.1749-6632.1977.tb19270.x] [PMID: 212992]

[31] Brown, E.R.; Hazdra, J.J.; Keith, L.; Greenspan, I.; Kwapinski, J.B.; Beamer, P. Frequency of fish tumors found in a polluted watershed as compared to nonpolluted Canadian waters. *Cancer Res.,* **1973**, *33*(2), 189-198.
[PMID: 4569787]

[32] Baumann, P.C. Epizootics of cancer in fish associated with genotoxins in sediment and water. *Mutat. Res.,* **1998**, *411*(3), 227-233.
[http://dx.doi.org/10.1016/S1383-5742(98)00015-5] [PMID: 9804959]

[33] Baumann, P.C.; Smith, I.R.; Metcalfe, C.D. Linkages between chemical contaminants and tumors benthic Great Lakes fish. *J. Gt. Lakes Res.,* **1996**, *22*, 22.
[http://dx.doi.org/10.1016/S0380-1330(96)70946-2]

[34] Coffee, L.L.; Casey, J.W.; Bowser, P.R. Pathology of tumors in fish associated with retroviruses: a review. *Vet. Pathol.,* **2013**, *50*(3), 390-403.
[http://dx.doi.org/10.1177/0300985813480529] [PMID: 23456970]

[35] Selkirk, J.K. Benzo[*a*]pyrene carcinogenesis: a biochemical selection mechanism. *J. Toxicol. Environ. Health,* **1977**, *2*(6), 1245-1258.
[http://dx.doi.org/10.1080/15287397709529527] [PMID: 328915]

[36] Slaga, T.J. Cancer: etiology, mechanisms, and prevention summary. *Carcinog. Compr. Surv.,* **1980**, *5*, 243-262.
[PMID: 6992999]

[37] Harvey, R.G. Polycyclic hydrocarbons and cancer. *Am. Sci.,* **1982**, *70*(4), 386-393.
[PMID: 7149435]

[38] Hendricks, J.D.; Meyers, T.R.; Shelton, D.W.; Casteel, J.L.; Bailey, G.S. Hepatocarcinogenicity of

benzo[*a*]pyrene to rainbow trout by dietary exposure and intraperitoneal injection. *J. Natl. Cancer Inst.,* **1985**, *74*(4), 839-851.
[PMID: 2985858]

[39] Hawkins, W.E.; Walker, W.W.; Overstreet, R.M.; Lytle, J.S.; Lytle, T.F. Carcinogenic effects of some polycyclic aromatic hydrocarbons on the Japanese medaka and guppy in waterborne exposures. *Sci. Total Environ.,* **1990**, *94*(1-2), 155-167.
[http://dx.doi.org/10.1016/0048-9697(90)90370-A] [PMID: 2163106]

[40] Melius, P. Comparative benzo[*a*]pyrene metabolite patterns in fish and rodents. *Natl. Cancer Inst. Monogr.,* **1984**, *65*, 387-390.
[PMID: 6087145]

[41] Iyer, R.P.; Lyga, J.W.; Secrist, J.A., III; Daub, G.H.; Slaga, T.J. Comparative tumor-initiating activity of methylated benzo(*a*)pyrene derivatives in mouse skin. *Cancer Res.,* **1980**, *40*(4), 1073-1076.
[PMID: 7357537]

[42] Varanasi, U.; Stein, J.E.; Nishimoto, M.; Reichert, W.L.; Collier, T.K. Chemical carcinogenesis in feral fish: uptake, activation, and detoxication of organic xenobiotics. *Environ. Health Perspect.,* **1987**, *71*, 155-170.
[http://dx.doi.org/10.1289/ehp.8771155] [PMID: 3297658]

[43] Vethaak, A.D.; Jol, J.G.; Meijboom, A.; Eggens, M.L.; Rheinallt, T.; Wester, P.W.; van de Zande, T.; Bergman, A.; Dankers, N.; Ariese, F.; Baan, R.A.; Everts, J.M.; Opperhuizen, A.; Marquenie, J.M. Skin and liver diseases induced in flounder (*Platichthys flesus*) after long-term exposure to contaminated sediments in large-scale mesocosms. *Environ. Health Perspect.,* **1996**, *104*(11), 1218-1229.
[http://dx.doi.org/10.1289/ehp.961041218] [PMID: 8959412]

[44] Larcher, T.; Perrichon, P.; Vignet, C.; Ledevin, M.; Le Menach, K.; Lyphout, L.; Landi, L.; Clerandeau, C.; Lebihanic, F.; Ménard, D.; Burgeot, T.; Budzinski, H.; Akcha, F.; Cachot, J.; Cousin, X. Chronic dietary exposure of zebrafish to PAH mixtures results in carcinogenic but not genotoxic effects. *Environ. Sci. Pollut. Res. Int.,* **2014**, *21*(24), 13833-13849.
[http://dx.doi.org/10.1007/s11356-014-2923-7] [PMID: 24777325]

[45] Malins, D.C.; McCain, B.B.; Myers, M.S.; Brown, D.W.; Krahn, M.M.; Roubal, W.T.; Schiewe, M.H.; Landahl, J.T.; Chan, S.L. Field and laboratory studies of the etiology of liver neoplasms in marine fish from Puget Sound. *Environ. Health Perspect.,* **1987**, *71*, 5-16.
[http://dx.doi.org/10.1289/ehp.87715] [PMID: 3297664]

[46] Malins, D.C.; McCain, B.B.; Landahl, J.T.; Myers, M.S.; Krahn, M.M.; Brown, D.W.; Chan, S-L.; Roubal, W.T. Neoplastic and other diseases in fish in relation to toxic chemicals: an overview. *Aquat. Toxicol.,* **1988**, *11*, 43-67.
[http://dx.doi.org/10.1016/0166-445X(88)90006-9]

[47] Paniagua-Michel, J.; Rosales, A. Marine bioremediation - a sustainable biotechnology of petroleum hydrocarbons biodegradation in coastal and marine environments. *J. Bioremediation Biodegrad.,* **2015**, *6*, 273.

[48] Baumann, P.C.; Harshbarger, J.C. Decline in liver neoplasms in wild brown bullhead catfish after coking plant closes and environmental PAHs plummet. *Environ. Health Perspect.,* **1995**, *103*(2), 168-170.
[http://dx.doi.org/10.1289/ehp.95103168] [PMID: 7737065]

[49]   De Flora, S.; Viganò, L.; DAgostini, F.; Camoirano, A.; Bagnasco, M.; Bennicelli, C.; Melodia, F.; Arillo, A. Multiple genotoxicity biomarkers in fish exposed *in situ* to polluted river water. *Mutat. Res.,* **1993**, *319*(3), 167-177.
[http://dx.doi.org/10.1016/0165-1218(93)90076-P] [PMID: 7694138]

[50]   Varanasi, U.; Reichert, W.L.; Stein, J.E. ^{32}P-postlabeling analysis of DNA adducts in liver of wild English sole (*Parophrys vetulus*) and winter flounder (*Pseudopleuronectes americanus*). *Cancer Res.,* **1989**, *49*(5), 1171-1177.
[PMID: 2917348]

[51]   al-Sabti, K.; Metcalfe, C.D. Fish micronuclei for assessing genotoxicity in water. *Mutat. Res.,* **1995**, *343*(2-3), 121-135.
[http://dx.doi.org/10.1016/0165-1218(95)90078-0] [PMID: 7791806]

[52]   Carballal, M.J.; Barber, B.J.; Iglesias, D.; Villalba, A. Neoplastic diseases of marine bivalves. *J. Invertebr. Pathol.,* **2015**, *131*, 83-106.
[http://dx.doi.org/10.1016/j.jip.2015.06.004] [PMID: 26146225]

[53]   Stegeman, J.J.; Lech, J.J. Cytochrome P-450 monooxygenase systems in aquatic species: carcinogen metabolism and biomarkers for carcinogen and pollutant exposure. *Environ. Health Perspect.,* **1991**, *90*, 101-109.
[http://dx.doi.org/10.2307/3430851] [PMID: 2050047]

[54]   Smolarz, K.; Renault, T.; Soletchnik, P.; Wolowicz, M. Neoplasia detection in *Macoma balthica* from the Gulf of Gdansk: comparison of flow cytometry, histology and chromosome analysis. *Dis. Aquat. Organ.,* **2005**, *65*(3), 187-195.
[http://dx.doi.org/10.3354/dao065187] [PMID: 16119887]

[55]   Krishnakumar, P.K.; Casillas, E.; Snider, R.G.; Kagley, A.N.; Varanasi, U. Environmental contaminants and the prevalence of hemic neoplasia (leukemia) in the common mussel (*Mytilus edulis* complex) from Puget Sound, Washington, U.S.A. *J. Invertebr. Pathol.,* **1999**, *73*(2), 135-146.
[http://dx.doi.org/10.1006/jipa.1998.4798] [PMID: 10066393]

[56]   Gardner, G.R.; Yevich, P.P.; Harshbarger, J.C.; Malcolm, A.R. Carcinogenicity of Black Rock Harbor sediment to the eastern oyster and trophic transfer of Black Rock Harbor carcinogens from the blue mussel to the winter flounder. *Environ. Health Perspect.,* **1991**, *90*, 53-66.
[http://dx.doi.org/10.2307/3430845] [PMID: 2050083]

[57]   Di, Y.; Schroeder, D.C.; Highfield, A.; Readman, J.W.; Jha, A.N. Tissue-specific expression of p53 and ras genes in response to the environmental genotoxicant benzo(α)pyrene in marine mussels. *Environ. Sci. Technol.,* **2011**, *45*(20), 8974-8981.
[http://dx.doi.org/10.1021/es201547x] [PMID: 21899289]

[58]   Browning, H.M.; Gulland, F.M.; Hammond, J.A.; Colegrove, K.M.; Hall, A.J. Common cancer in a wild animal: the California sea lion (*Zalophus californianus*) as an emerging model for carcinogenesis. *Philos. Trans. R. Soc. Lond. B Biol. Sci.,* **2015**, *370*(1673), 1673.
[http://dx.doi.org/10.1098/rstb.2014.0228] [PMID: 26056370]

[59]   Simeone, C.A.; Gulland, F.M.; Norris, T.; Rowles, T.K. A systematic review of changes in marine

mammal health in North America, 19722012: the need for a novel integrated approach. *PLoS One,* **2015**, *10*(11), e0142105.
[http://dx.doi.org/10.1371/journal.pone.0142105] [PMID: 26579715]

[60] Muir, D.C.; Ford, C.A.; Rosenberg, B.; Norstrom, R.J.; Simon, M.; Béland, P. Persistent organochlorines in beluga whales (*Delphinapterus leucas*) from the St Lawrence River estuaryI. Concentrations and patterns of specific PCBs, chlorinated pesticides and polychlorinated dibenzo--dioxins and dibenzofurans. *Environ. Pollut.,* **1996**, *93*(2), 219-234.
[http://dx.doi.org/10.1016/0269-7491(96)00006-1] [PMID: 15091361]

[61] Hanahan, D.; Weinberg, R.A. The hallmarks of cancer. *Cell,* **2000**, *100*(1), 57-70.
[http://dx.doi.org/10.1016/S0092-8674(00)81683-9] [PMID: 10647931]

[62] Sarasin, A. An overview of the mechanisms of mutagenesis and carcinogenesis. *Mutat. Res.,* **2003**, *544*(2-3), 99-106.
[http://dx.doi.org/10.1016/j.mrrev.2003.06.024] [PMID: 14644312]

[63] Pietras, K.; Ostman, A. Hallmarks of cancer: interactions with the tumor stroma. *Exp. Cell Res.,* **2010**, *316*(8), 1324-1331.
[http://dx.doi.org/10.1016/j.yexcr.2010.02.045] [PMID: 20211171]

[64] Vineis, P.; Schatzkin, A.; Potter, J.D. Models of carcinogenesis: an overview. *Carcinogenesis,* **2010**, *31*(10), 1703-1709.
[http://dx.doi.org/10.1093/carcin/bgq087] [PMID: 20430846]

[65] Hill, A.B. The environment and disease: association or causation? 1965. *J. R. Soc. Med.,* **2015**, *108*(1), 32-37.
[http://dx.doi.org/10.1177/0141076814562718] [PMID: 25572993]

[66] Fedak, K.M.; Bernal, A.; Capshaw, Z.A.; Gross, S. Applying the Bradford Hill criteria in the 21st century: how data integration has changed causal inference in molecular epidemiology. *Emerg. Themes Epidemiol.,* **2015**, *12*, 14.
[http://dx.doi.org/10.1186/s12982-015-0037-4] [PMID: 26425136]

[67] Angerer, J.; Ewers, U.; Wilhelm, M. Human biomonitoring: state of the art. *Int. J. Hyg. Environ. Health,* **2007**, *210*(3-4), 201-228.
[http://dx.doi.org/10.1016/j.ijheh.2007.01.024] [PMID: 17376741]

[68] Grandjean, P.; Brown, S.S.; Reavey, P.; Young, D.S. Biomarkers in environmental toxicology: state of the art. *Clin. Chem.,* **1995**, *41*(12 Pt 2), 1902-1904.
[PMID: 7497652]

[69] Sarkar, A.; Ray, D.; Shrivastava, A.N.; Sarker, S. Molecular Biomarkers: their significance and application in marine pollution monitoring. *Ecotoxicology,* **2006**, *15*(4), 333-340.
[http://dx.doi.org/10.1007/s10646-006-0069-1] [PMID: 16676218]

[70] Ohe, T.; Watanabe, T.; Wakabayashi, K. Mutagens in surface waters: a review. *Mutat. Res.,* **2004**, *567*(2-3), 109-149.
[http://dx.doi.org/10.1016/j.mrrev.2004.08.003] [PMID: 15572284]

[71] Ames, B.N.; Lee, F.D.; Durston, W.E. An improved bacterial test system for the detection and classification of mutagens and carcinogens. *Proc. Natl. Acad. Sci. USA,* **1973**, *70*(3), 782-786.
[http://dx.doi.org/10.1073/pnas.70.3.782] [PMID: 4577135]

[72]   Møller, M.; Hagen, I.; Ramdahl, T. Mutagenicity of polycyclic aromatic compounds (PAC) identified in source emissions and ambient air. *Mutat. Res.,* **1985**, *157*(2-3), 149-156.
[http://dx.doi.org/10.1016/0165-1218(85)90110-7] [PMID: 3894960]

[73]   Pahlman, R.; Pelkonen, O. Mutagenicity studies of different polycyclic aromatic hydrocarbons: the significance of enzymatic factors and molecular structure. *Carcinogenesis,* **1987**, *8*(6), 773-778.
[http://dx.doi.org/10.1093/carcin/8.6.773] [PMID: 3301044]

[74]   Mersch-Sundermann, V.; Mochayedi, S.; Kevekordes, S.; Kern, S.; Wintermann, F. The genotoxicity of unsubstituted and nitrated polycyclic aromatic hydrocarbons. *Anticancer Res.,* **1993**, *13*(6A), 2037-2043.
[PMID: 8297112]

[75]   Carver, J.H.; Machado, M.L.; MacGregor, J.A. Application of modified Salmonella/microsome prescreen to petroleum-derived complex mixtures and polynuclear aromatic hydrocarbons (PAH). *Mutat. Res.,* **1986**, *174*(4), 247-253.
[http://dx.doi.org/10.1016/0165-7992(86)90042-4] [PMID: 3526138]

[76]   Hermann, M. Synergistic effects of individual polycyclic aromatic hydrocarbons on the mutagenicity of their mixtures. *Mutat. Res.,* **1981**, *90*(4), 399-409.
[http://dx.doi.org/10.1016/0165-1218(81)90062-8] [PMID: 7038461]

[77]   White, P.A. The genotoxicity of priority polycyclic aromatic hydrocarbons in complex mixtures. *Mutat. Res.,* **2002**, *515*(1-2), 85-98.
[http://dx.doi.org/10.1016/S1383-5718(02)00017-7] [PMID: 11909757]

[78]   Kuo, C.Y.; Cheng, Y.W.; Chen, C.Y.; Lee, H. Correlation between the amounts of polycyclic aromatic hydrocarbons and mutagenicity of airborne particulate samples from Taichung City, Taiwan. *Environ. Res.,* **1998**, *78*(1), 43-49.
[http://dx.doi.org/10.1006/enrs.1998.3838] [PMID: 9630444]

[79]   De Flora, S.; Bagnasco, M.; Izzotti, A.; DAgostini, F.; Pala, M.; Valerio, F. Mutagenicity of polycyclic aromatic hydrocarbon fractions extracted from urban air particulates. *Mutat. Res.,* **1989**, *224*(2), 305-318.
[http://dx.doi.org/10.1016/0165-1218(89)90169-9] [PMID: 2677712]

[80]   Bostrøm, E.; Engen, S.; Eide, I. Mutagenicity testing of organic extracts of diesel exhaust particles after spiking with polycyclic aromatic hydrocarbons (PAH). *Arch. Toxicol.,* **1998**, *72*(10), 645-649.
[http://dx.doi.org/10.1007/s002040050555] [PMID: 9851680]

[81]   Brinkmann, M.; Blenkle, H.; Salowsky, H.; Bluhm, K.; Schiwy, S.; Tiehm, A.; Hollert, H. Genotoxicity of heterocyclic PAHs in the micronucleus assay with the fish liver cell line RTL-W1. *PLoS One,* **2014**, *9*(1), e85692.
[http://dx.doi.org/10.1371/journal.pone.0085692] [PMID: 24416442]

[82]   Kienzler, A.; Mahler, B.J.; Van Metre, P.C.; Schweigert, N.; Devaux, A.; Bony, S. Exposure to runoff from coal-tar-sealed pavement induces genotoxicity and impairment of DNA repair capacity in the RTL-W1 fish liver cell line. *Sci. Total Environ.,* **2015**, *520*, 73-80.
[http://dx.doi.org/10.1016/j.scitotenv.2015.03.005] [PMID: 25795989]

[83] Chen, G.; White, P.A. The mutagenic hazards of aquatic sediments: a review. *Mutat. Res.,* **2004**, *567*(2-3), 151-225.
[http://dx.doi.org/10.1016/j.mrrev.2004.08.005] [PMID: 15572285]

[84] Cachot, J.; Geffard, O.; Augagneur, S.; Lacroix, S.; Le Menach, K.; Peluhet, L.; Couteau, J.; Denier, X.; Devier, M.H.; Pottier, D.; Budzinski, H. Evidence of genotoxicity related to high PAH content of sediments in the upper part of the Seine estuary (Normandy, France). *Aquat. Toxicol.,* **2006**, *79*(3), 257-267.
[http://dx.doi.org/10.1016/j.aquatox.2006.06.014] [PMID: 16887205]

[85] Yang, F.; Zhang, Q.; Guo, H.; Zhang, S. Evaluation of cytotoxicity, genotoxicity and teratogenicity of marine sediments from Qingdao coastal areas using *in vitro* fish cell assay, comet assay and zebrafish embryo test. *Toxicol. In Vitro,* **2010**, *24*(7), 2003-2011.
[http://dx.doi.org/10.1016/j.tiv.2010.07.019] [PMID: 20656009]

[86] Thomas, K.V.; Balaam, J.; Barnard, N.; Dyer, R.; Jones, C.; Lavender, J.; McHugh, M. Characterisation of potentially genotoxic compounds in sediments collected from United Kingdom estuaries. *Chemosphere,* **2002**, *49*(3), 247-258.
[http://dx.doi.org/10.1016/S0045-6535(02)00316-8] [PMID: 12363302]

[87] Ji, K.; Seo, J.; Liu, X.; Lee, J.; Lee, S.; Lee, W.; Park, J.; Khim, J.S.; Hong, S.; Choi, Y.; Shim, W.J.; Takeda, S.; Giesy, J.P.; Choi, K. Genotoxicity and endocrine-disruption potentials of sediment near an oil spill site: two years after the Hebei Spirit oil spill. *Environ. Sci. Technol.,* **2011**, *45*(17), 7481-7488.
[http://dx.doi.org/10.1021/es200724x] [PMID: 21786741]

[88] Goldstein, B.; Gibson, J.; Henderson, R.; Hobbie, J.; Landrigan, P.; Mattison, D.; Perera, F.; Pfitzer, E.; Silbergeld, E.; Wogan, G.; Davis, D.; Thomas, R.; Wagener, D.; Peter, F.; Wakefield, L. Biological markers in environmental health research. *Environ. Health Perspect.,* **1987**, *74*, 3-9.
[PMID: 3691432]

[89] McCarthy, J.F.; Shugart, L.R. *Biomarkers of Environmental Contamination*; Lewis Publishers: Chelsea, **1990**.

[90] Reichert, W.L.; Myers, M.S.; Peck-Miller, K.; French, B.; Anulacion, B.F.; Collier, T.K.; Stein, J.E.; Varanasi, U. Molecular epizootiology of genotoxic events in marine fish: linking contaminant exposure, DNA damage, and tissue-level alterations. *Mutat. Res.,* **1998**, *411*(3), 215-225.
[http://dx.doi.org/10.1016/S1383-5742(98)00014-3] [PMID: 9804956]

[91] Le Dû-Lacoste, M.; Akcha, F.; Dévier, M-H.; Morin, B.; Burgeot, T.; Budzinski, H. Comparative study of different exposure routes on the biotransformation and genotoxicity of PAHs in the flatfish species, *Scophthalmus maximus*. *Environ. Sci. Pollut. Res. Int.,* **2013**, *20*(2), 690-707.
[http://dx.doi.org/10.1007/s11356-012-1388-9] [PMID: 23247530]

[92] Grøsvik, B.E.; Jonsson, H.; Rodríguez-Ortega, M.J.; Roepstorff, P.; Goksøyr, A. CYP1A-immunopositive proteins in bivalves identified as cytoskeletal and major vault proteins. *Aquat. Toxicol.,* **2006**, *79*(4), 334-340.
[http://dx.doi.org/10.1016/j.aquatox.2006.07.003] [PMID: 16949163]

[93] Livingstone, D.R.; Mitchelmore, C.L.; Peters, L.D.; OHara, S.C.; Shaw, J.P.; Chesman, B.S.; Doyotte, A.; McEvoy, J.; Ronisz, D.; Larsson, D.G.; Förlin, L. Development of hepatic CYP1A and blood

vitellogenin in eel (*Anguilla anguilla*) for use as biomarkers in the Thames Estuary, UK. *Mar. Environ. Res.,* **2000**, *50*(1-5), 367-371.
[http://dx.doi.org/10.1016/S0141-1136(00)00060-X] [PMID: 11460720]

[94] Nahrgang, J.; Camus, L.; Gonzalez, P.; Goksøyr, A.; Christiansen, J.S.; Hop, H. PAH biomarker responses in polar cod (*Boreogadus saida*) exposed to benzo(*a*)pyrene. *Aquat. Toxicol.,* **2009**, *94*(4), 309-319.
[http://dx.doi.org/10.1016/j.aquatox.2009.07.017] [PMID: 19709761]

[95] Fragoso, N.M.; Hodson, P.V.; Zambon, S. Evaluation of an exposure assay to measure uptake of sediment PAH by fish. *Environ. Monit. Assess.,* **2006**, *116*(1-3), 481-511.
[http://dx.doi.org/10.1007/s10661-006-7667-5] [PMID: 16779608]

[96] Oliva, M.; Gravato, C.; Guilhermino, L.; Galindo-Riaño, M.D.; Perales, J.A. EROD activity and cytochrome P4501A induction in liver and gills of Senegal sole Solea senegalensis from a polluted Huelva estuary (SW Spain). *Comp. Biochem. Physiol. C Toxicol. Pharmacol.,* **2014**, *166*, 134-144.
[http://dx.doi.org/10.1016/j.cbpc.2014.07.010] [PMID: 25110325]

[97] Kammann, U.; Lang, T.; Berkau, A-J.; Klempt, M. Biological effect monitoring in dab (*Limanda limanda*) using gene transcript of CYP1A1 or EROD-a comparison. *Environ. Sci. Pollut. Res. Int.,* **2008**, *15*(7), 600-605.
[http://dx.doi.org/10.1007/s11356-008-0048-6] [PMID: 18853211]

[98] Trisciani, A.; Corsi, I.; Torre, C.D.; Perra, G.; Focardi, S. Hepatic biotransformation genes and enzymes and PAH metabolites in bile of common sole (*Solea solea*, Linnaeus, 1758) from an oil-contaminated site in the Mediterranean Sea: a field study. *Mar. Pollut. Bull.,* **2011**, *62*(4), 806-814.
[http://dx.doi.org/10.1016/j.marpolbul.2011.01.001] [PMID: 21276988]

[99] Jewett, S.C.; Dean, T.A.; Woodin, B.R.; Hoberg, M.K.; Stegeman, J.J. Exposure to hydrocarbons 10 years after the Exxon Valdez oil spill: evidence from cytochrome P4501A expression and biliary FACs in nearshore demersal fishes. *Mar. Environ. Res.,* **2002**, *54*(1), 21-48.
[http://dx.doi.org/10.1016/S0141-1136(02)00093-4] [PMID: 12148943]

[100] Collier, T.K.; Singh, S.V.; Awasthi, Y.C.; Varanasi, U. Hepatic xenobiotic metabolizing enzymes in two species of benthic fish showing different prevalences of contaminant-associated liver neoplasms. *Toxicol. Appl. Pharmacol.,* **1992**, *113*(2), 319-324.
[http://dx.doi.org/10.1016/0041-008X(92)90131-B] [PMID: 1561641]

[101] Wessel, N.; Santos, R.; Menard, D.; Le Menach, K.; Buchet, V.; Lebayon, N.; Loizeau, V.; Burgeot, T.; Budzinski, H.; Akcha, F. Relationship between PAH biotransformation as measured by biliary metabolites and EROD activity, and genotoxicity in juveniles of sole (*Solea solea*). *Mar. Environ. Res.,* **2010**, *69* Suppl., S71-S73.
[http://dx.doi.org/10.1016/j.marenvres.2010.03.004] [PMID: 20417553]

[102] Pottenger, L.H.; Carmichael, N.; Banton, M.I.; Boogaard, P.J.; Kim, J.; Kirkland, D.; Phillips, R.D.; van Benthem, J.; Williams, G.M.; Castrovinci, A. ECETOC workshop on the biological significance of DNA adducts: summary of follow-up from an expert panel meeting. *Mutat. Res.,* **2009**, *678*(2), 152-157.
[http://dx.doi.org/10.1016/j.mrgentox.2009.07.006] [PMID: 19628052]

[103] Dunn, B.P.; Black, J.J.; Maccubbin, A. ^{32}P-postlabeling analysis of aromatic DNA adducts in fish from

polluted areas. *Cancer Res.,* **1987**, *47*(24 Pt 1), 6543-6548.
[PMID: 3677092]

[104] Baan, R.A.; Steenwinkel, M.J.; van den Berg, P.T.; Roggeband, R.; van Delft, J.H. Molecular dosimetry of DNA damage induced by polycyclic aromatic hydrocarbons; relevance for exposure monitoring and risk assessment. *Hum. Exp. Toxicol.,* **1994**, *13*(12), 880-887.
[http://dx.doi.org/10.1177/096032719401301211] [PMID: 7718309]

[105] Harvey, J.S.; Lyons, B.P.; Waldock, M.; Parry, J.M. The application of the 32P-postlabelling assay to aquatic biomonitoring. *Mutat. Res.,* **1997**, *378*(1-2), 77-88.
[http://dx.doi.org/10.1016/S0027-5107(97)00099-7] [PMID: 9288887]

[106] Reynolds, W.J.; Feist, S.W.; Jones, G.J.; Lyons, B.P.; Sheahan, D.A.; Stentiford, G.D. Comparison of biomarker and pathological responses in flounder (*Platichthys flesus* L.) induced by ingested polycyclic aromatic hydrocarbon (PAH) contamination. *Chemosphere,* **2003**, *52*(7), 1135-1145.
[http://dx.doi.org/10.1016/S0045-6535(03)00332-1] [PMID: 12820994]

[107] Malmström, C.; Konn, M.; Bogovski, S.; Lang, T.; Lönnström, L-G.; Bylund, G. Screening of hydrophobic DNA adducts in flounder (*Platichthys flesus*) from the Baltic Sea. *Chemosphere,* **2009**, *77*(11), 1514-1519.
[http://dx.doi.org/10.1016/j.chemosphere.2009.09.055] [PMID: 19846194]

[108] Varanasi, U.; Reichert, W.L.; Le Eberhart, B.T.; Stein, J.E. Formation and persistence of benzo[*a*]pyrene-diolepoxide-DNA adducts in liver of English sole (*Parophrys vetulus*). *Chem. Biol. Interact.,* **1989**, *69*(2-3), 203-216.
[http://dx.doi.org/10.1016/0009-2797(89)90078-1] [PMID: 2495192]

[109] Aas, E.; Baussant, T.; Balk, L.; Liewenborg, B.; Andersen, O.K. PAH metabolites in bile, cytochrome P4501A and DNA adducts as environmental risk parameters for chronic oil exposure: a laboratory experiment with Atlantic cod. *Aquat. Toxicol.,* **2000**, *51*(2), 241-258.
[http://dx.doi.org/10.1016/S0166-445X(00)00108-9] [PMID: 11064127]

[110] Balk, L.; Hylland, K.; Hansson, T.; Berntssen, M.H.; Beyer, J.; Jonsson, G.; Melbye, A.; Grung, M.; Torstensen, B.E.; Børseth, J.F.; Skarphedinsdottir, H.; Klungsøyr, J. Biomarkers in natural fish populations indicate adverse biological effects of offshore oil production. *PLoS One,* **2011**, *6*(5), e19735.
[http://dx.doi.org/10.1371/journal.pone.0019735] [PMID: 21625421]

[111] Cachot, J.; Cherel, Y.; Larcher, T.; Pfohl-Leszkowicz, A.; Laroche, J.; Quiniou, L.; Morin, J.; Schmitz, J.; Burgeot, T.; Pottier, D. Histopathological lesions and DNA adducts in the liver of European flounder (*Platichthys flesus*) collected in the Seine estuary *versus* two reference estuarine systems on the French Atlantic coast. *Environ. Sci. Pollut. Res. Int.,* **2013**, *20*(2), 723-737.
[http://dx.doi.org/10.1007/s11356-012-1287-0] [PMID: 23161498]

[112] Pinkney, A.E.; Harshbarger, J.C.; Karouna-Renier, N.K.; Jenko, K.; Balk, L.; Skarphéðinsdóttir, H.; Liewenborg, B.; Rutter, M.A. Tumor prevalence and biomarkers of genotoxicity in brown bullhead (*Ameiurus nebulosus*) in Chesapeake Bay tributaries. *Sci. Total Environ.,* **2011**, *410-411*, 248-257.
[http://dx.doi.org/10.1016/j.scitotenv.2011.09.035] [PMID: 21995877]

[113] Wirgin, I.; Waldman, J.R. Altered gene expression and genetic damage in North American fish populations. *Mutat. Res.,* **1998**, *399*(2), 193-219.
[http://dx.doi.org/10.1016/S0027-5107(97)00256-X] [PMID: 9672660]

[114] Myers, M.S.; French, B.L.; Reichert, W.L.; Willis, M.L.; Anulacion, B.F.; Collier, T.K.; Stein, J.E. Reductions in CYP1A expression and hydrophobic DNA adducts in liver neoplasms of English sole (*Pleuronectes vetulus*): further support for the resistant hepatocyte model of hepatocarcinogenesis. *Mar. Environ. Res.,* **1998**, *46*, 197-202. [http://dx.doi.org/10.1016/S0141-1136(98)00010-5]

[115] *Report of the Study Group on Integrated Monitoring of Contaminants and Biological Effects (SGIMC),*; Copenhagen, Denmark, **2011**, pp. March;14-18.

[116] Mathieu, A.; Payne, J.F.; Fancey, L.L.; Santella, R.M.; Young, T.L. Polycyclic aromatic hydrocarbon-DNA adducts in beluga whales from the Arctic. *J. Toxicol. Environ. Health,* **1997**, *51*(1), 1-4. [http://dx.doi.org/10.1080/00984109708984006] [PMID: 9169056]

[117] Martineau, D.; De Guise, S.; Fournier, M.; Shugart, L.; Girard, C.; Lagacé, A.; Béland, P. Pathology and toxicology of beluga whales from the St. Lawrence Estuary, Quebec, Canada. Past, present and future. *Sci. Total Environ.,* **1994**, *154*(2-3), 201-215. [http://dx.doi.org/10.1016/0048-9697(94)90088-4] [PMID: 7973607]

[118] Reichert, W.L.; French, B.L.; Stein, J.E. Exposure of marine mammals to genotoxic environmental contaminants: application of the 32P-postlabeling assay for measuring DNA-xenobiotic adducts. *Environ. Monit. Assess.,* **1999**, *56*, 225-239. [http://dx.doi.org/10.1023/A:1005991116993]

[119] Frenzilli, G.; Nigro, M.; Lyons, B.P. The Comet assay for the evaluation of genotoxic impact in aquatic environments. *Mutat. Res.,* **2009**, *681*(1), 80-92. [http://dx.doi.org/10.1016/j.mrrev.2008.03.001] [PMID: 18439870]

[120] Martins, M.; Costa, P.M.; Ferreira, A.M.; Costa, M.H. Comparative DNA damage and oxidative effects of carcinogenic and non-carcinogenic sediment-bound PAHs in the gills of a bivalve. *Aquat. Toxicol.,* **2013**, *142-143*, 85-95. [http://dx.doi.org/10.1016/j.aquatox.2013.07.019] [PMID: 23969285]

[121] Large, A.T.; Shaw, J.P.; Peters, L.D.; McIntosh, A.D.; Webster, L.; Mally, A.; Chipman, J.K. Different levels of mussel (*Mytilus edulis*) DNA strand breaks following chronic field and acute laboratory exposure to polycyclic aromatic hydrocarbons. *Mar. Environ. Res.,* **2002**, *54*(3-5), 493-497. [http://dx.doi.org/10.1016/S0141-1136(02)00139-3] [PMID: 12408607]

[122] Martins, M.; Ferreira, A.M.; Costa, M.H.; Costa, P.M. Comparing the genotoxicity of a potentially carcinogenic and a noncarcinogenic PAH, singly, and in binary combination, on peripheral blood cells of the European sea bass. *Environ. Toxicol.,* **2016**, *31*(11), 1307-1318. [http://dx.doi.org/10.1002/tox.22135] [PMID: 25728603]

[123] Dasgupta, S.; Cao, A.; Mauer, B.; Yan, B.; Uno, S.; McElroy, A. Genotoxicity of oxy-PAHs to Japanese medaka (*Oryzias latipes*) embryos assessed using the comet assay. *Environ. Sci. Pollut. Res. Int.,* **2014**, *21*(24), 13867-13876. [http://dx.doi.org/10.1007/s11356-014-2586-4] [PMID: 24510601]

[124] Williams, A.B.; Schumacher, B. p53 in the DNA-Damage-Repair Process. *Cold Spring Harb. Perspect. Med.,* **2016**, *6*(5), a026070. [http://dx.doi.org/10.1101/cshperspect.a026070] [PMID: 27048304]

[125] Duerksen-Hughes, P.J.; Yang, J.; Ozcan, O. p53 induction as a genotoxic test for twenty-five chemicals undergoing *in vivo* carcinogenicity testing. *Environ. Health Perspect.,* **1999**, *107*(10), 805-812.
[http://dx.doi.org/10.1289/ehp.99107805] [PMID: 10504146]

[126] Brzuzan, P.; Jurczyk, L.; Luczyński, M.K.; Góra, M. Real-time PCR analysis of p53 mRNA levels in tissues of whitefish (*Coregonus lavaretus*) exposed to benzo[*a*]pyrene. *Pol. J. Vet. Sci.,* **2006**, *9*(2), 139-143.
[PMID: 16780182]

[127] Rau Embry, M.; Billiard, S.M.; Di Giulio, R.T. Lack of p53 induction in fish cells by model chemotherapeutics. *Oncogene,* **2006**, *25*(14), 2004-2010.
[http://dx.doi.org/10.1038/sj.onc.1209238] [PMID: 16434976]

[128] Brzuzan, P.; Woźny, M.; Góra, M.; Łuczyński, M.K.; Jabłońska, A. Benzo[*a*]pyrene and cyclopenta[*c*]phenanthrene suppress expression of p53 in head kidney of rainbow trout. *Environ. Biotechnol.,* **2001**, *7*, 41-45.

[129] Cheng, C-H.; Yang, F-F.; Liao, S-A.; Miao, Y-T.; Ye, C-X.; Wang, A-L.; Tan, J-W.; Chen, X-Y. High temperature induces apoptosis and oxidative stress in pufferfish (*Takifugu obscurus*) blood cells. *J. Therm. Biol.,* **2015**, *53*, 172-179.
[http://dx.doi.org/10.1016/j.jtherbio.2015.08.002] [PMID: 26590470]

[130] Rivlin, N.; Brosh, R.; Oren, M.; Rotter, V. Mutations in the p53 tumor suppressor gene: important milestones at the various steps of tumorigenesis. *Genes Cancer,* **2011**, *2*(4), 466-474.
[http://dx.doi.org/10.1177/1947601911408889] [PMID: 21779514]

[131] Henkler, F.; Stolpmann, K.; Luch, A. Exposure to polycyclic aromatic hydrocarbons: bulky DNA adducts and cellular responses. *EXS,* **2012**, *101*, 107-131.
[http://dx.doi.org/10.1007/978-3-7643-8340-4_5] [PMID: 22945568]

[132] Cachot, J.; Cherel, Y.; Galgani, F.; Vincent, F. Evidence of p53 mutation in an early stage of liver cancer in European flounder, *Platichthys flesus* (L.). *Mutat. Res.,* **2000**, *464*(2), 279-287.
[http://dx.doi.org/10.1016/S1383-5718(99)00205-3] [PMID: 10648915]

[133] Bhaskaran, A.; May, D.; Rand-Weaver, M.; Tyler, C.R. Fish p53 as a possible biomarker for genotoxins in the aquatic environment. *Environ. Mol. Mutagen.,* **1999**, *33*(3), 177-184.
[http://dx.doi.org/10.1002/(SICI)1098-2280(1999)33:3<177::AID-EM1>3.0.CO;2-X] [PMID: 10334619]

[134] Santos, M.M.; Solé, M.; Lima, D.; Hambach, B.; Ferreira, A.M.; Reis-Henriques, M.A. Validating a multi-biomarker approach with the shanny Lipophrys pholis to monitor oil spills in European marine ecosystems. *Chemosphere,* **2010**, *81*(6), 685-691.
[http://dx.doi.org/10.1016/j.chemosphere.2010.07.065] [PMID: 20797766]

[135] Suárez-Ulloa, V.; Fernández-Tajes, J.; Manfrin, C.; Gerdol, M.; Venier, P.; Eirín-López, J.M. Bivalve omics: state of the art and potential applications for the biomonitoring of harmful marine compounds. *Mar. Drugs,* **2013**, *11*(11), 4370-4389.
[http://dx.doi.org/10.3390/md11114370] [PMID: 24189277]

CHAPTER 5

PAH Metabolites in Fish and Invertebrates: Analysis and Endocrine Disruptive Potential

Denise Fernandes[1], Anna Marqueño[1], Cinta Porte[1,*] and Montserrat Solé[2]

[1] *Environmental Chemistry Department, IDAEA-CSIC, 08034 Barcelona, Spain*

[2] *Marine Science Institute (ICM-CSIC), 08003 Barcelona, Spain*

Abstract: This chapter describes the metabolism of PAHs to oxidized and conjugated metabolites by both fish and invertebrates. So far, a wide range of metabolites have been described and a number of studies have alerted for the potential of phenol and quinone derivatives, as well as other reactive intermediates, to exert toxicity in early life stages of fish and reproductive disorders in adults. The endocrine disruptive properties have been mainly related to action at the receptor level, due to structural similarities of PAH and their metabolites with natural hormones, and also to interaction with key steroidogenic enzymes, which are present in vertebrate and invertebrate groups. The fast development of new analytical techniques, based on the use of high resolution mass spectrometry, will lead in the near future to the detection of a wider range of oxidized PAHs in abiotic matrices (*i.e.* water, sediment) and hopefully in biota, bringing along new research challenges.

Keywords: Analysis, Endocrine alteration, Metabolism, Oxidized PAHs.

PAH METABOLITE FORMATION IN AQUATIC SPECIES

In the last decade, the analysis of PAH metabolites has gained importance in front of the more traditional analysis of parent compounds. It was wisely pointed out that the analysis of only parent compounds in fish may underestimate exposure to PAHs [1]. Since then, PAH metabolites have been screened for in muscle, gut and gills from different fish species [2]. Although bile remains the

---

* **Corresponding author Cinta Porte**: Environmental Chemistry Department, IDAEA-CSIC, 08034 Barcelona, Spain; Tel/Fax: +34 93 4006100; E-mail: cpvqam@cid.csic.es

Daniela M. Pampanin and Magne O. Sydnes (Eds.)
All rights reserved-© 2017 Bentham Science Publishers

most frequently analyzed matrix, and the presence of PAH metabolites in bile has been successfully used as a biomarker of PAH exposure in field studies [3]. PAH metabolites have also been successfully determined in the urine of crabs [4, 5]. For other invertebrate species, the absence of an easily accessible biofluid makes the tissue analysis necessary in order to determine PAH exposure. Often the whole animal [6, 7], or specific tissues, such as muscle or visceral mass, are analyzed [8 - 10]. Recently, Layshock *et al.* detected the presence of PAHs substituted with keto- or quinone functional groups [11], namely, 9-fluorenone, 9,10- anthraquinone, benzofluorenone, and 7,12-benz[*a*]anthracenequinone in mussels, alerting that levels of oxygenated PAHs (OPAHs) could be similar or even exceed those of parental PAHs. This is one of the first studies reporting the presence of PAH metabolites in molluscs, as usually, OPAH are generated in fish and other aquatic organisms possessing well-developed enzymatic detoxification systems that efficiently convert PAHs to epoxide and hydroxylated derivatives, among others, during phase I metabolism (*e.g.* cytochrome P450 isoenzymes). These derivatives are further converted by phase II enzymes into highly water-soluble conjugates, such as glucuronides or sulphates that are readily excreted [7, 8].

Two of the main commonly phase I metabolites formed by vertebrates, and some invertebrates, (*e.g.* 1-hydroxypyrene and 9-hydroxyphenanthrene) share a structural resemblance to the natural hormone estradiol and this characteristic confers them estrogenic properties in humans and fish [12]. In addition, the conjugated forms of these phase I metabolites could also be deconjugated in the aquatic environment by microbial processes [13]. Thus, in the last years, concern has been raised about the disruptive potential of not only PAHs but also their metabolites over the endocrine reproductive regulation in fish and invertebrates. In this context, the scope of this review is to report on the different metabolic capacities of fish and invertebrates exposed to either parental or substituted PAHs and/or some of their main metabolites. Biotransformation of PAHs will be placed in the context of metabolites identified in vertebrates and invertebrates. It is thanks to recent advances in analytical technologies that progress on the identification of PAH metabolite formation in different aquatic species has been made. Tissue distribution of PAH metabolites will be described for fish and invertebrates collected from different field environments or exposed to crude oil and/or specific PAHs in the laboratory. In addition, already reported evidences of endocrine reproductive disturbances in aquatic species of both groups exposed to either parental PAHs and/or their respective metabolites will be considered.

PAH Metabolites in Fish

Fish have long been considered to possess a well developed enzymatic system able to biotransform PAHs into more polar metabolites and in this way eliminate them from their body compartments. Nonetheless, during these biotransformation reactions and due to some particular metabolite formation, toxicity and other endocrine disrupting events may arise [3]. The metabolic profile originated may also differ due to several factors even within fish. Species, age, physiological status, tissue distribution, way of exposure, pre-exposure conditions may be some factors that account for these differences and they have often been explored (see Table **1** for a summary).

Table 1. PAH metabolites identified in vertebrate and invertebrate tissues using different liquid- or gas-chromatography analytical methods.

Metabolites analyzed	Matrix	Sample treatment	Analytical method	Reference
1-, 2-Hydroxynaphthalene.	Fish bile	SPE + NH_2 cleanup + derivatization (BSTFA)	GC-MS (SIM)	[14]
	Fish bile	SPE + derivatization (BSTFA)	GC-MS (SIM)	[15]
1-Hydroxynaphthalene	Octopus	MeOH/ethyl acetate (1:1)	SPE-LC-FD	[7]
($1R,2R$)-1,2-dihydronaphthalene-1,2-Diol.	Fish bile	Dilution (MeOH:H_2O; 1:1)	F and SFS	[16]
1-Naphtoic acid.	Mussels	Soxhlet (DCM) (A) + saponification (B)	HPLC-UV (A) and GC-MS (B)	[17]
1,4-Naphthoquinone; 1-naphthaldehyde; 1,2-acenaphthenequinone; 1,8-naphthalic anhydride; naphthene-5,12-dione.	Fish (muscle and guts and gills)	ASE + filtration + silica gel column.	GC-MS (SIM)	[2]
2-Hydroxybiphenyl.	Fish bile	SPE + NH_2 cleanup + derivatization (BSTFA)	GC-MS (SIM)	[14]

(Table 1) contd.....

Metabolites analyzed	Matrix	Sample treatment	Analytical method	Reference
	Fish bile	SPE + derivatization (BSTFA)	GC-MS (SIM)	[15]
2-Biphenylcarboxaldehyde.	Fish (muscle and guts and gills)	ASE + filtration + silica gel column.	GC-MS (SIM)	[2]
9-Hydroxyfluorene.	Fish bile	SPE + NH₂ cleanup + derivatization (BSTFA)	GC-MS (SIM)	[14]
9-Fluorenone; benzo[*a*]fluorenone.	Fish (muscle, guts and gills)	ASE + filtration + silica gel column.	GC-MS (SIM)	[2]
9,10-Anthraquinone.	Mussels	Soxhlet (DCM) (A) + saponification (B)	HPLC-UV (A) and GC-MS (B)	[17]
9,10-Anthraquinone; 2-methyl-9,10-anthraquinone; 7H-benzo[*d,e*]anthracene-7-one; benzo[*a*]anthracene-7,12-dione; 1-indanone.	Fish (muscle and guts and gills)	ASE + filtration + silica gel column.	GC-MS (SIM)	[2]
Phenanthrene glucose; phenanthrene glucosesulphate; phenanthrene epoxide; phenanthrene monolepoxide; phenanthrene diolepoxide; phenanthrene diolepoxide; phenanthrene glucoside; phenanthrene diol; trihydroxy phenanthrene; phenanthrene orthoquinone; hidroxy phenanthrene; phenanthrene tetrol.	Fish bile and/or crab urine	SPE + filtration (0.22 μm)	LC-APCI(+)-IT (SRM)	[5]
1-, 2-, 3-Hydroxyphenanthrene.	Fish bile	SPE + NH₂ cleanup + derivatization (BSTFA)	GC-MS (SIM)	[14]
1-, 2-, 3-,4-,9-Hydroxyphenanthrene; 1,2-dihydrophenanthrene-1,2-diol; 9,10-dihidrophenanthrenee-9,10-diol; 3,4-dihidrophenanthrene-3,4-diol.	Fish liver microsomes	Liquid extraction (ethyl acetate)	HPLC-UV/Vis and radioactivity counting	[18]
9-Hydroxyphenanthrene.	Fish bile	SPE + derivatization (BSTFA)	GC-MS (SIM)	[15]
1,2-Dihydrophenanthrene-1,2-diol.	Fish bile	Dilution (methanol:water; 1:1)	F and SFS	[16]
1,2-Dihydrophenanthrene-1,2-diol;9,10-dihydrophenanthrene-9,10-diol or 9-hydroxyphenanthrene.	Fish bile and hepatic microsomes	Liquid extraction (ethyl acetate) + derivatizaton (TMSI)	GC-MS (SIM)	[19]
4H-Cyclopenta[*d,e,f*]phenanthrenone.	Fish (muscle and guts and gills)	ASE + filtration + silica gel column	GC-MS (SIM)	[2]
Pyrene glucose sulphate	Fish bile	SPE + filtration (0.22 μm)	LC-APCI(+)-IT (SRM)	[5]
1-Hydroxypyrene; pyrene glucose sulphate; pyrene-1-sulfate; pyrene-1-glucuronide; pyrenediol sulfate; pyrenediol glucuronide sulfate.	Marine whelk (muscle and visceral mass)	SPE + second SPE cleanup	LC-FLD and HPLC-ES--QqQ (MRM)	[8,10]

(Table 1) contd.....

Metabolites analyzed	Matrix	Sample treatment	Analytical method	Reference
1-Hydroxypyrene.	Fish bile	SPE + NH$_2$ cleanup + derivatization (BSTFA)	GC-MS (SIM)	[14]
1-Hydroxypyrene; 3-hydroxybenzo[*a*]pyrene.	Fish bile	SPE + Al$_2$O$_3$ and SiO$_2$ cleanup+ derivatization (BSTFA)	GC-MS (SIM)	[15]
1-Hydroxypyrene; pyrene-1-sulfate; pyrene-1-glucuronide; 1-pyrenecarboxylic acid; 1-methylpyrene glucuronide 1; 1-methylpyrene glucuronide 2; 1-carbonylpyrene glycine; glucoside–sulfate pyrene.	Annelid	Liquid extraction (methanol) + filtration (0.45 μm)	UHPLC-UV/FLR/QTOF-MS	[6]
1-Hydroxypyrene; pyrene-1-sulfate; pyrene-1-glucuronide.	Annelid	Protein precipitation (ethanol) + filtration (0.22 μm)	SFS, HPLC-DAD, HPLC-F	[20]
Pyrene-1-sulfate.	Crab urine	Dilution (EtOH:H$_2$O; 1:1)	F, SFS, HPLC-F	[4]
6H-Benzo[*c,d*]pyren-6-one.	Fish (muscle, guts, gills)	ASE + filtration + silica gel column	GC-MS (SIM)	[2]
1-Hydroxychrysene	Fish bile	SPE + derivatization (BSTFA)	GC-MS (SIM)	[15]
1-, 2-, 3-, 4-, 6-Hydroxychrysene; 1,2-dihydroxy-1,2-dihydrochrysene; 3,4-dihydroxy-3,4-dihydrochrysene.	Fish bile	Liquid extraction (ethyl acetate) + derivatizaton (TMSI)	GC-MS (SIM)	[1]
1-, 2-, 3-, 4-, 6-Hydroxychrysene; 1,2-dihydroxy-1,2-dihydrochrysene; 3,4-dihydroxy-3,4-dihydrochrysene; 5,6-dihidrochrysene-5,6-diol; 5,6-quinone chrysene.	Fish liver microsomes	Liquid extraction (ethyl acetate)	HPLC-UV/Vis and radioactivity counting	[18]

Bandowe *et al*. recently reported concentrations of 28 PAHs and 15 OPAHs in muscle, gut, and gill tissues of demersal fish (*Drae africana, Cynoglossus senegalensis,* and *Pomadasys peroteti*) finding that concentrations of the Σ15 OPAHs in fish muscle doubled the average concentration of the Σ28 PAHs [2]. The authors ascribed this difference to the fact that the OPAHs are more water-soluble and hence more bioavailable. The composition of the OPAHs mixture was dominated by 1,2-acenaphthenequinone (45%), 1-indanone (26%) 1,4-naphthoquinone (13%) and 6H-benzo[*c,d*]pyren-6-one (7%) whereas in guts and gills, the composition slightly differed: 1,2-acenaphthenequinone (41%), 1-indanone (23%) 1,8-naphthalic anhydride (15%), and 2-methyl-9,-0-anthraquinone (10%). Another field study, analyzing phase I metabolites of PAHs, identified 1- and 2-hydroxynaphthalene, 2-hydroxybiphenyl, 9-hydroxyfluorene, 1-hydroxypyrene, 1-, 2-, and 3-hydroxyphenanthrene as the main metabolites in bile of flounder (*Platichthys flesus*) collected from the Seine Bay [14]. Advances on fish metabolism after alkyl-PAH exposures have also derived from recent field studies. Sette *et al.* analysed the presence of alkylated-PAHs in bile of different fish species such as catfish (*Genidens genidens*), croaker (*Micropogonias furnieri*) and mullet (*Mugil liza*) by high performance liquid chromatography-atmospheric pressure chemical ionization/tandem mass spectrometry (HPLC-APCI/MS^2) and identified conjugated metabolites as pyrene glucosesulphate, phenanthrene glucoside and phenanthrene glucosesulphate in non-hydrolysed samples [5]. These same samples once hydrolysed using β-glucuronidase/arylsulfatase, produced phenanthrene epoxide, phenanthrene monolepoxide, phenanthrene diolepoxide, phenanthrene glucoside and their respective isomers. Overall, these studies highlight that the metabolizing system of different species may produce different metabolites, which some authors seem to relate with species feeding habits or to properties of the particular enzymatic system of each species.

Several laboratory experiments have been carried out in order to come to a better understanding of the fate of PAHs in the marine environment and especially on their bioaccumulation and biotransformation effects in fish. Dû-Lacoste *et al.* exposed juveniles of turbot (*Scophthalmus maximus*) to a water solution of a mixture of PAHs (naphthalene, phenanthrene, anthracene, fluoranthene, pyrene, chrysene, and B[*a*]P), a PAH-polluted sediment and an oil fuel elutriate for 4 days followed by a 6-day depuration period [15]. In all three experiments, a rapid production of biliary metabolites was observed, although the formation rate was slower in the sediment exposure group and higher with the dissolved mixture. The metabolites detected by gas chromatography-mass spectrometry in selected ion monitoring mode (GC-MS-SIM) were 1-, 2-hydroxynaphtalene, 2-hydroxybiphenyl, 9-hydroxyfluorene, 9-hydroxyphenanthrene and isomers, 1-hydroxypyrene, 1-hydroxychrysene, and 3-hydroxybenzo[*a*]pyrene. Biotransformation was also shown to be more efficient for pyrene than for phenanthrene. Benzo[*a*]pyrene was initially oxidised to several arene oxides that may rearrange spontaneously to phenols (3-OH-, 6-OH-, 7-OH- and 9-hydroxybenzo[*a*]pyrene). Regardless of the way of exposure, 1-hydroxypyrene was the major metabolite formed. Others have reported the presence of glucuronide conjugates or phenols in fish bile when exposing to phase I metabolites of PAHs, *viz.* 1-, 2-hydroxynaphthalene, 2-hydroxybiphenyl, 9-hydroxyfluorene, 1-hydroxypyrene, 1-, 2- and 3-hydroxyphenanthrene [21]. In another laboratory study, Atlantic cod (*Gadus morhua*) was exposed to chrysene (1 mg/kg) *via* intra-peritoneal (IP) and muscular injections showing that chrysene metabolites were three times higher after IP injection, even though the relative distribution of metabolites formed was very similar. The use of HPLC coupled to fluorescence detection in this study allowed the identification of chrysene and up to seven metabolites (1-hydroxychrysene, 2-hydroxychrysene, 3-hydroxy-chrysene, 4-hydroxychrysene and/or 6-hydroxychrysene, 1,2-dihydroxy-1-2-dihydrochrysene and 3,4-dihydroxy-3,4-dihydrochrysene) [22]. These meta-bolites were further confirmed by GC-MS-SIM mode, except for 3,4-dihydrochrysene, for which the method was not sensitive enough.

Although concentrations of alkylated PAHs in oil-contaminated sediments are much higher than those of unsubstituted PAHs, only little attention has been given to metabolism of alkyl-PAHs. There is increasing evidence that not only parental but mostly substituted PAHs, as main components of heavy crude oil, and their metabolites are responsible for toxicity and endocrine alterations in early life stages of fish [23, 24]. However, while PAH metabolism by CYP1A enzymes for compounds such are pyrene and chrysene is well characterized (Table **1**), knowledge on the metabolism and the resulting metabolites of alkyl-PAHs is still limited. The *in vitro* and *in vivo* CYP1A metabolism of alkyl-PAH such as retene (7-isopropyl-1-methylphenanthrene) and 1-methylphenanthrene generates a mixture of mono and di-hydroxylated metabolites. In trout exposed to alkyl-PAH, most metabolites are mono-hydroxylated (OH-ring) and chain hydroxylated (OH-chain) derivatives, as evidenced by HPLC analysis of tissue or bile. However, early eluting metabolites are presumed to be conjugated di-hydroxylated derivatives based on their UV and mass spectra analysis [24, 25]. Another study with alkyl-PAHs was performed by Turcotte *et al.* [23], who exposed rainbow trout (*Oncorhynchus mykiss*) through independent IP injections to anthracene, 1-methylanthracene, 9-methylanthracene, 9,10-dimethylanthracene, 2,3-dimethylanthracene and 2,3,6,7-tetramethylanthracene. Bile metabolites formed were analyzed by HPLC-diode array detection (DAD) and indicated that alkyl-anthracenes were mainly metabolized to benzene rings. When some of the fish were co-injected with the same compounds and β-naphthoflavone (BNF) (enzymatic induction), the metabolites produced were identical although at higher concentrations.

A number of fish studies have compared metabolites produced *in vivo versus* those produced after exposure of liver microsomes. When Atlantic cod (*Gadus morhua*) was intragastrically exposed to phenanthrene, 1,2-dihydrophenanthren--1,2-diol was the most abundant metabolite formed in bile after hydrolysis of the glucuronide and sulfate conjugates [19]. This pattern, however, changed when liver microsomes of cod, previously IP injected to phenanthrene and BNF independently, were incubated with phenanthrene (substrate) and NADPH (co-factor) *in vitro*. Thus, hepatic microsomes converted phenanthrene to 9,10-dihidrophenanthrene-9,10-diol, whereas those treated with BNF showed a dominating oxygenation of phenanthrene at the 1,2-position, although to a less extent to that previously observed *in vivo*. This difference could be due to a low regioselectivity of the cytochrome P450 in a reconstituted system, such as microsomal incubations. In fact, preincubations *in vitro* with anti-cytP450-serum did not change the phenanthrene metabolite pattern to any extent. Interestingly, this study reports a lack of agreement between microsomal and reconstituted cytochrome P450 metabolite patterns, suggesting that other control systems in addition to the P450-forms are involved in the *in vivo* transformation of phenanthrene by Atlantic cod to the 1,2-dihydrodiol metabolite [19]. In line with this former research, Pangrekar *et al.* also incubated liver microsomes (of brown Bullhead; *Ameriurus nebulosus*) of non-injected (control) and injected (IP) fish with 3-methylcholanthrene (3-MC) using [3H]phenanthrene as substrate [18]. All phenanthrene metabolites formed in microsomes were qualitatively similar in both control and 3-MC-treated fish; being the most predominant 1,2-dihydrophenanthrene-1,2-diol (dihydrodiol with a bay-region double bond) and phenanthrene phenols (1-,2-,3-,4-,9-hydroxyphenanthrene), whereas phenanthrene 3,4-dihidrophenanthrene-3,4-diol and 9,10-dihidrophenanthrene-9,10-diol (K-region dihydrodiol) were minor metabolites. Moreover, the relative proportions of the individual metabolites formed were markedly different. A notable aspect of phenanthrene metabolism by microsomes from 3-MC-treated fish was a significant decrease in the proportion of phenanthrene dihydrodiols in respect to control fish. Nonetheless, comparable metabolites were obtained when liver microsomes (control and 3-MC treated) were incubated with chrysene. In this last case, benzo-ring diols (1,2-diol and 3,4-diol) represented the major chrysene metabolite formed in both situations although the relative proportions of the two benzo ring diols differed. The 3-MC-treated microsomes produced a lower proportion of chrysene 1,2-diol compared to chrysene 3,4-diol, being the reverse observed for control microsomes.

PAH Metabolites in Invertebrates

It is well established that, within aquatic phyla, invertebrates express CYP activities to a lesser extent than vertebrates [26, 27]. In this context and due to the lower metabolism that characterizes this group, less research has been conducted on PAH metabolites. Within invertebrates, biotransformation of PAHs has been mainly investigated on annelids, molluscs and crustaceans.

Annelids and in particular, *Nereis diversicolor*, has often been used as model invertebrate for PAH metabolism [28, 29]. It has recently been demonstrated that this worm metabolized alkyl-PAHs primarily to polycyclic aromatic acids (PAAs) [6, 30]. These studies reported on the metabolism of 1-methyl pyrene, 1- methyl phenanthrene, 3,6-dimethylphenanthrene, and 1-, 2-, 3-, and 6-methyl chrysene in respect to their unsubstituted parent PAHs. Revealing that body burdens and production of PAAs was related to the position of the methyl group, showing in the annelid the same isomer specific preferences as in microbial degradation and a larger metabolism of alkyl-PAHs than their parent forms.

Molluscs is an invertebrate group that also has received notable attention on PAH metabolism. Accumulation of PAHs and alkyl-PAHs in a dose dependent manner was observed in oysters (*Crassostrea brasiliana*) when exposed to the water accommodated fraction (WAF) of diesel oil [31]. Others, using HPLC and GC/MS analysis, have reported the biotransformation of 1- methylnaphthalene and anthracene to 1-naphthoic acid and 9-, 10-anthraquinone by mussels (*Mytilus galloprovincialis*), evidencing that oxygenation of anthracene by these organisms occurs only on carbon number 9 and 10 [17]. Beach *et al.* exposed marine whelks

(*Buccinum undatum*) to pyrene and 1-hydroxypyrene through the diet over 15 days [10], and muscle and visceral mass were separately analyzed by HPLC-fluorescence and HPLC-electrospray ionization (ESI)/triple quadrupole mass spectrometer (QqQ) in multiple reaction monitoring (MRM) mode. They detected nine biotransformation products, namely 1-hydroxypyrene, pyrene-1-sulfate, pyrene-1-glucuronide, pyrene glucose sulfate, two isomers of pyrenediol sulfate and pyrenediol disulfate, and one isomer of pyrenediol glucuronide sulfate. The phase II metabolites formed did not differ between muscle and visceral mass. Interestingly, diconjugated metabolites were equally important in whelks exposed to pyrene as in those exposed to 1-hydroxypyrene, suggesting that both biotransformations take place through similar pathways, being 1-hydroxypyrene the main phase I metabolite of pyrene and further converted to phase II metabolites. The amount of pyrene metabolites formed represented 90% of the initial concentration of parent compound. Interestingly, the tissue distribution of [^{14}C]pyrene was also studied by autoradiography and revealed that activity was primarily present in the digestive and excretory system of the whelks and not in the gonads or muscle tissue. Likewise, Beach and Hellou exposed another marine whelk (*Neptunea lyrata*) to 1-hydroxypyrene through the diet over 35 days, and by using HPLC-fluorescence the authors could identify the formation of 1-hydroxypyrene, pyrene sulphate, pyrene glucuronide and one isomer of pyrenediol disulphate [8]. These results were taken as evidence that the biotransformation of PAHs in this invertebrate group is also species specific. Moreover, in these gastropods, the formation of di-conjugated metabolites such as pyrene disulphate that can originate more than one isomer complicates metabolic profiles, if compared to the single conjugated form typically analysed as metabolites of pyrene. Enzymatic hydrolysis of these diconjugates, a common practice used to simplify metabolic profiles of PAH, would produce pyrenediol instead of 1-hydroxypyrene, the latter being currently obtained in hydrolysed extracts. Pyrenediol has a reduced stability under laboratory conditions, as it degrades to pyrenequinone which has different fluorescence properties than single conjugated pyrene metabolites [8].

Regarding crustaceans, crabs (*Carcinus maenas*) were exposed to pyrene and phenanthrene and after 48 h, urine was analysed by HPLC-fluorescence and synchronous fluorescence spectrometry (SFS) [4]. Both techniques evidenced the production of pyrene-1-sulfate and other conjugate, as the main phase II pyrene metabolites while the peak of the parent compound was almost negligible, being glucosidation more efficient than sulfation, with a considerable interindividual variability that was related to moulting differences and/or pre-exposure to xenobiotics. For phenanthrene exposed crabs, urine analysis produced such a broad fluorescence that did not allow the identification of the metabolites. However, the use of LC-APCI(+)-MS/MS in SRM acquisition mode allowed the detection of phenanthrene diol, trihydroxy phenanthrene, phenanthrene orthoquinone, hydroxyphenanthrene; and phenanthrene epoxide, phenanthrene monolepoxide, phenanthrene diolepoxide, phenanthrene glucoside, and phenanthrene tetrol, after hydrolysis in urine of crabs (*Ucides cordatus*) [5].

Sexual Hormone Disruption

First studies to alert on PAHs reproductive disorders in aquatic wildlife were conducted in fish exposed to diesel oil (WSF), naphthalene [32], and to sediments collected from natural petroleum seeps [33], and in mussels exposed to WAF of different types of crude oil [34], and in the bivalve *Mya arenaria* inhabiting PAH chronically polluted sites [35, 36]. A link between endocrine disruption and PAHs was firstly related to the xenobiotic action over the cytochrome P450 metabolic pathway (CYP induction) mediated though the nuclear aryl hydrocarbon receptor (AhR). This CYP activity induction could also increase the clearance of sex hormones which are metabolised the same way as xenobiotics [37]. Another proposed action was a cross talk mechanism between receptors: the formerly mentioned AhR and the estrogen receptor (ER) causing anti-estrogenicity [38]. A number of studies in vertebrates have already demonstrated reduction of circulating plasmatic estradiol in the presence of PAHs both in field and laboratory conditions, associated with an increase of the excretion of estradiol metabolites in the bile and a reduction in vitellogenin (Vtg) synthesis in female fish (see review by Nicolas [39]). On the other hand, Tollefsen reported the existence of significant amounts of AhR antagonists in North Sea produced water (complex mixture of seawater with oil, PAHs and alkylphenols (AP)) [40]. Although the identity of the AhR antagonists remained unknown, the authors indicated that blocking of the AhR during sensitive windows of development would disrupt the masculinizing action of androgens and lead to feminization.

Endocrine Effects in Fish

As the number of studies progresses and the analytical technologies evolve, more refined approaches have allowed identifying which reproductive pathways are susceptible to be altered and the nature of the compound responsible for the alteration. Sex steroids are crucial in fish development since they act as morphogenic factors during sex differentiation [41], and as activation factors during sexual maturation that is regulated by maturation-inducing hormones, *e.g.* $17\alpha,20\alpha/\beta$-dihydroxyprogesterone ($17\alpha,20\alpha/\beta P$) [42]. Moreover, factors such as sex, maturation stage, and the degree of expression of target enzymes should be considered when assessing disruptive effects since they will likely modulate the sensitivity/susceptibility of the organisms to suffer endocrine alterations.

A number of *in vivo* and *in vitro* studies have focused on target key enzymes of the steroidogenic pathway and their potential modulation by PAHs or their metabolites. In particular, the activity of the isoenzyme CYP19 (P450aromatase), that converts androgens to estrogens is a frequently assessed enzyme as it is affected by PAH exposures [43 - 45]. Nonetheless, there are other enzymes from the steroidogenic pathway potentially targeted by this class of chemicals (Fig. **1**). An *in vivo* study with zebrafish (*Danio rerio*) reported a statistically significant increase in the expression of 20β-HSD mRNA in both the heads and ovaries of females exposed to B[*a*]P for 56 days [46]. In gonads, 20β-HSD is responsible for the conversion of 17α-hydroxyprogesterone (17P4) to the final maturation-inducing hormone $17\alpha,20\alpha/\beta P$ (Fig. **1**), whereas in the gills it is likely present to produce $17,20\alpha/\beta P$ for release into the water to function as a pheromone [47]. Therefore, these results alert on the possibility that B[*a*]P may alter reproductive behaviours by affecting the production of both hormones and pheromones. Ovarian tissue of flounder (*Platichthys flesus*) exposed to 15 µM of phenanthrene, chrysene and B[*a*]P showed an inhibition of CYP17 (C17,20-lyase) and CYP19, leading to reduced secretion of androstenedione (AD) and estradiol (E_2) in vitellogenic ovarian tissue [48]. C17,20-lyase plays a key role in the conversion of 17P4 to AD, a precursor of testosterone (T), whereas CYP19 is the enzyme responsible for converting AD and T into estrone and E_2, thus regulating local and systemic levels of estrogens in the body (Fig. **1**) [49]. Phenanthrene and chrysene also inhibited the activity of 17β-HSD; the enzyme responsible for the reduction of AD to T and estrone to E_2 (Fig. **1**) [50]. In addition, both conjugated T and E_2 were significantly decreased in phenanthrene-treated incubations, suggesting potential alterations in the levels of free active endogenous steroids in fish [50]. Meanwhile, an increase in hepatic UGT-T (increased clearance of T) (Fig. **1**), a significant depletion on plasmatic T levels and a trend towards higher P450aromatase in gonads was detected in juvenile turbot exposed to 0.5 μg/g of dispersed North Sea crude oil for 21 days [51]. Similarly, exposure to the Prestige fuel oil (2.5-50 mg fuel/g food, containing 3.6-19.8 μg PAHs/g food) sharply reduced the circulating levels of T in plasma of juvenile turbot [52]. Meanwhile, exposure to 0.06-6.0 μg/g of phenanthrene for 50 days significantly decreased testicular levels of E_2 and inhibited spermatogenesis in male *Sebastiscus marmoratus* [53]. Recently, Polar cod (*Boreogadus saida*) exposed for 28 days to a mixture of PAH, alkylated-PAH, and AP simulating the composition of North Sea produced water, reported a decrease in oocyte number and an inhibition of spermatogenesis [54].

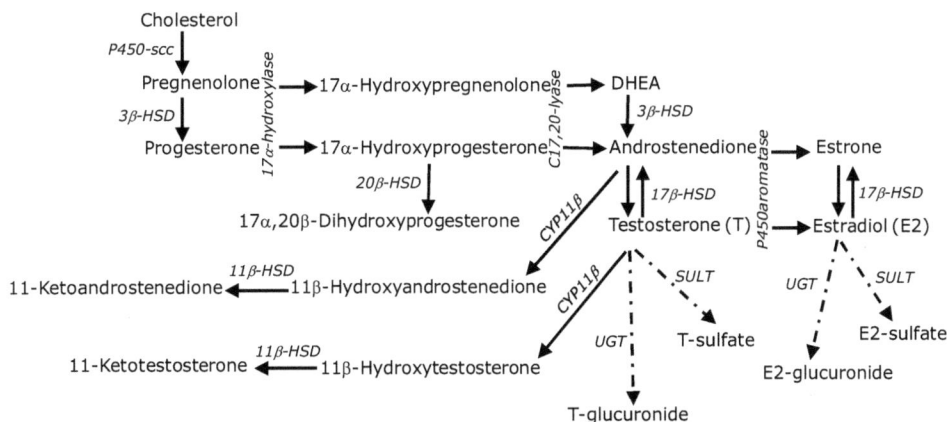

Fig. (1). Scheme illustrating the main enzymatic pathways involved in the synthesis and clearance of key sex steroids in teleost fish gonads. *P450scc*: P450 side chain cleavage; *17α-hydroxylase*: CYP17A1; *C17,20-lyase*: CYP17; *CYP11β*: 11β- hydroxylase; *P450aromatase*: CYP19; *HSD*: hydroxysteroid dehydrogenases; *UGT*: UDP-glucuronosyltransferase *SULT*: sulfotransferase; DHEA: Dihydroxyepiandrosterone ($-\cdot-\!\!\!\succ$) phase II metabolism of steroids.

Despite this evidence, the mechanisms behind these changes are not fully elucidated as most studies relating endocrine disturbances in fish to PAHs are usually based on exposure to the parental form. Considering that chemicals suffer *in vivo* transformation reactions, a relationship of cause-effect is not so straightforward and new associations using parental and metabolite forms should be addressed. First attempts on this direction revealed that several hydroxylated-PAH metabolites (*e.g.* 1-hydroxynaphthalene, 9-hydroxyphenanthrene, 1-hydroxypyrene, 1-hydroxychrysene), inhibited the mitochondrial activity of C17,20-lyase and CYP11β enzymes in carp male gonads, in particularly 9-hydroxyphenanthrene (IC50s: 11 and 31 μM for C17,20-lyase and CYP11β, respectively) [55]. Furthermore, an enhancement in the microsomal formation of 11-ketandrostenedione was also reported in the presence of 9-hydroxyphenanthrene, indicating the stimulation of 11β-HSD2 activity [55]. In male gonads, both T and AD are transformed into their respective 11-hydroxylated metabolites by CYP11β activity (11β-hydroxylases) and further metabolized into their 11-ketoandrogen form by 11β-hydroxysteorid dehydrogenase (11β-HSD) catalyzed reaction (Fig. **1**) [56, 57]. 11-Ketotestosterone is considered to be the main androgen in male teleost fish, being more effective than T in stimulating secondary sexual characters, influencing spermatogenesis and stimulating reproductive behavior [58]. In addition, 9-hydroxyphenanthrene also inhibited carp ovarian microsomal CYP19aromatase activity with a rather low IC50 (4.3 μM) when compared with other environmental endocrine disruptors [55]. Meanwhile, the parental forms tested (*i.e.* B[*a*]P, pyrene, chrysene, phenanthrene and naphthalene) had no effect on the above mentioned steroidogenic enzymes neither in carp or in sea bass gonads [55, 59]. Similarly, incubations with rainbow trout (*Onchorhynchus mykiss*) microsomes and 10 μM of chrysene and B[*a*]P had no significant effect on CYP19aromatase activity in both brain and ovaries [43]. Thus, these evidences suggest that PAH metabolites rather than their parental form are likely responsible for endocrine disrupting properties in fish by interfering with the activity of steroidogenic enzymes. In fact, it is the phenolic moiety of hydroxylated-PAH (and of other emerging endocrine disrupting chemicals) that makes them structural analogues to sex steroids, and there are strong evidences that hydroxylated-PAHs are estrogenic as they bind to the ER in reporter gene assays

[12, 60]. Moreover, estrogenic and antiestrogenic properties have been attributed to quinoid PAH derivates by the use of the yeast-two hybrid assay [61].

Endocrine Effects in Invertebrates

Further support to the action of the metabolite rather than the parental form can be found in the differences in metabolic activities between vertebrates and invertebrates. Quantitative and qualitative differences exist in different phyla in terms of xenobiotic metabolising enzymes capacities, including CYPs and other detoxifying mechanisms; therefore, the severity of the endocrine disturbances is also likely to differ. It was formerly accepted that fish were able to transform PAHs whereas invertebrates only accumulate them, so signs of disturbances were more evident in the vertebrate group. However the strictness of the former statement is not so clear as in each phylogenic group there is a balance between both processes and the final outcome depends on many factors (*e.g.* species, age, size, dose and length of the exposures and targeted tissue) regulating them [62].

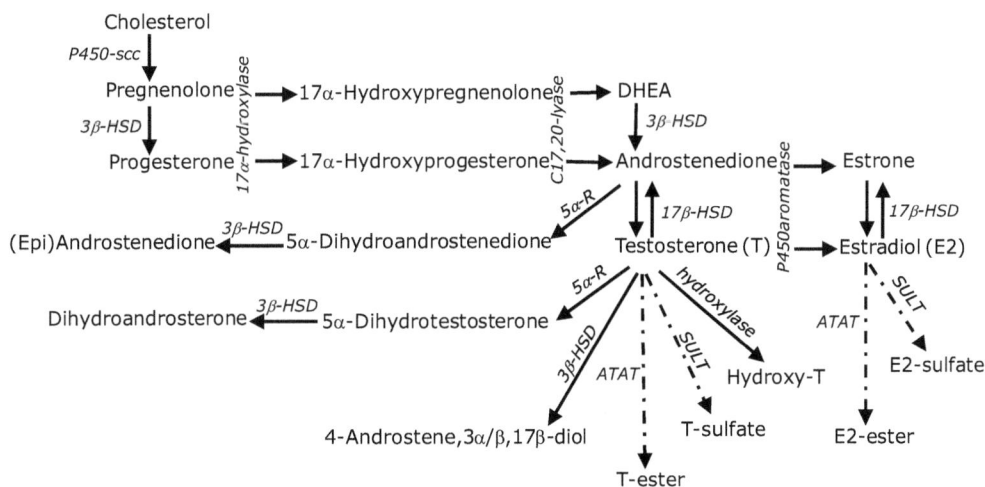

Fig. (2). Principal steroid synthesis and metabolic pathways that are presently described in invertebrates. *P450scc*: P450 side chain cleavage; *17α-hydroxylase*: CYP17A1; *C17,20-lyase*: CYP17; *HSD*: hydroxysteroid dehydrogenases; *5α-R*: 5α-reductase; *ATAT*: fatty acid acyl-CoA acyltransferase; *SULT*: sulfotransferase; DHEA: Dihydroxyepiandrosterone; 5α-Androstane-3α,17β-diol: Dihydroandrosterone; (— · — ➤) phase II metabolism of steroids. Adapted from Fernandes *et al.* [67].

Most key steps of steroidogenesis and steroid metabolism formerly described for fish have also been reported in different invertebrate species (Fig. **2**). In particular, 3β-HSD, various types of CYP17s and 17β-HSDs have been described in different invertebrate species, either by directly exposing animals to steroid precursors or by incubating homogenates with these precursors (Fig. **2**). Furthermore, the potency of progesterone (P4), E_2 and T to promote gamete development or its related energy metabolism (induction of vitellogenic process by E_2) has been demonstrated in molluscs [63 - 65]. Research on the mechanisms of endocrine disruption by PAHs on invertebrates is still much needed. Up to now, there are few reports relating changes in sex hormone levels and steroidogenesis by PAHs in molluscs or any other species of invertebrate. In this context, the mussels *Mytilus edulis* and the scallop *Chlamys farreri* have shown to be suitable species for studying endocrine-disrupting effects of PAHs in marine invertebrates over the metabolic (CYPs) and steroidogenic pathways. Mussels (*M. edulis*) exposed to a mixture of 0.5 μg/g of North Sea crude oil together with 0.1 μg/g of AP reported a significant increase in the microsomal P450 aromatase activity in both digestive glands and gonads [66]. An increase in esterified T and E2 levels in gonad tissue of mussels was reported, with no effects in steroid free levels [66]. It has been suggested that esterification is the major biotransformation pathway for T and E_2 in molluscs (Fig. **2**), regulating the levels of free steroids [67]. In addition, the sulfation of E_2 was also investigated as a conjugation pathway (Fig. **2**) and increased activities up to 2.8-fold were observed in the digestive gland cytosol of *M. edulis* exposed to both crude oil and crude oil + AP from produced water [66]. Meanwhile, a 10 day B[*a*]P exposure severely disrupted the normal fluctuation of P4 levels and also inhibited E_2 and T secretion in ovary of sexually mature *Chlamys farreri* [68]. The same study showed that 3β-HSD, CYP17a 17β-HSD1 (Fig. **2**) are potential target genes of B[*a*]P disruption to ovary steroidogenesis, and B[*a*]P can induce up-regulation of ER-Vtg mRNA expression and AhR action. The latest was suggested to be involved in invertebrate CYP regulation and ER transcription [68]. Similarly, B[*a*]P caused a significant increase in mRNA expression of ER and Vtg at 0.4 and 2 μg/L but down regulated them at 10 μg/L in *C. farreri* [69]. Altogether, these evidences suggest that PAHs and chemicals present in produced water can lead to alterations in key biochemical pathways of invertebrates that could turn into physiological consequences for the organism.

CONCLUDING REMARKS

In summary, this chapter describes the metabolism of PAHs to oxidised and conjugated metabolites in both fish and invertebrates. A wide range of metabolites have been described by using different analytical techniques and different approaches (*e.g.* field and laboratory exposures, *in vitro* methods). However, the metabolites detected so far might just be the tip of the iceberg. The fast development of new analytical techniques based on the use of high resolution MS will probably lead to the detection in the next coming years of a wider range of oxidized PAHs in abiotic matrices (water, sediment) and hopefully, in biota. It's also important to highlight that some studies have evidenced stronger endocrine disrupting properties of PAH metabolites in comparison to parental compounds. These metabolites act, not only at the receptor level due to structural similarities with natural hormones, but also through direct interaction with key steroidogenic enzymes. Moreover, some studies, dealing with invertebrates, alert of the importance of carboxylic acid metabolites of alkyl-PAHs. Additionally, it should be considered that even if invertebrates do not actively transform PAHs to the same extent than fish, they can take up oxidised PAHs from the environment [8, 70], and this can lead to the transfer of oxidised PAH through the food chain [62, 71 - 73]. Finally, it should be mentioned that in a context of climate change, increased abundance of oxidised PAH forms would be expected, due to increased UV radiation, microbial activity, *etc.*, and this might have severe consequences for aquatic ecosystems.

CONFLICT OF INTEREST

The authors confirm that they have no conflict of interest to declare for this publication.

ACKNOWLEDGEMENTS

Anna Marqueño acknowledges a pre-doctoral fellowship BES-2015-074842. This work has been carried out under project CGL2014-52144-P.

REFERENCES

[1]   Jonsson, G.; Taban, I.C.; Jørgensen, K.B.; Sundt, R.C. Quantitative determination of de-conjugated

chrysene metabolites in fish bile by HPLC-fluorescence and GCMS. *Chemosphere,* **2004**, *54*(8), 1085-1097.
[http://dx.doi.org/10.1016/j.chemosphere.2003.09.026] [PMID: 14664837]

[2]   Bandowe, B.A.; Bigalke, M.; Boamah, L.; Nyarko, E.; Saalia, F.K.; Wilcke, W. Polycyclic aromatic compounds (PAHs and oxygenated PAHs) and trace metals in fish species from Ghana (*West Africa*): bioaccumulation and health risk assessment. *Environ. Int.,* **2014**, *65*, 135-146.
[http://dx.doi.org/10.1016/j.envint.2013.12.018] [PMID: 24486971]

[3]   Beyer, J.; Jonsson, G.; Porte, C.; Krahn, M.M.; Ariese, F. Analytical methods for determining metabolites of polycyclic aromatic hydrocarbon (PAH) pollutants in fish bile: A review. *Environ. Toxicol. Pharmacol.,* **2010**, *30*(3), 224-244.
[http://dx.doi.org/10.1016/j.etap.2010.08.004] [PMID: 21787655]

[4]   Fillmann, G.; Watson, G.M.; Howsam, M.; Francioni, E.; Depledge, M.H.; Readman, J.W. Urinary PAH metabolites as biomarkers of exposure in aquatic environments. *Environ. Sci. Technol.,* **2004**, *38*(9), 2649-2656.
[http://dx.doi.org/10.1021/es0350839] [PMID: 15180061]

[5]   Sette, C.B.; Pedrete, Tde.A.; Felizzola, J.; Nudi, A.H.; Scofield, Ade.L.; Wagener, Ade.L. Formation and identification of PAHs metabolites in marine organisms. *Mar. Environ. Res.,* **2013**, *91*, 2-13.
[http://dx.doi.org/10.1016/j.marenvres.2013.02.004] [PMID: 23518369]

[6]   Malmquist, L.M.; Selck, H.; Jørgensen, K.B.; Christensen, J.H. Polycyclic aromatic acids are primary metabolites of alkyl-PAHs-A case study with *Nereis diversicolor*. *Environ. Sci. Technol.,* **2015**, *49*(9), 5713-5721.
[http://dx.doi.org/10.1021/acs.est.5b01453] [PMID: 25827176]

[7]   Lourenço, D.; Silva, L.J.; Lino, C.M.; Morais, S.; Pena, A. SPE-LC-FD determination of polycyclic aromatic hydrocarbon monohydroxy derivatives in cephalopods. *J. Agric. Food Chem.,* **2014**, *62*(12), 2685-2691.
[http://dx.doi.org/10.1021/jf4056852] [PMID: 24588515]

[8]   Beach, D.G.; Hellou, J. Bioaccumulation and biotransformation of 1-hydroxypyrene by the marine whelk *Neptunea lyrata*. *Int. J. Environ. Anal. Chem.,* **2011**, *91*, 1227-1243.
[http://dx.doi.org/10.1080/03067310903524830]

[9]   Beach, D.G.; Quilliam, M.A.; Hellou, J. Analysis of pyrene metabolites in marine snails by liquid chromatography using fluorescence and mass spectrometry detection. *J. Chromatogr. B Analyt. Technol. Biomed. Life Sci.,* **2009**, *877*(22), 2142-2152.
[http://dx.doi.org/10.1016/j.jchromb.2009.06.006] [PMID: 19553165]

[10]   Beach, D.G.; Quilliam, M.A.; Rouleau, C.; Croll, R.P.; Hellou, J. Bioaccumulation and biotransformation of pyrene and 1-hydroxypyrene by the marine whelk *Buccinum undatum*. *Environ. Toxicol. Chem.,* **2010**, *29*(4), 779-788.
[http://dx.doi.org/10.1002/etc.112] [PMID: 20821506]

[11]   Layshock, J.A.; Wilson, G.; Anderson, K.A. Ketone and quinone-substituted polycyclic aromatic hydrocarbons in mussel tissue, sediment, urban dust, and diesel particulate matrices. *Environ. Toxicol. Chem.,* **2010**, *29*(11), 2450-2460.
[http://dx.doi.org/10.1002/etc.301] [PMID: 20830751]

[12] Wenger, D.; Gerecke, A.C.; Heeb, N.V.; Naegeli, H.; Zenobi, R. Catalytic diesel particulate filters reduce the *in vitro* estrogenic activity of diesel exhaust. *Anal. Bioanal. Chem.,* **2008**, *390*(8), 2021-2029.
[http://dx.doi.org/10.1007/s00216-008-1872-8] [PMID: 18264702]

[13] Gomes, R.L.; Scrimshaw, M.D.; Lester, J.N. Fate of conjugated natural and synthetic steroid estrogens in crude sewage and activated sludge batch studies. *Environ. Sci. Technol.,* **2009**, *43*(10), 3612-3618.
[http://dx.doi.org/10.1021/es801952h] [PMID: 19544862]

[14] Mazéas, O.; Budzinski, H. Solid-phase extraction and purification for the quantification of polycyclic aromatic hydrocarbon metabolites in fish bile. *Anal. Bioanal. Chem.,* **2005**, *383*(6), 985-990.
[http://dx.doi.org/10.1007/s00216-005-0096-4] [PMID: 16244859]

[15] Le Dû-Lacoste, M.; Akcha, F.; Dévier, M.H.; Morin, B.; Burgeot, T.; Budzinski, H. Comparative study of different exposure routes on the biotransformation and genotoxicity of PAHs in the flatfish species, Scophthalmus maximus. *Environ. Sci. Pollut. Res. Int.,* **2013**, *20*(2), 690-707.
[http://dx.doi.org/10.1007/s11356-012-1388-9] [PMID: 23247530]

[16] Pampanin, D.M.; Kemppainen, E.K.; Skogland, K.; Jørgensen, K.B.; Sydnes, M.O. Investigation of fixed wavelength fluorescence results for biliary metabolites of polycyclic aromatic hydrocarbons formed in Atlantic cod (*Gadus morhua*). *Chemosphere,* **2016**, *144*, 1372-1376.
[http://dx.doi.org/10.1016/j.chemosphere.2015.10.013] [PMID: 26492423]

[17] Tomruk, A.; Güven, K.C. Biotransformation of 1- methylnaphthalene and anthracene in mussels (*Mytilus galloprovincialis* Lamarck, 1819). *Freseniu Environ. Bull.,* **2008**, *17*, 256-259.

[18] Pangrekar, J.; Kole, P.L.; Honey, S.A.; Kumar, S.; Sikka, H.C. Metabolism of phenanthrene by brown bullhead liver microsomes. *Aquat. Toxicol.,* **2003**, *64*(4), 407-418.
[http://dx.doi.org/10.1016/S0166-445X(03)00075-4] [PMID: 12878411]

[19] Goksøyr, A.; Solbakken, J.E.; Klungsøyr, J. Regioselective metabolism of phenanthrene in Atlantic cod (*Gadusmorhua*): studies on the effects of monooxygenase inducers and role of cytochromes P-450. *Chem. Biol. Interact.,* **1986**, *60*(3), 247-263.
[http://dx.doi.org/10.1016/0009-2797(86)90056-6] [PMID: 3791491]

[20] Giessing, A.M.; Mayer, L.M.; Forbes, T.L. Synchronous fluorescence spectrometry of 1-hydroxypyrene: a rapid screening method for identification of PAH exposure in tissue from marine polychaetes. *Mar. Environ. Res.,* **2003**, *56*(5), 599-615.
[http://dx.doi.org/10.1016/S0141-1136(03)00045-X] [PMID: 12927741]

[21] Leonard, J.D.; Hellou, J. Separation and characterization of gall bladder bile metabolites from speckled trout, Salvelinus fontinalis, exposed to individual polycyclic aromatic compounds. *Environ. Toxicol. Chem.,* **2001**, *20*(3), 618-623.
[http://dx.doi.org/10.1002/etc.5620200322] [PMID: 11349864]

[22] Jonsson, G.; Sundt, R.C.; Aas, E.; Beyer, J. An evaluation of two fluorescence screening methods for the determination of chrysene metabolites in fish bile. *Chemosphere,* **2004**, *56*(1), 81-90.
[http://dx.doi.org/10.1016/j.chemosphere.2004.02.026] [PMID: 15109882]

[23] Turcotte, D.; Headley, J.V.; Abudulai, N.L.; Hodson, P.V.; Brown, R.S. Identification of phase II *in vivo* metabolites of alkyl-anthracenes in rainbow trout (*Oncorhynchusmykiss*). *Chemosphere,* **2011**, *85*(10), 1585-1591.
[http://dx.doi.org/10.1016/j.chemosphere.2011.08.005] [PMID: 21907385]

[24]　Fallahtafti, S.; Rantanen, T.; Brown, R.S.; Snieckus, V.; Hodson, P.V. Toxicity of hydroxylated alkyl-phenanthrenes to the early life stages of Japanese medaka (*Oryzias latipes*). *Aquat. Toxicol.,* **2012**, *106-107*, 56-64.
　　　　[http://dx.doi.org/10.1016/j.aquatox.2011.10.007] [PMID: 22071127]

[25]　Hodson, P.V.; Qureshi, K.; Noble, C.A.; Akhtar, P.; Brown, R.S. Inhibition of CYP1A enzymes by α-naphthoflavone causes both synergism and antagonism of retene toxicity to rainbow trout (*Oncorhynchus mykiss*). *Aquat. Toxicol.,* **2007**, *81*(3), 275-285.
　　　　[http://dx.doi.org/10.1016/j.aquatox.2006.12.012] [PMID: 17257690]

[26]　Livingstone, D.R. The fate of organic xenobiotics in aquatic ecosystems: quantitative and qualitative differences in biotransformation by invertebrates and fish. *Comp. Biochem. Physiol. A Mol. Integr. Physiol.,* **1998**, *120*(1), 43-49.
　　　　[http://dx.doi.org/10.1016/S1095-6433(98)10008-9] [PMID: 9773498]

[27]　Rewitz, K.F.; Styrishave, B.; Løbner-Olsen, A.; Andersen, O. Marine invertebrate cytochrome P450: emerging insights from vertebrate and insects analogies. *Comp. Biochem. Physiol. C Toxicol. Pharmacol.,* **2006**, *143*(4), 363-381.
　　　　[http://dx.doi.org/10.1016/j.cbpc.2006.04.001] [PMID: 16769251]

[28]　Jørgensen, A.; Giessing, A.M.; Rasmussen, L.J.; Andersen, O. Biotransformation of polycyclic aromatic hydrocarbons in marine polychaetes. *Mar. Environ. Res.,* **2008**, *65*(2), 171-186.
　　　　[http://dx.doi.org/10.1016/j.marenvres.2007.10.001] [PMID: 18023473]

[29]　Dejong, C.A.; Wilson, J.Y. The Cytochrome P450 superfamily complement (CYPome) in the annelid *Capitella teleta. PLoS One,* **2014**, *9*(11), e107728.
　　　　[http://dx.doi.org/10.1371/journal.pone.0107728] [PMID: 25390889]

[30]　Malmquist, L.M.; Selck, H.; Jørgensen, K.B.; Christensen, J.H. Polycyclic aromatic acids are primary metabolites of Alkyl-PAHs-A case study with *Nereis diversicolor. Environ. Sci. Technol.,* **2015**, *49*(9), 5713-5721.
　　　　[http://dx.doi.org/10.1021/acs.est.5b01453] [PMID: 25827176]

[31]　Lüchmann, K.H.; Mattos, J.J.; Siebert, M.N.; Granucci, N.; Dorrington, T.S.; Bícego, M.C.; Taniguchi, S.; Sasaki, S.T.; Daura-Jorge, F.G.; Bainy, A.C. Biochemical biomarkers and hydrocarbons concentrations in the mangrove oyster Crassostrea brasiliana following exposure to diesel fuel water-accommodated fraction. *Aquat. Toxicol.,* **2011**, *105*(3-4), 652-660.
　　　　[http://dx.doi.org/10.1016/j.aquatox.2011.09.003] [PMID: 21963596]

[32]　Thomas, P.; Budiantara, L. Reproductive life-history stages sensitive to oil and naphthalene in Atlantic croaker. *Mar. Environ. Res.,* **1995**, *39*, 147-150.
　　　　[http://dx.doi.org/10.1016/0141-1136(94)00072-W]

[33]　Roy, L.A.; Steinert, S.; Bay, S.M.; Greenstein, D.; Sapozhnikova, Y.; Bawardi, O.; Leifer, I.; Schlenk, D. Biochemical effects of petroleum exposure in hornyhead turbot (*Pleuronichthys verticalis*) exposed to a gradient of sediments collected from a natural petroleum seep in CA, USA. *Aquat. Toxicol.,* **2003**, *65*(2), 159-169.
　　　　[http://dx.doi.org/10.1016/S0166-445X(03)00135-8] [PMID: 12946616]

[34]　Cajaraville, M.P.; Marigómez, J.A.; Angulo, E. Comparative effects of the water accommodated fraction of three oils on mussels. 1. Survival, growth and gonad development. *Comp. Biochem. Physiol. C. Comp. Pharmacol. Toxicol.,* **1992**, *102*(1), 103-112.
[http://dx.doi.org/10.1016/0742-8413(92)90051-8] [PMID: 1358514]

[35]　Gagné, F.; Blaise, C.; Pellerin, J.; Gauthier-Clerc, S. Alteration of the biochemical properties of female gonads and vitellins in the clam *Mya arenaria* at contaminated sites in the Saguenay Fjord. *Mar. Environ. Res.,* **2002**, *53*(3), 295-310.
[http://dx.doi.org/10.1016/S0141-1136(01)00122-2] [PMID: 11939294]

[36]　Gauthier-Clerc, S.; Pellerin, J.; Blaise, C.; Gagné, F. Delayed gametogenesis of *Mya arenaria* in the Saguenay fjord (Canada): a consequence of endocrine disruptors? *Comp. Biochem. Physiol. C Toxicol. Pharmacol.,* **2002**, *131*(4), 457-467.
[http://dx.doi.org/10.1016/S1532-0456(02)00041-8] [PMID: 11976061]

[37]　Tintos, A.; Gesto, M.; Míguez, J.M.; Soengas, J.L. Naphthalene treatment alters liver intermediary metabolism and levels of steroid hormones in plasma of rainbow trout (*Oncorhynchus mykiss*). *Ecotoxicol. Environ. Saf.,* **2007**, *66*(2), 139-147.
[http://dx.doi.org/10.1016/j.ecoenv.2005.11.008] [PMID: 16466791]

[38]　Navas, J.M.; Segner, H. Antiestrogenicity of beta-naphthoflavone and PAHs in cultured rainbow trout hepatocytes: evidence for a role of the arylhydrocarbon receptor. *Aquat. Toxicol.,* **2000**, *51*(1), 79-92.
[http://dx.doi.org/10.1016/S0166-445X(00)00100-4] [PMID: 10998501]

[39]　Nicolas, J.M. Vitellogenesis in fish and the effects of polycyclic aromatic hydrocarbon contaminants. *Aquat. Toxicol.,* **1999**, *45*, 77-90.
[http://dx.doi.org/10.1016/S0166-445X(98)00095-2]

[40]　Tollefsen, K.E. Binding of alkylphenols and alkylated non-phenolics to the rainbow trout (*Oncorhynchus mykiss*) plasma sex steroid-binding protein. *Ecotoxicol. Environ. Saf.,* **2007**, *68*(1), 40-48.
[http://dx.doi.org/10.1016/j.ecoenv.2006.07.002] [PMID: 16945415]

[41]　Piferrer, F. Endocrine sex control strategies for the feminization of teleost fish. *Aquaculture,* **2001**, *197*, 229-281.
[http://dx.doi.org/10.1016/S0044-8486(01)00589-0]

[42]　Scott, A.P.; Sumpter, J.P.; Stacey, N. The role of the maturation-inducing steroid, 17,20beta-dihydroxypregn-4-en-3-one, in male fishes: a review. *J. Fish Biol.,* **2010**, *76*(1), 183-224.
[http://dx.doi.org/10.1111/j.1095-8649.2009.02483.x] [PMID: 20738705]

[43]　Hinfray, N.; Palluel, O.; Turies, C.; Cousin, C.; Porcher, J.M.; Brion, F. Brain and gonadal aromatase as potential targets of endocrine disrupting chemicals in a model species, the zebrafish (*Danio rerio*). *Environ. Toxicol.,* **2006**, *21*(4), 332-337.
[http://dx.doi.org/10.1002/tox.20203] [PMID: 16841311]

[44]　Patel, M.R.; Scheffler, B.E.; Wang, L.; Willett, K.L. Effects of benzo(a)pyrene exposure on killifish (*Fundulus heteroclitus*) aromatase activities and mRNA. *Aquat. Toxicol.,* **2006**, *77*(3), 267-278.
[http://dx.doi.org/10.1016/j.aquatox.2005.12.009] [PMID: 16458981]

[45]　Dong, W.; Wang, L.; Thornton, C.; Scheffler, B.E.; Willett, K.L. Benzo(a)pyrene decreases brain and

ovarian aromatase mRNA expression in *Fundulus heteroclitus. Aquat. Toxicol.,* **2008**, *88*(4), 289-300.
[http://dx.doi.org/10.1016/j.aquatox.2008.05.006] [PMID: 18571745]

[46]   Hoffmann, J.L.; Oris, J.T. Altered gene expression: a mechanism for reproductive toxicity in zebrafish exposed to benzo[*a*]pyrene. *Aquat. Toxicol.,* **2006**, *78*(4), 332-340.
[http://dx.doi.org/10.1016/j.aquatox.2006.04.007] [PMID: 16765461]

[47]   Pankhurst, N.W. Gonadal steroids: functions and patterns of change. In: *Fish Reproduction*; Rocha, M.J.; Arukwe, A.; Kapoor, B.G., Eds.; Science Publishers, **2008**; pp. 67-111.

[48]   Reis-Henriques, M.A.; Coimbra, J.; Coimbra, J. Polycyclic aromatic hydrocarbons inhibit *in vitro* ovarian steroidogenesis in the flounder (*Platichthys flesus* L.). *Aquat. Toxicol.,* **2000**, *48*(4), 549-559.
[http://dx.doi.org/10.1016/S0166-445X(99)00055-7] [PMID: 10794836]

[49]   Cheshenko, K.; Pakdel, F.; Segner, H.; Kah, O.; Eggen, R.I. Interference of endocrine disrupting chemicals with aromatase CYP19 expression or activity, and consequences for reproduction of teleost fish. *Gen. Comp. Endocrinol.,* **2008**, *155*(1), 31-62.
[http://dx.doi.org/10.1016/j.ygcen.2007.03.005] [PMID: 17459383]

[50]   Monteiro, P.R.; Reis-Henriques, M.A.; Coimbra, J. Plasma steroid levels in female flounder (*Platichthys flesus*) after chronic dietary exposure to single polycyclic aromatic hydrocarbons. *Mar. Environ. Res.,* **2000**, *49*(5), 453-467.
[http://dx.doi.org/10.1016/S0141-1136(99)00085-9] [PMID: 11285723]

[51]   Martin-Skilton, R.; Coughtrie, M.W.; Porte, C. Sulfotransferase activities towards xenobiotics and estradiol in two marine fish species (*Mullus barbatus* and *Lepidorhombus boscii*): characterization and inhibition by endocrine disrupters. *Aquat. Toxicol.,* **2006**, *79*(1), 24-30.
[http://dx.doi.org/10.1016/j.aquatox.2006.04.012] [PMID: 16806523]

[52]   Martin-Skilton, R.; Saborido-Rey, F.; Porte, C. Endocrine alteration and other biochemical responses in juvenile turbot exposed to the Prestige fuel oil. *Sci. Total Environ.,* **2008**, *404*(1), 68-76.
[http://dx.doi.org/10.1016/j.scitotenv.2008.06.006] [PMID: 18625515]

[53]   Sun, L.; Zuo, Z.; Luo, H.; Chen, M.; Zhong, Y.; Chen, Y.; Wang, C. Chronic exposure to phenanthrene influences the spermatogenesis of male Sebastiscus marmoratus: U-shaped effects and the reason for them. *Environ. Sci. Technol.,* **2011**, *45*(23), 10212-10218.
[http://dx.doi.org/10.1021/es202684w] [PMID: 22029749]

[54]   Geraudie, P.; Nahrgang, J.; Forget-Leray, J.; Minier, C.; Camus, L. *In vivo* effects of environmental concentrations of produced water on the reproductive function of polar cod (*Boreogadus saida*). *J. Toxicol. Environ. Health A,* **2014**, *77*(9-11), 557-573.
[http://dx.doi.org/10.1080/15287394.2014.887420] [PMID: 24754392]

[55]   Fernandes, D.; Porte, C. Hydroxylated PAHs alter the synthesis of androgens and estrogens in subcellular fractions of carp gonads. *Sci. Total Environ.,* **2013**, *447*, 152-159.
[http://dx.doi.org/10.1016/j.scitotenv.2012.12.068] [PMID: 23376527]

[56]   Kime, D. Classical and non-classical reproductive steroids in fish. *Rev. Fish Biol. Fish.,* **1993**, *3*, 160-180.
[http://dx.doi.org/10.1007/BF00045230]

[57] Kusakabe, M.; Nakamura, I.; Young, G. 11β-hydroxysteroid dehydrogenase complementary deoxyribonucleic acid in rainbow trout: cloning, sites of expression, and seasonal changes in gonads. *Endocrinology,* **2003**, *144*(6), 2534-2545.
[http://dx.doi.org/10.1210/en.2002-220446] [PMID: 12746316]

[58] Borg, B. Androgens in teleost fishes. *Comp. Biochem. Physiol.,* **1994**, *109C*, 219-245.

[59] Fernandes, D.; Bebianno, M.J.; Porte, C. Mitochondrial metabolism of 17alpha-hydroxyprogesterone in male sea bass (*Dicentrarchus labrax*): a potential target for endocrine disruptors. *Aquat. Toxicol.,* **2007**, *85*(4), 258-266.
[http://dx.doi.org/10.1016/j.aquatox.2007.09.010] [PMID: 17977610]

[60] Noguchi, K.; Toriba, A.; Chung, S.W.; Kizu, R.; Hayakawa, K. Identification of estrogenic/anti-estrogenic compounds in diesel exhaust particulate extract. *Biomed. Chromatogr.,* **2007**, *21*(11), 1135-1142.
[http://dx.doi.org/10.1002/bmc.861] [PMID: 17583877]

[61] Hayakawa, K.; Bekki, K.; Yoshita, M.; Tachikawa, C.; Kameda, T.; Tang, N.; Toriba, A.; Hosoi, S. Estrogenic/antiestrogenic activities of quinoid polycyclic aromatic hydrocarbons. *J. Health Sci.,* **2011**, *57*, 274-280.
[http://dx.doi.org/10.1248/jhs.57.274]

[62] Hellou, J.; Beach, D.G.; Leonard, J.; Banoub, J.H. Integrating field analyses with laboratory exposures to assess ecosystems health. *Polycycl. Aromat. Compd.,* **2012**, *32*, 97-132.
[http://dx.doi.org/10.1080/10406638.2011.651681]

[63] Matsumoto, T.; Osada, M.; Osawa, Y.; Mori, K. Gonadal estrogen profile and immunohistochemical localization of steroidogenic enzymes in the oyster and scallop during sexual maturation. *Comp. Biochem. Physiol.,* **1997**, *118B*, 811-817.
[http://dx.doi.org/10.1016/S0305-0491(97)00233-2]

[64] Li, Q.; Osada, M.; Suzuki, T.; Mori, K. Changes in vitellin during oogenesis and effect of estradiol-17β on vitellogenesis in the Pacific oyster *Crassostrea gigas. Invertebr. Reprod. Dev.,* **1998**, *33*, 87-93.
[http://dx.doi.org/10.1080/07924259.1998.9652345]

[65] Wang, C.; Croll, P.R. Effects of sex steroids on gonadal development and gender determination in the sea scallop, *Placopecten magellanicus. Aquaculture,* **2004**, *238*, 483-498.
[http://dx.doi.org/10.1016/j.aquaculture.2004.05.024]

[66] Lavado, R.; Janer, G.; Porte, C. Steroid levels and steroid metabolism in the mussel *Mytilus edulis*: the modulating effect of dispersed crude oil and alkylphenols. *Aquat. Toxicol.,* **2006**, *78* Suppl. 1, S65-S72.
[http://dx.doi.org/10.1016/j.aquatox.2006.02.018] [PMID: 16600398]

[67] Fernandes, D.; Loi, B.; Porte, C. Biosynthesis and metabolism of steroids in molluscs. *J. Steroid Biochem. Mol. Biol.,* **2011**, *127*(3-5), 189-195.
[http://dx.doi.org/10.1016/j.jsbmb.2010.12.009] [PMID: 21184826]

[68] Tian, S.; Pan, L.; Sun, X. An investigation of endocrine disrupting effects and toxic mechanisms modulated by benzo[a]pyrene in female scallop *Chlamys farreri. Aquat. Toxicol.,* **2013**, *144-145*, 162-171.
[http://dx.doi.org/10.1016/j.aquatox.2013.09.031] [PMID: 24185101]

[69]   Zhang, H.; Pan, L.; Zhang, L. Molecular cloning and characterization of estrogen receptor gene in the scallop *Chlamys farreri*: expression profiles in response to endocrine disrupting chemicals. *Comp. Biochem. Physiol. C Toxicol. Pharmacol.,* **2012,** *156*(1), 51-57.
[http://dx.doi.org/10.1016/j.cbpc.2012.03.007] [PMID: 22507668]

[70]   Beyer, J.; Aarab, N.; Tandberg, A.H.; Ingvarsdottir, A.; Bamber, S.; Børseth, J.F.; Camus, L.; Velvin, R. Environmental harm assessment of a wastewater discharge from Hammerfest LNG: a study with biomarkers in mussels (*Mytilus* sp.) and Atlantic cod (*Gadus morhua*). *Mar. Pollut. Bull.,* **2013,** *69*(1-2), 28-37.
[http://dx.doi.org/10.1016/j.marpolbul.2013.01.001] [PMID: 23419752]

[71]   Lazartigues, A.; Thomas, M.; Grandclaudon, C.; Brun-Bellut, J.; Feidt, C. Polycyclic aromatic hydrocarbons and hydroxylated metabolites in the muscle tissue of Eurasian perch (*Perca fluviatilis*) through dietary exposure during a 56-day period. *Chemosphere,* **2011,** *84*(10), 1489-1494.
[http://dx.doi.org/10.1016/j.chemosphere.2011.04.037] [PMID: 21546054]

[72]   Carrasco Navarro, V.; Leppänen, M.T.; Honkanen, J.O.; Kukkonen, J.V. Trophic transfer of pyrene metabolites and nonextractable fraction from Oligochaete (*Lumbriculus variegatus*) to juvenile brown trout (*Salmo trutta*). *Chemosphere,* **2012,** *88*(1), 55-61.
[http://dx.doi.org/10.1016/j.chemosphere.2012.02.060] [PMID: 22475154]

[73]   Carrasco Navarro, V.; Leppänen, M.T.; Kukkonen, J.V.; Godoy Olmos, S. Trophic transfer of pyrene metabolites between aquatic invertebrates. *Environ. Pollut.,* **2013,** *173*, 61-67.
[http://dx.doi.org/10.1016/j.envpol.2012.09.023] [PMID: 23202283]

CHAPTER 6

Synthesis of Environmental Relevant Metabolites

Emil Lindbäck and **Magne O. Sydnes**[*]

Faculty of Science and Technology, Department of Mathematics and Natural Science, University of Stavanger, Stavanger, Norway

Abstract: Polycyclic aromatic hydrocarbons are metabolized *in vivo* resulting in the formation a range of oxidized products – metabolites. The metabolites generated are more water soluable and therefore easier to excrete from the system, but these compounds are also more toxic for the organism. A range of PAH metabolites have been synthesized and used in order to study their toxicity and further faith *in vivo*. Different synthetic strategies have been used in order to prepare the metabolites. Herein, the synthetic strategies utilized for the formation of environmental relevant metabolites of naphthalene, acenaphthene, fluorine, phenanthrene, and chrysene, the PAHs found with the highest concentration in crude oil, are described in detail.

Keywords: Acenaphthene, Chrysene, Diols, Fluorine, Naphthalene, Oxidation, Phenanthrene, Phenols.

INTRODUCTION

Polycyclic aromatic hydrocarbons (PAHs) are enzymatically metabolized *in vivo* in organisms in order to make these compounds more water soluble. A range of oxidation products are formed in these processes with a large variation of products and product distribution varying from organism to organism due to species specific metabolism [1, 2]. In order to evaluate the toxicity of the generated metabolites and study their further faith *in vivo* it is necessary to obtain the metabolites in pure form.

[*] **Corresponding author Magne O. Sydnes**: Faculty of Science and Technology, Department of Mathematics and Natural Science, University of Stavanger, Stavanger, Norway; Tel: +47 51831761; E-mail: magne.o.sydnes@uis.no

Daniela M. Pampanin and Magne O. Sydnes (Eds.)
All rights reserved-© 2017 Bentham Science Publishers

During metabolism of PAHs a range of phenols, diols, and epoxides are formed. In the case of phenols different regioisomers are generated, and diols and epoxides are formed as different regioisomers with specific stereochemistry. For the chiral metabolites the toxicity of the compounds is very dependent on their stereochemistry. For biological studies it is therefore desirable to utilize metabolites with high enantiomerical excess (*ee*) in order to assure that an observed biological effect can be linked to a specific enantiomer.

The best method to generate sufficient quantities of these specific metabolites required for biological studies is through total synthesis. There are two main strategies to consider when designing a total synthesis of diols and epoxides. Either an asymmetric synthesis, or a racemic synthesis. The latter strategy requires a final separation in order to isolate the two enantiomers formed, and by such means obtain the enantiomerically pure compounds.

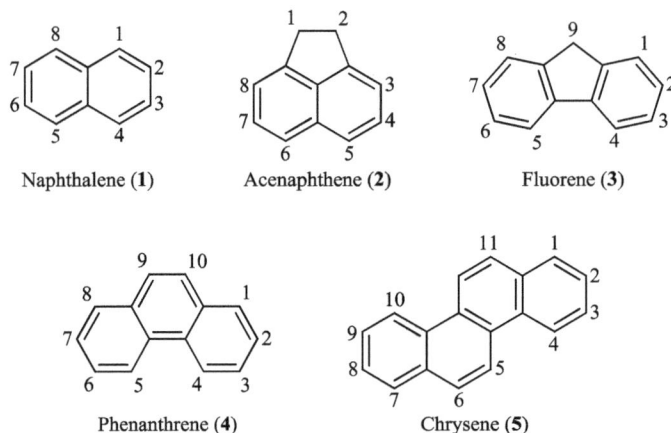

Fig. (1). The structure of naphthalene (**1**), acenaphthene (**2**), fluorine (**3**), phenanthrene (**4**), and chrysene (**5**) with indication of how the carbons are numbered.

Many of the relevant metabolites generated *in vivo* from the EPA 16 PAHs [3] have been prepared synthetically. These syntheses have been presented in a large amount of publications over the last 50-60 years. In preparing this chapter, it was necessary to decide which parent PAHs should be focus. In crude oil, there is a great variation in concentration of the various PAHs. However, there are a few PAHs that are generally present in higher concentrations than other PAHs found in crude oil. It was therefore decided to focus on the few PAHs that generally make up the majority of the concentration of PAHs found in crude oil. Therefore, the focus herein will be on the reported synthesis of metabolites that can be formed *in vivo* by oxidation of naphthalene (**1**), acenaphthene (**2**), fluorine (**3**), phenanthrene (**4**), and chrysene (**5**) (Fig. **1**).

Naphthols

Formation of Naphthols by Hydroxylation of Halonaphthalenes

In the recent years, metal catalyzed, in particular copper catalyzed, direct hydroxylation of halonaphthalenes has emerged as an attractive methodology for the synthesis of naphthols. As outlined in Scheme **1** and Scheme **2**, several examples exist where both 1-naphthol (**6**) and 2-naphthol (**7**) have been obtained by copper catalyzed direct hydroxylation of the corresponding halonaphthalenes in the presence of various hydroxide salts. For instance, You and co-workers concluded that 1-naphthol (**6**) could be obtained in very good yield when 1-iodonaphthalene (**8**) was coupled with the hydroxide ion of potassium hydroxide employing copper iodide as a precatalyst in the presence of 1,10-phenanthroline as a ligand (Scheme **1**, Conditions A) [4]. Likewise, Wang and colleagues also used copper iodide as a precatalyst for the direct hydroxylation of 2-iodonaphthalene (**9**) with potassium hydroxide as the hydroxyl source and tetrabutylammonium bromide (TBAB) as phase-transfer catalyst [5] using triethanolamine as ligand (Scheme **1**, Conditions B) [6].

In contrast to the examples presented in Scheme **1**, Conditions A-B, the Chae research-group utilized copper(II) as the copper source for the copper catalyzed direct hydroxylation of both 1- (**8**) and 2-iodonaphthalene (**9**) [7]. In fact, the use of copper(II) hydroxide in combination with glycolic acid as ligand afforded 1-naphthol (**6**) and 2-naphthol (**7**) in 98 and 77% yield, respectively (Scheme **1**, Conditions C). Another efficient copper source for the direct hydroxylation of 1- (**8**) and 2-iodonaphthalene (**9**) is $Cu_3(btc)_2$ (btc = 1,3,5-benzenetricarboxylate (Scheme **1**, Conditions D) [8].

From substrate **8**
A) CuI (10 mol%)
1,10-phenanthroline
(20 mol%)
KOH (2-6 equiv)
DMSO/H$_2$O (1:1)
100 °C, 89%
C) CuOH$_2$ (5 mol%)
glycolic acid (30 mol%)
NaOH 6 (equiv)
DMSO/H$_2$O (1:1)
120 °C, 98%
D) Cu$_3$(btc)$_2$
(10 mol% Cu)
H$_2$O/DMSO (1:1)
125 °C, 85%

From substrate **9**
B) CuI (5 mol%)
N(CH$_2$CH$_2$OH)$_3$
(40 mol%)
TBAB (20 mol%)
KOH (4 equiv)
H$_2$O, 120 °C, 77%
C) 77%
D) 90%

6

8: X = I, Y = H
9: X = H, Y = I

7

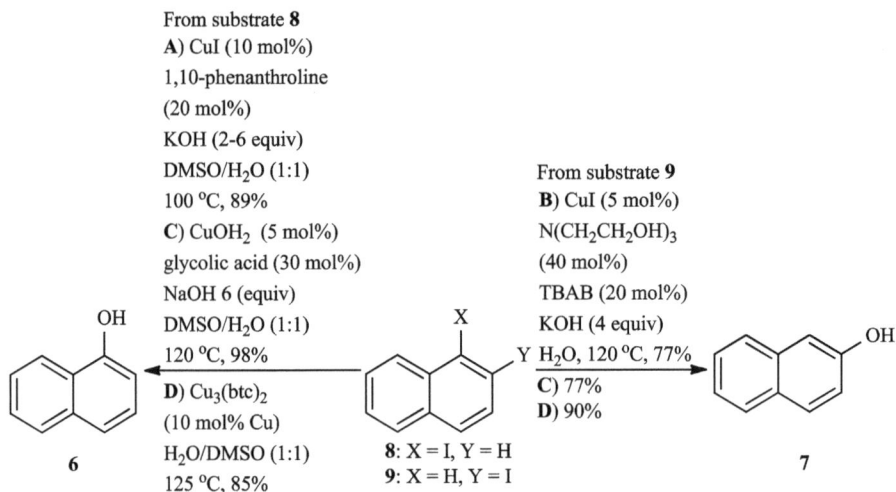

Scheme 1. Copper catalyzed direct hydroxylation of iodonaphthalenes.

In the literature it is also reported that 1- (**6**) and 2-naphthol (**7**) can be obtained *via* copper catalyzed direct hydroxylation of 1- (**10**) and 2-bromonaphthalene (**11**), which are less reactive electrophiles than the corresponding iodonaphthalenes **8** and **9**. Jiang and co-workers reported that 1- (**6**) and 2-naphthol (**7**) were obtained when 8-hydroxyquinoline-*N*-oxide served as ligand in the copper catalyzed direct hydroxylation of 1- (**10**) and 2-bromonaphthalene (**11**) utilizing copper iodide as precatalyst and cesium hydroxide as the hydroxyl source (Scheme **2**, Conditions A) [9]. Likewise, Chen and co-workers also made use of copper iodide as precatalyst in direct hydroxylation of 1- (**10**) and 2-bromonaphthalene (**11**) (Scheme **2**, Conditions B) [10]. The reactions were performed in a H$_2$O/polyethylene glycol (PEG)-400 (1:4) mixture where potassium hydroxide was the hydroxyl source. Feng's research group explored copper iodide nanoparticles as catalyst for direct hydroxylation of aryl bromides and iodides in water under ligand free conditions using tetrabutylammonium hydroxide as base [11]. The conditions worked excellently when 1- (**10**) and 2-bromonaphthalene (**11**) was hydroxylated into 1- (**6**) and 2-naphthol (**7**), respectively (Scheme **2**, Conditions C). Copper iodide was also used by Taillefer and co-workers as precatalyst in order to achieve *indirect* hydroxylation of 2-bromonaphthalene (**11**) in the presence of N^1,N^2-dimethylethane-1,2-diamine and sodium iodide (Scheme **2**, Conditions D) [12]. The hydroxylation took place in two steps: 1) copper catalyzed production of 2-iodonaphthalene (**9**) from 2-bromonaphthalene (**11**) and sodium iodide, and 2) addition of the hydroxyl source (cesium hydroxide) that was trapped by the 2-iodonaphthalene (**9**) intermediate with the aid of the copper catalyst.

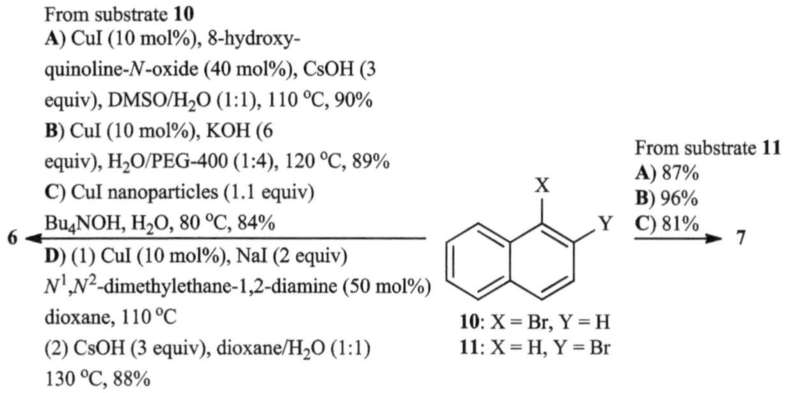

From substrate **10**
A) CuI (10 mol%), 8-hydroxy-quinoline-*N*-oxide (40 mol%), CsOH (3 equiv), DMSO/H$_2$O (1:1), 110 °C, 90%
B) CuI (10 mol%), KOH (6 equiv), H$_2$O/PEG-400 (1:4), 120 °C, 89%
C) CuI nanoparticles (1.1 equiv) Bu$_4$NOH, H$_2$O, 80 °C, 84%
D) (1) CuI (10 mol%), NaI (2 equiv) *N*^{1},*N*^{2}-dimethylethane-1,2-diamine (50 mol%) dioxane, 110 °C
(2) CsOH (3 equiv), dioxane/H$_2$O (1:1) 130 °C, 88%

From substrate **11**
A) 87%
B) 96%
C) 81%

10: X = Br, Y = H
11: X = H, Y = Br

Scheme 2. Copper catalyzed direct hydroxylation of bromonaphthalenes.

As described above, copper catalysis has frequently been employed for hydroxylation of halonaphthalenes in order to afford the corresponding naphthols. Iron catalysis on the other hand has been less utilized for this type of transformations. Wang and co-workers introduced such a protocol when they reported that iron(III) chloride behaved as precatalyst for the direct hydroxylation of 1-iodonaphthalene (**8**), 1- (**10**), and 2-bromonaphthalene (**11**) (Scheme 3) [13]. The reactions were conducted in water with TBAB as phase-transfer catalyst due to the poor solubility of the substrates in water. *N*,*N*'-Dimethylethylenediamine (DMEDA) was used as ligand and potassium phosphate as base resulting in the formation of the products in good to excellent yields.

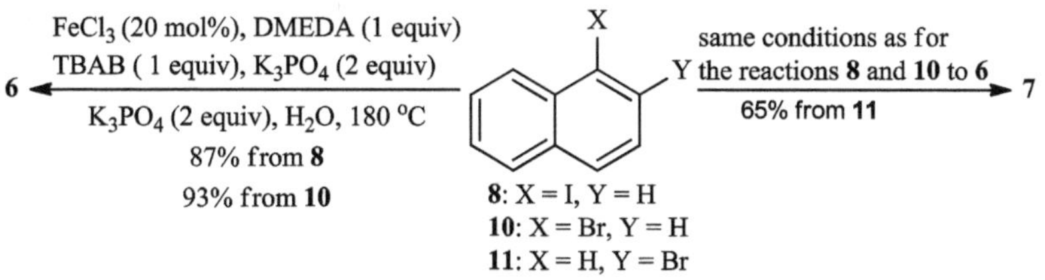

FeCl$_3$ (20 mol%), DMEDA (1 equiv)
TBAB (1 equiv), K$_3$PO$_4$ (2 equiv)
K$_3$PO$_4$ (2 equiv), H$_2$O, 180 °C
87% from **8**
93% from **10**

same conditions as for the reactions **8** and **10** to **6**
65% from **11**

8: X = I, Y = H
10: X = Br, Y = H
11: X = H, Y = Br

Scheme 3. Iron catalyzed direct hydroxylation of halonaphthalenes.

Formation of Naphthols by Hydroxylation of Naphthylboronic Acids

Copper catalysis has not only been used for the conversion of halonaphthalenes into their corresponding naphthols, it has also been utilized for oxidative hydroxylation of naphthylboronic acids into their corresponding naphthols. Hu and co-workers reported such a strategy when they showed that 1- (**12**) and 2-naphthylboronic acid (**13**) undergo oxidative hydroxylation in the presence of a catalytic amount of copper sulfate, 1,10-phenanthroline, and air (Scheme **4**, Conditions A) [14]. Kaboudin and co-workers produced a Cu_2-β-CD (CD = cyclodextrin) complex, which behaved as a supramolecular catalyst, for the conversion of naphthylboronic acids under ligand- and base-free conditions [15]. Interestingly, this complex exhibited very different reactivity with 1-napthylboronic acid (**12**) compared to 2-napthylboronic acid (**13**). In fact, compound **13** was hydroxylated into 2-naphthol (**7**) in 91% yield (Scheme **4**, Conditions B), whereas 1-napthylboronic acid (**12**) was converted to naphthalene (**1**) in 90% yield with no trace of 1-naphthol (**6**) being formed (Scheme **4**, Conditions B). The latter result is possibly due to steric effects.

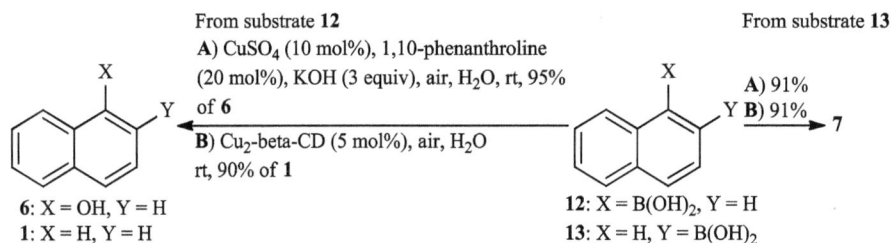

From substrate **12**
A) CuSO$_4$ (10 mol%), 1,10-phenanthroline (20 mol%), KOH (3 equiv), air, H$_2$O, rt, 95% of **6**
B) Cu$_2$-beta-CD (5 mol%), air, H$_2$O rt, 90% of **1**

From substrate **13**
A) 91%
B) 91%

6: X = OH, Y = H
1: X = H, Y = H

12: X = B(OH)$_2$, Y = H
13: X = H, Y = B(OH)$_2$

Scheme 4. Copper catalyzed oxidative hydroxylation of naphthylboronic acids.

Sawant and colleagues reported an example of oxidative hydroxylation of 1- (**12**) and 2-naphthylboronic acid (**13**) into the corresponding naphthols **6** and **7** (Scheme **5**, Conditions A), respectively, employing iron(III) oxide as a precatalyst in solar VIS-light (VIS = visible) irradiation under base- and ligand-free conditions [16]. When conducting the reaction under inert conditions no reaction took place indicating that the reaction requires dioxygen as an oxidant in order to occur. In another strategy, Wang and co-workers utilized a catalytic amount of strongly magnetic CuFe$_2$O$_4$ without any ligand present under an atmosphere of air for the oxidative hydroxylation of 2-naphthylboronic acid (**13**), thus affording 2-

naphthol (**7**) in 54% yield (Scheme **5**, Conditions B) [17].

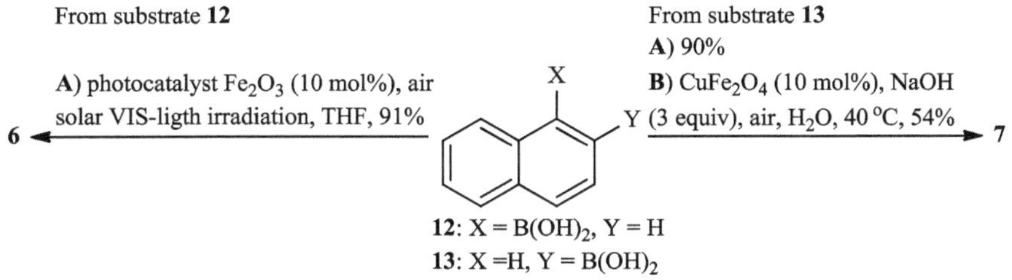

From substrate **12**

A) photocatalyst Fe_2O_3 (10 mol%), air
solar VIS-ligth irradiation, THF, 91%

From substrate **13**
A) 90%
B) $CuFe_2O_4$ (10 mol%), NaOH
Y (3 equiv), air, H_2O, 40 °C, 54%

12: X = $B(OH)_2$, Y = H
13: X =H, Y = $B(OH)_2$

6 ← → **7**

Scheme 5. Iron catalyzed oxidative hydroxylation of naphthylboronic acids.

It is also worth noting that the catalytic oxidative hydroxylation of naphthylboronic acids to their corresponding naphthols has not only been restricted to transition metal-catalysis. In fact, both Amberlite IR-400 resin [18], a deep eutectic mixture based on choline chloride together with 1,1,1,3,3,3-hexafluoro-2-propanol (HFIP) [19], and choline chloride together with urea [20] have also been used as catalytic systems together with hydrogen peroxide as the oxidant.

Perhaps more remarkable is the fact that oxidative hydroxylation of naphthylboronic acids have been performed in the absence of catalyst. For instance, Zhu and co-workers recognized that the oxidative hydroxylation of 1-(**12**) and 2-naphthylboronic acid (**13**) into their corresponding naphthols **6** and **7**, respectively, took place in methylene chloride in the presence of a stoichiometric amount of *N,N*-dimethylaniline oxide (Scheme **6**, Conditions A) [21]. Another catalyst free oxidative hydroxylation method for naphthylboronic acids was reported by Gogoi and colleagues who utilized sodium chlorite ($NaClO_2$) as oxidation agent when the reactions were performed in water (Scheme **6**, Conditions B) [22]. Hydrogen peroxide has also been employed as an oxidation agent without any use of catalyst for the formation of naphthol **6** from the corresponding naphthylboronic acid **12** (Scheme **6**, Conditions C) [23]. The reaction was carried out in PEG-400, which was proposed to behave as a hydrogen bond acceptor for hydrogen peroxide, and thereby promoting it to be nucleophilic enough for the addition to the naphthylboronic acid without the aid of a base for deprotonation prior to the addition reaction.

From substrate **12**
A) *N,N*-dimethylaniline oxide (1.2 equiv)
air, CH$_2$Cl$_2$, rt, 94%
B) NaClO$_2$ (1.2 equiv), H$_2$O, rt, 86%
C) H$_2$O$_2$ (1.5 equiv), PEG-400, rt, 97%

From substrate **13**
A) 92%
B) 91%

6 ◄——————————————————

——————————► **7**

12: X = B(OH)$_2$, Y = H
13: X = H, Y = B(OH)$_2$

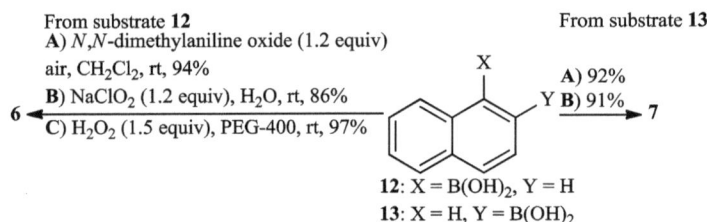

Scheme 6. Catalyst free oxidative hydroxylation of naphthylboronic acids.

Feng and co-workers applied a two-step palladium catalyzed coupling sequence in order to prepare 2-napthanol (**7**) from 2-bromobenzyl bromide (**14**) (Scheme 7) [24]. The first step constituted a carbonylative Stille cross-coupling on to compound **14** with vinyltributylstannane to afford 2-bromobezyl α,β-unsaturated ketone **15**, which in the subsequent step underwent an intramolecular Heck reaction [25, 26] to furnish naphthol **7**.

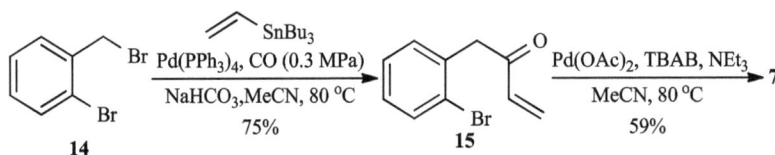

SnBu$_3$

Br   Pd(PPh$_3$)$_4$, CO (0.3 MPa)
—————————————————————
NaHCO$_3$, MeCN, 80 °C
Br   75%

14

O   Pd(OAc)$_2$, TBAB, NEt$_3$
—————————————————————► **7**
Br   MeCN, 80 °C
59%

15

Scheme 7. 2-Napthol (**7**) was obtained *via* two subsequent palladium-catalyzed steps from 2-bromobenzyl bromide (**14**).

1-Naphthol (**6**) has been obtained *via* a metal catalyzed isomerization of 1,4-epoxy-1,4-dihydronaphthalene (**16**). For instance, in 1,2-dichloroethene (DCE), copper(II) triflate behaves as a Lewis acid catalyst for the isomerization of substrate **16** into naphthol **6** (Scheme 8, Conditions A) [27]. In contrast to copper catalysis, Sawama and co-workers demonstrated that in the presence of a gold catalyst (HAuCl$_4$·H$_2$O) 1,4-epoxy-1,4-dihydronaphthalene (**16**) undergo isomerization into naphthol **6** when treated with TMSCl (TMS = trimethylsilyl) (Scheme 8, Conditions B) [28]. Tam and colleagues have investigated various ruthenium catalysts for the isomerization of 1,4-epoxy-1,4-dihydronaphthalenes into naphthols [29, 30]. Among the catalysts tested, (RuCl$_2$(CO)$_3$)$_2$ was found to give the desired product (**6**) in quantitative (quant) yield in a range of solvents (*e.g.* DCE, THF, and toluene) (Scheme 8, Conditions C).

A) Cu(OTf)$_2$ (5 mol%), DCE, rt, 98%
B) HAuCl$_4$·3H$_2$O (5 mol%), TMSCl (4 equiv)
CH$_2$Cl$_2$, -40 °C, 79%
C) [RuCl$_2$(CO)$_3$]$_2$ (5 mol%), solvent, 60 °C
quant

16 → **6**

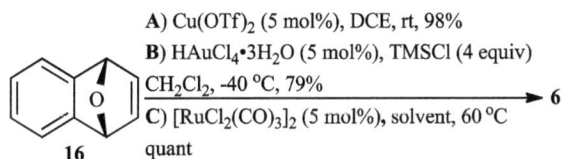

Scheme 8. Metal catalyzed isomerization of 1,4-epoxy-1,4-dihydronaphthalene (**16**) into 1-naphthol (**6**) (solvent = acetone, DCE, dioxane, hexanes, THF, or toluene).

Synthesis of Naphthalene Diols

A methodology publication concerning the reduction of non-K-region *ortho*-quinones to the corresponding dihydrodiols included the synthesis of *trans*-(±--1,2-dihydroxy-1,2-dihydronaphthalene **17(±)** along with the *cis*-(±)-counterpart **18(±)** [31, 32]. Thus, bromination of 1,2-naphthoquinone (**19**) with bromine in benzene (PhH) gave the corresponding dibromo intermediate **20(±)**, which could not be isolated due to rapid decomposition. Compound **20(±)** was therefore immediately used in the following step, hence, the crude product of compound **20(±)** was reduced with sodium borohydride (NaBH$_4$) in ethanol to give *trans*-**17(±)** and *cis*-dihydrodiol **18(±)** in a combined yield of 34% (the diastereoselectivity was not given) (Scheme **9**). Beneficially, diastereomer **17(±)** precipitated out from the reaction mixture leaving behind diastereomer **18(±)**.

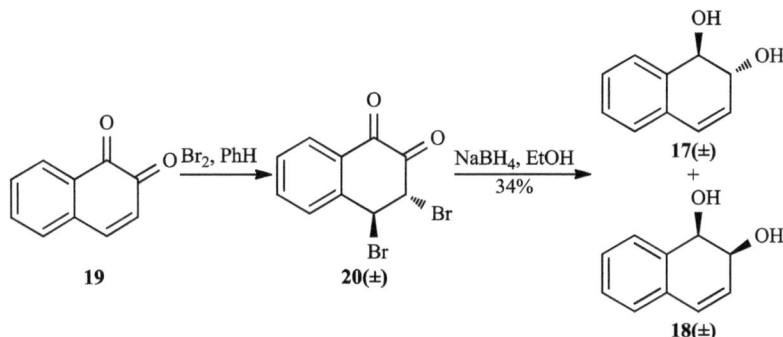

Scheme 9. A two-step procedure for the preparation of *trans*- (**17(±)**) and *cis*-1,2-dihydroxy-1,2-dihydronaphthalene (**18(±)**) from 1,2-naphthoquinone (**19**).

Platt and Oesch reported that an atmosphere of air over the reaction mixture was required in order to achieve efficient stereoselective reduction of 1,2-naphthoquinone (**19**) with sodium borohydride to *trans*-(±)-dihydrodiol **17(±)**

(Scheme **10**) [33]. The need for an atmosphere of air over the reaction mixture was supported by the fact that when air was replaced by argon catechol **21** was formed instead of *trans*-(±)-dihydrodiol **17**(±) under otherwise identical conditions (Scheme **10**). The authors proposed a mechanism to rationalize the role of air in the reduction as outlined in Scheme **11**. Thus, 1,2-naphthoquinone (**19**) was reduced to form α-hydroxy ketone **22**, which either isomerized into catechol **21** or was further reduced to furnish the *trans*-(±)-dihydrodiol **17**(±). In the presence of air, catechol **21** (the product formed under an argon atmosphere) was faster oxidized into the starting material **19** than diol **17**(±) was reoxidized back to α-hydroxy ketone **22**. Hence, α-hydroxy ketone **22**, regenerated from catechol **21** *via* compound **19**, could once again be reduced leading to *trans*-dihydrodiol **17**(±) as the end product in this reduction-oxidation process.

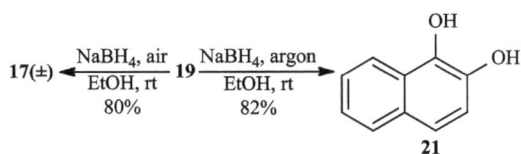

Scheme 10. Reduction of 1,2-naphthoquinone (**19**) with sodium borohydride under air and argon furnished product **17**(±) and **21**, respectively.

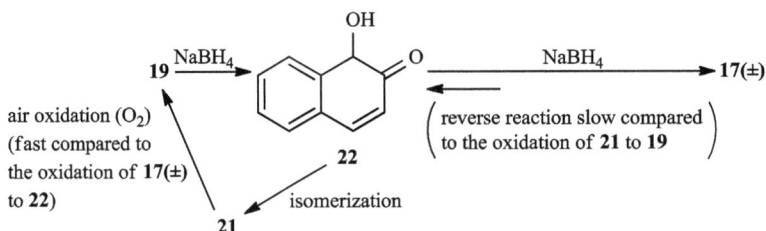

Scheme 11. Proposed mechanism for the role of air in the reduction of 1,2-quinone **19** with sodium borohydride into *trans*-(±)-dihydrodiol **17**(±).

Jeffrey and co-workers launched a synthetic protocol that led to the formation of both *cis*-(±)-1,2-dihydroxy-1,2-dihydronaphthalene (**18**(±)) and *cis*-1,4-dihydroxy-1,4-dihydronaphthalene (**23**) (Scheme **12**) [34]. For target compounds **18**(±) and **23**, 1,4-naphthoquinone (**24**) constituted the starting material and was converted into *cis*-1,4-dihydroxy-2,3-epoxynaphthalene intermediate **25**, using a previously reported epoxidation/reduction sequence [35]. In order to reach the target compound **18**(±), the epoxide moiety within substrate **25** was ring-opened with sodium iodide with the aid of zinc dust in acetic acid forming an iodohydrin intermediate, which was immediately treated with aqueous sodium carbonate furnishing the first target diol **18(±)**. In order to obtain the second target compound **23**, epoxide **25** was transformed into thioepoxide **26** upon treatment with potassium thiocyanate in ethanol, which underwent desulfurisation in the presence of triphenylphosphine to furnish desired compound **23**. It is noteworthy that 1,4-naphthoquinone (**24**) could be reduced immediately into compound **23**, albeit in low yield (20%), with diisobutylaluminium hydride (DIBAL) in toluene at room temperature.

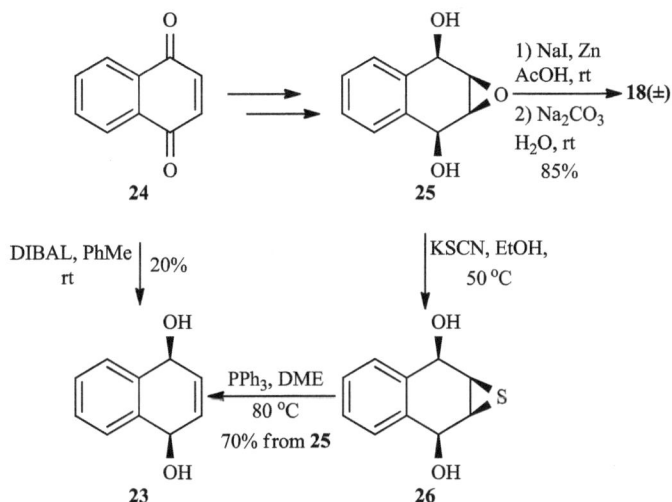

Scheme 12. Synthesis of *cis*-(±)-1,2-dihydroxy-1,2-dihydronaphthalene (**18(±)**) and *cis*-1,4-dihydroxy-1,4-dihydronaphthalene (**23**).

Jerina and co-workers made use of *trans*-(±)-1,2-dihydroxy-1,2-dihydronaphthalene (**17(±)**) as a building block in a stereoselective synthesis leading to both *anti*- and *syn*-(±)-tetrahydrodiol epoxide **27(±)** and **28(±)**, respectively [36]. The synthesis of the *anti*-(±)-tetrahydrodiol epoxide **27(±)** was readily achieved when compound **17(±)** was stereoselectively epoxidized with 3-chloroperbenzoic acid (*m*CPBA) in methylene chloride. In order to achieve the diastereomeric *syn*-(±)-counterpart **28(±)** a two-step protocol was utilized (Scheme **13**, Method 1). The first step constituted bromination of compound **17(±)** with *N*-bromoacetamide (NBA) in the presence of water, which acted as nucleophiles, to stereoselectively furnish bromotriol **29(±)**. In the final step, hydrogen bromide was eliminated from the bromotriol intermediate **29(±)** upon treatment with Amberlite-IRA-400 (hydroxylic form) to give the desired product **28(±)**. The authors stated that since the benzylic hydroxy group was *trans* to the bromide substituent in bromotriol **29(±)** in contrast to the adjacent non-benzylic hydroxy group which was *cis* only the former hydroxy group was able to displace bromide, and thus form compound **28(±)** upon treatment with base. Noteworthy, the methodology (in Scheme **13**, Method 1) developed by Jerina and co-workers in order to prepare compound **28(±)** from building block **17(±)** [36] did not work in the hands of Bruice and co-workers [37]. Thus, they launched a slightly modified procedure (Scheme **13**, Method 2) including bromination of compound **17(±)** with NBA together with water as a nucleophile in the presence of a catalytic (cat) amount of hydrogen chloride to furnish bromotriol **29(±)**, which underwent cyclization forming *syn*-tetrahydrodiol epoxide **28(±)** in the presence of potassium *tert*-butoxide.

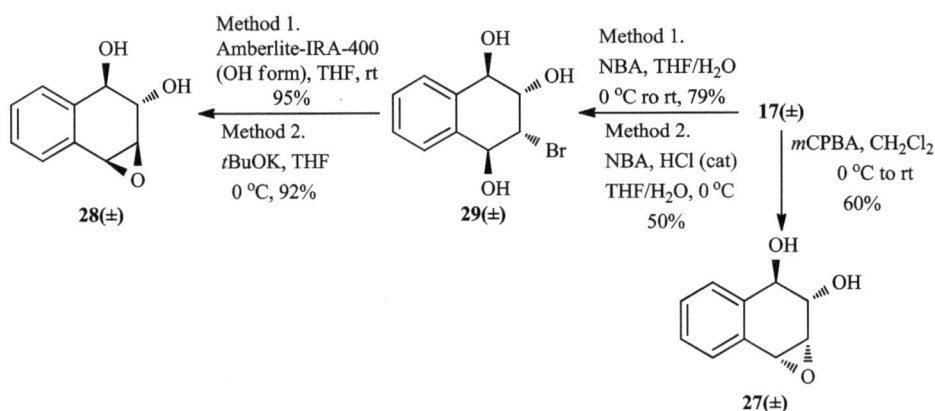

Scheme 13. Synthesis of *anti*- and *syn*-(±)-tetrahydrodiol epoxides **28(±)** and **29(±)**, respectively.

Angerbauer and Schmidt came up with another approach in order to achieve the preparation of *syn*-(±)-tetrahydrodiol epoxide **28(±)** (Scheme **14**) [38]. The methodology was initiated by a Diels-Alder reaction between benzenediazonium-2-carboxylate (**30**), which constitutes a benzyne precursor, and diene **31** to furnish the cyclization product **32**. Substrate **32** underwent a *cis*-dihydroxylation with hydrogen peroxide using osmium tetroxide as catalyst to generate diol **33**. In the following step, diol **33** was subjected to mono-tosylation with *p*-toluenesulfonyl chloride (TsCl) in pyridine to give tosylate **34(±)**. In the final step, when tosylate **34(±)** was subjected to sodium carbonate in methanol the acetyl groups were removed *via* transesterification along with intramolecular cyclization to generate the *syn*-tetrahydrodiol epoxide **28(±)**. Notably this methodology also gave easy access to *cis*-1,4-dihydroxy-1,4-dihydronaphthalene (**23**), since this diol was formed when the *O*-acetylated counterpart **32** was subjected to methanolysis.

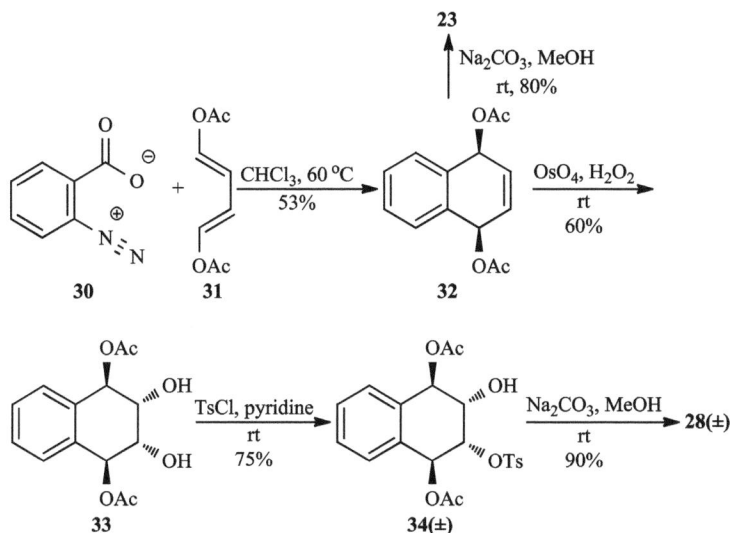

Scheme 14. Synthesis of *syn*-(±)-tetrahydrodiol epoxide **28(±)**.

Recently, Tsui and Lautens introduced a rhodium(I) catalyzed strategy in order to achieve asymmetric ring-opening (ARO) of 1,4-epoxy-1,4-dihydronaphthalenes with water to furnish the corresponding enantioenriched *trans*-1,2-diols [39]. In their work they found that ARO of 1,4-epoxy-1,4-dihydronaphthalenes (**16**) in aqueous THF employing 1 mol% of (Rh(cod)Cl)$_2$ (cod = cyclooctadiene) as catalyst in the presence of the chiral ligand (*R,S*)-PPF-P*t*Bu$_2$ (2.2 mol%) gave *trans*-(+)-1,2-dihydroxy-1,2-dihydronaphthalene (**17(+)**) in >99% enantiomeric ratio (e.r.) and 89% yield (Scheme **15**). The authors also identified α-hydroxy ketone **35** along with compound **36** as minor side products.

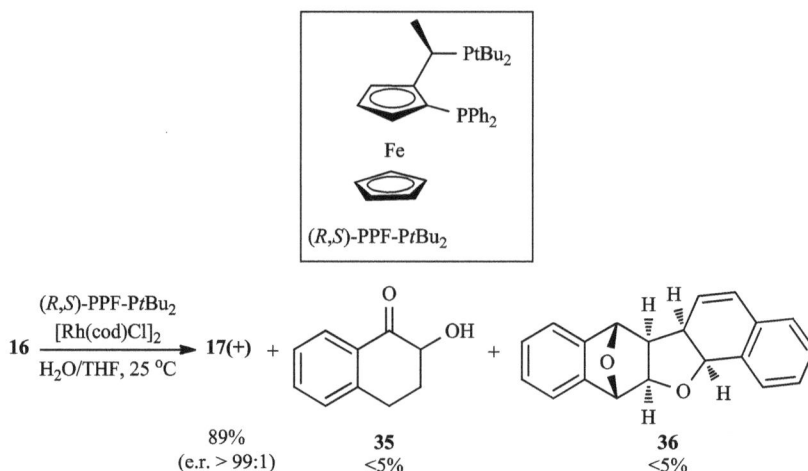

Scheme 15. Rhodium(I) catalyzed ARO of 1,4-epoxy-1,4-dihydronaphthalene (**16**) to furnish enantioenriched *trans*-(+)-1,2-dihydroxy-1,2-dihydronaphthalene (**17(+)**).

In 1977, Hamilton and co-workers reported that various arene oxides could be obtained when the corresponding arenes were subjected to aqueous sodium hypochlorite in the presence of tetrabutylammonium hydrogen sulfate as phase transfer catalyst [40]. The authors pointed out that it was essential to perform the reaction at pH 8-9 (pH was adjusted by addition of concentrated HCl) since at lower pH other oxidation products were observed and at higher pH only modest yields of the desired product could be obtained. The reaction conditions that worked well for a range of arenes was used in order to convert naphthalene (**1**) into *syn*-1,2:3,4-naphthalene dioxide (**37**) (Scheme **16**), however, the yield was poor (19%).

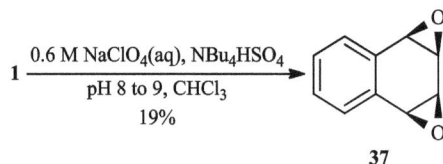

Scheme 16. Epoxidation of naphthalene (**1**) into *syn*-1,2:3,4-naphthalene dioxide (**37**).

Anti-(±)-1,2:3,4-Naphthalene dioxide (**38(±)**) has been generated, albeit in poor yield (15-20%), when a solution of naphthalene (**1**) (0.04 M) in methylene chloride was epoxidized with *m*CPBA (Scheme **17**, Conditions A) [41].

Methyl(trifluoromethyl)dioxirane on the other hand has been found to be a significantly more efficient oxidation reagent than *m*CPBA in order to epoxidize naphthalene (**1**) into *anti*-($\pm$)1,2:3,4-naphthalene dioxide (**38($\pm$)**) (90% yield) (Scheme **17**, Conditions B) [42].

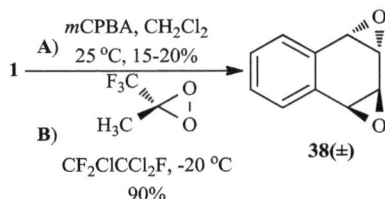

Scheme 17. Diepoxidation of naphthalene (**1**) with *m*CPBA **A**) and mehyl(trifluoromethyl)dioxirane **B**).

Demirtas and co-workers reported a three step synthesis of *anti*-($\pm$)-1,2:3-4-naphthalene dioxide (**38($\pm$)**) from naphthalene (**1**) (Scheme **18**) [43]. Thus, naphthalene (**1**) was subjected to light-induced tetrabromination to furnish substrate **39($\pm$)**, which in aqueous acetone in the presence of silver perchlorate as promoter underwent a substitution reaction with water in both benzylic positions to furnish compound **40($\pm$)**. The desired compound **38($\pm$)** was generated upon treatment of substrate **40($\pm$)** with sodium methoxide in THF.

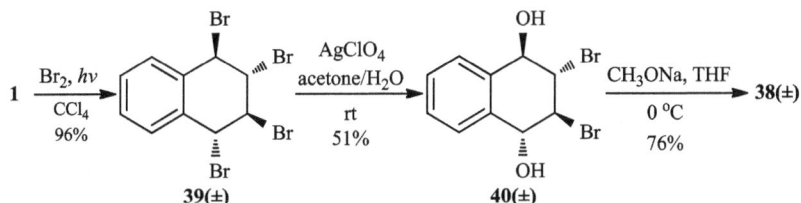

Scheme 18. Three step synthesis of *anti*-($\pm$)-1,2:3,4-naphthalene dioxide (**38($\pm$)**) from naphthalene (**1**).

Boyd and co-workers converted *cis*-(+)-1,2-dihydroxy-1,2-dihydronaphthalene **18(+)** to enantiopure *anti*-(+)-1,2:3,4-naphthalene dioxide (**38(+)**) by first subjecting substrate **18(+)** to a dihydroxylation reaction with osmium tetraoxide employing trimethylamine *N*-oxide (TMAO) as a terminal oxidant, thus forming 1,2,3,4-tetraol **41**. Subsequent treatment of 1,2,3,4-tetraol **41** with α-acetoxyisobutyryl bromide promoted bromination in the benzylic positions along with *O*-acetylation of the remaining hydroxyl groups to give compound **42**, which underwent de-*O*-acetylation and cyclization upon treatment with sodium methoxide to afford the desired compound **38(+)** (Scheme **19**) [44].

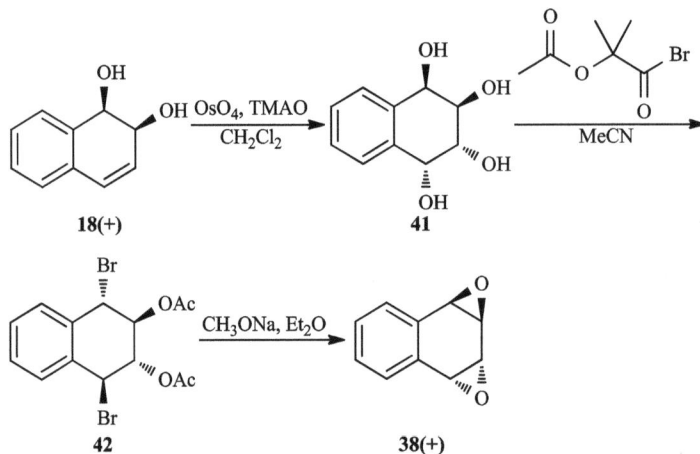

Scheme 19. Synthesis of compound **38(+)**.

Metabolites of Acenaphthene and Acenaphthylene

Acenaphthenols

In order to obtain enantioenriched (*R*)-(-)-1-acenaphthenol (**42(-)**), Hu and Ziffer resolved (±)-1-acenaphthenol (**42(±)**) by fractional crystallization of the corresponding camphanate ester followed by subjecting the crystallized species to ester hydrolysis [45].

A strategy that led to (*S*)-(+)-1- (**43(+)**) and (*R*)-(-)-1-acenaphthenol (**43(-)**) as enantioenriched species has been to utilize lipase-catalyzed hydrolytic kinetic resolution of (±)-1-acenaphthyl acetate (**44(±)**) [46, 47]. Likewise, enantiomerically enriched 1-acenaphthyl esters, which can be converted to the corresponding 1-acenaphthenols, can be achieved in the opposite way *i.e.* by lipase-catalyzed acylative kinetic resolution of (±)-1-acenaphthenol (**43(±)**) [46, 48, 49]. However, since a racemic mixture consists of 50% of each enantiomer no more than 50% of the desired enantiomer can be obtained. An approach to circumvent this limitation has been to couple lipase-catalyzed acylative kinetic resolution of (±)-1-acenaphthenol (**43(±)**) or lipase-catalyzed hydrolytic kinetic resolution of (±)-1-acenaphthyl acetate (**44(±)**) with a Mitsunobu inversion (for the Mitsunobu reaction see reference [50]). Scheme **20a** illustrates how (*R*)-(+)-1-acenaphthyl acetate **44(+)** was obtained in > 99% *ee* when (±)-1-acenaphthenol **43(±)** was deracemized in a two-step sequence, which includes: 1) *Candida Antarctica lipase* B (CAL B) catalyzed acylative kinetic resolution of racemate **43(±)** using isopropenyl acetate as acetylating reagent and 2) Mitsunobu inversion *in situ* of the unreacted enantiomer **43(+)** [51]. The opposite enantiomer **44(-)** was also obtained in > 99% *ee* when (±)-1-acenaphthyl acetate **44(±)** underwent deracemization *via* CAL B catalyzed hydrolytic resolution and then in the following step the unreacted enantiomer **44(-)** underwent Mitsunobu inversion *in situ* (Scheme **20b**) [52].

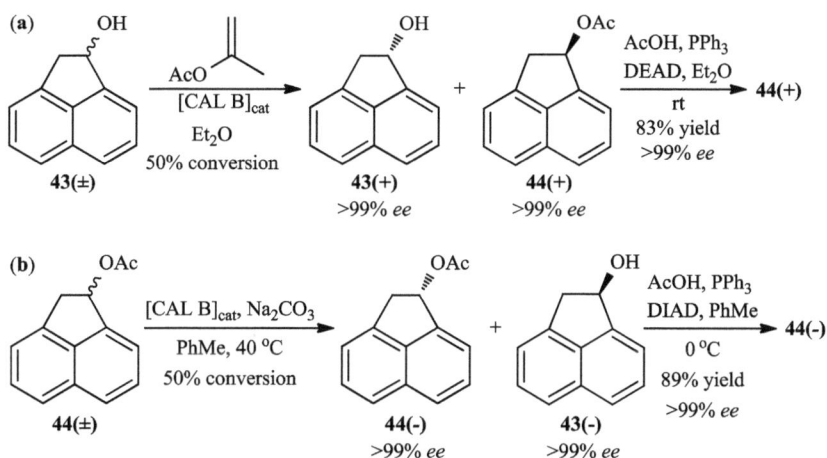

Scheme 20. (a) Coupled CAL B catalyzed acylative kinetic resolution of racemate **43(±)** and (b) CAL B catalyzed hydrolytic kinetic resolution of racemate **44(±)** followed by *in situ* Mitsunobu inversion in both cases (DEAD = diethyl azodicarboxylate, DIAD = diisopropyl azodicarboxylate).

In contrast to the enzyme catalyzed kinetic resolution of racemate (**43(±)**), which is described above, Mandal and Sigman employed palladium catalyzed aerobic oxidative kinetic resolution of racemate **43(±)** (Scheme **21**) [53]. The most reactive enantiomer out of (*S*)-(+)-1- (**43(+)**) and (*R*)-(-)-1-acenaphthenol (**43(-)**) was oxidized 13 times faster into 1-acenaphthenone (**45**) than the slower reacting enantiomer, which was obtained in 99% *ee* when 68% of racemate **43(±)** had been consumed. However, the authors did not point out which of the enantiomers that reacted fastest.

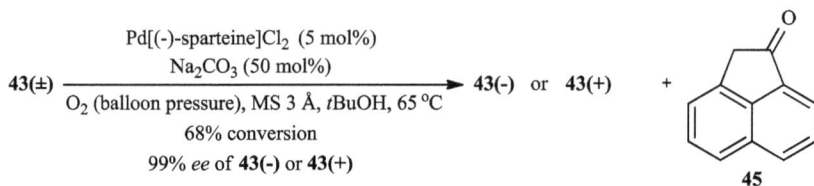

Scheme 21. Palladium catalyzed aerobic oxidative kinetic resolution of racemate (**43(±)**) (MS = molecular sieves).

Organocatalysis constitutes another method for the kinetic resolution of racemate **43(±)**. Wiskur and colleagues employed (-)-tetramisole as a nucleophilic chiral catalyst to afford silylation based kinetic resolution of racemic alcohols to selectively silylate one enantiomer over the other [54]. Application of this enantioselective silylation methodology on racemate **43(±)** provided a selectivity factor (s) of 5.1 when triphenylsilyl ether **46(+)** was formed in 74/26 e.r. leaving behind 80/20 e.r. of unreacted (*R*)-(-)-1-acenaphthenol (**43(-)**) (Scheme **22**).

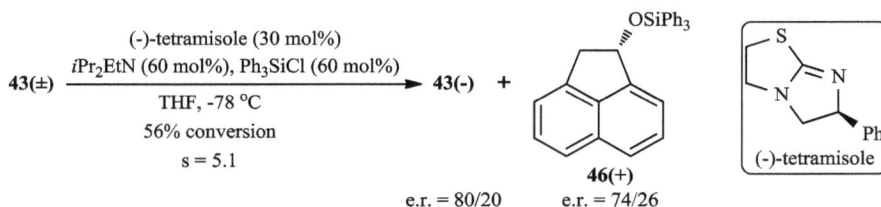

Scheme 22. Enantioselective silylation of (±)-1-acenaphthenol (**43(±)**) with triphenylsilyl chloride in the presence of (-)-tetramisole as a chiral organocatalyst.

In a protocol for catalytic enantioselective hydrosilylation of aromatic alkenes, Nishiyama *et al.* launched the chiral rhodium acetate complex **47** as an enantioselective catalyst [55]. The chiral complex **47** was for instance used as a catalyst for the enantioselective hydrosilylation of acenaphthylene (**48**) with triethoxysilane to furnish a chiral silane intermediate, which underwent oxidation with hydrogen peroxide to furnish enantiomer **43(+)** in 83% *ee* (Scheme **23**).

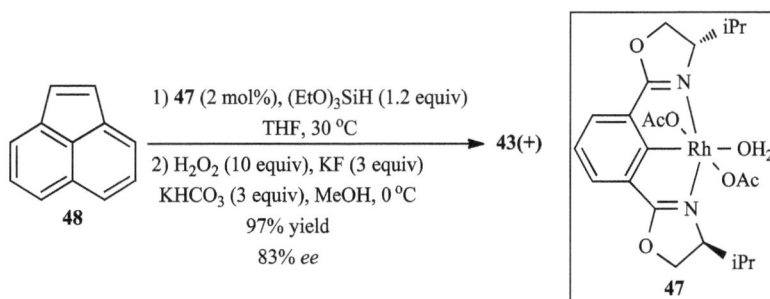

Scheme 23. Sequential enantioselective hydrosilylation/oxidation of acenaphthylene (**48**).

Both 2-hydroxy-1-acenaphthenones [56] and 1,2-acenaphthendiols [57] have been synthesized as racemic mixtures from 1,2-acenaphthenediones. However, baker's yeast mediated reduction of 1,2-acenaphthenediones constitutes a method to afford both 2-hydroxy-1-acenaphthenones [58, 59] and 1,2-acenaphthendiols [60] as enantioenriched species. For instance, Yusufoğli and co-workers reduced 1,2-acenaphthenedione (**49**) into (+)-(*R*)-1,2-dihydro-1-hydroxy-2-acenaphthyleneone **50(+)** in >99% *ee* with baker's yeast (Scheme **24**) [59]. Yeast cells contain some reductases that provide the *R*-enantiomer and others that give the *S*-enantiomer. Thus, prior to the reaction with substrate **49** baker's yeast was incubated with an inhibitor (allyl alcohol) in order to inhibit the reductase that provides the *S*-enantiomer. After inhibition the substrate together with heptane/EtOAc (1:1) as substrate co-solvent was added to the pre-incubated baker's yeast. The choice of substrate co-solvent markedly affected both the conversion and *ee*. In another baker's yeast mediated reduction of 1,2-acenaphthenedione (**49**) in the presence of dimethyl sulfoxide (DMSO) as a co-solvent (-)-(1*S*,2*S*)-1,2-acenaphthendiol (**51(-)**) was obtained in 97% *ee* and 74% yield (Scheme **24**) [60]. The authors explained that DMSO as substrate co-solvent was superior to dimethylformamide (DMF) in both yield and *ee*. In fact, it was proposed that the higher enantioselectivity obtained using DMSO as substrate co-solvent compared to DMF arose from the sulfur atom of DMSO that might affect the enantiofacial discrimination of the reductases in baker's yeast.

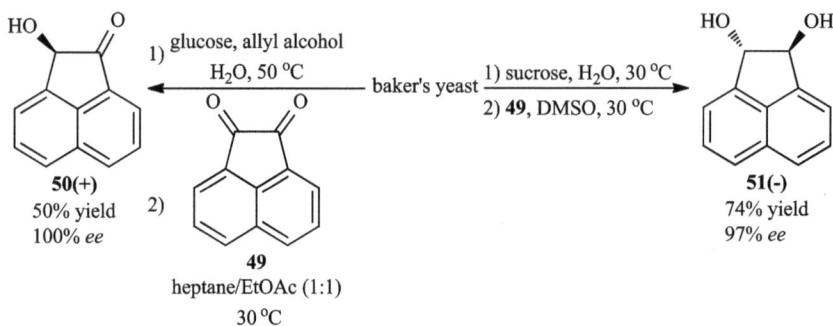

Scheme 24. Baker's yeast mediated reduction of 1,2-acenaphthenedione (**49**).

Acenaphthenones

This section describes synthetic methods used for the formation of acenaphthenones since El Ashry *et al*. published a review within this field in 1998 [61]. Only a few syntheses of 1-acenaphthenone (**45**) from (±)-1-acenaphthenol (**43(±)**) by oxidation has been reported since 1998 [62 - 64]. Another method to obtain 1-acenaphthenone (**45**) and 1,2-acenaphthenedione (**49**) has been to utilize metal catalysis in order to oxidize one or two benzylic methylene groups, respectively, of acenaphthene (**2**). For instance, Li and co-workers reported the synthesis of a heterogeneous silica supported cobalt(II) salen complex catalyst **52** (Fig. **2**), which in the presence of *N*-hydroxyphthalimide (NHPI) under an atmosphere of dioxygen functioned as an efficient catalyst for the oxidation of alkylaromatics containing benzylic C-H bonds to the corresponding benzylic ketones [65]. One of the investigated substrates was acenaphthene (**2**), which was oxidized to 1-acenaphthenone (**45**) in essentially complete selectivity (99%) (Scheme **25**, Conditions A). Likewise, Islam *et al*. synthesized chloromethylated polystyrene supported cobalt(II) complexes and employed them as efficient recyclable heterogeneous catalysts for the oxidation of alkylaromatics, including acenaphthene (**2**), into the corresponding benzylic ketones [66]. In order to generate another metal catalysts for the oxidation of alkylaromatics to the corresponding benzyl ketones a series of perovskite catalysts were synthesize of which LaCrO$_3$ turned out to be superior over all the others catalysts prepared in the series [67]. Under solvent free conditions and in the presence of *tert*-butyl hydroperoxide (TBHP) as oxidant, LaCrO$_3$ exhibited the highest conversion (84% ≤) and selectivity (93% ≤) of all catalysts investigated. One of the substrates included in the study was acenaphthene (**2**), which was oxidized to 1-acenaphthenone (**45**) in 100% selectivity at 87% conversion (Scheme **25**, Conditions B). Jayaram and co-workers prepared amorphous nano manganese(IV) oxide (ANMnO$_2$) as a heterogeneous catalyst for installing an oxo-group at the benzylic methylene groups using TBHP as oxidant [68]. The catalytic performance of ANMnO$_2$ was compared with commercial available manganese(IV) oxide for the oxidation of diphenylmethane into the corresponding ketone. The comparison revealed that ANMnO$_2$ exhibited 2.6 times higher activity than the commercial available manganese(IV) oxide. The catalytic performance of ANMO$_2$ was investigated on various alkylaromatic substrates including acenaphthene (**2**), which underwent 88% conversion and 100% selective oxidation into 1-acenaphthenone (**45**) (Scheme **25**, Conditions C). In contrast to the heterogeneous catalytic systems described in Scheme **25**, Conditions A-C, Maradani and Golchoubian utilized a manganese(III) salen complex **53** (Fig. **2**) as a homogeneous precatalyst for the oxidation of hydrocarbons with hydrogen peroxide [69]. The authors pointed out that the catalytic oxidation of benzylic methylene groups took place through a manganese(V) oxo salen intermediate. Acenaphthene (**2**) was one of the substrates investigated and was oxidized into 1,2-acenaphthenedione (**49**) with hydrogen peroxide in the presence of precatalyst **53** (Scheme **25**, Conditions D).

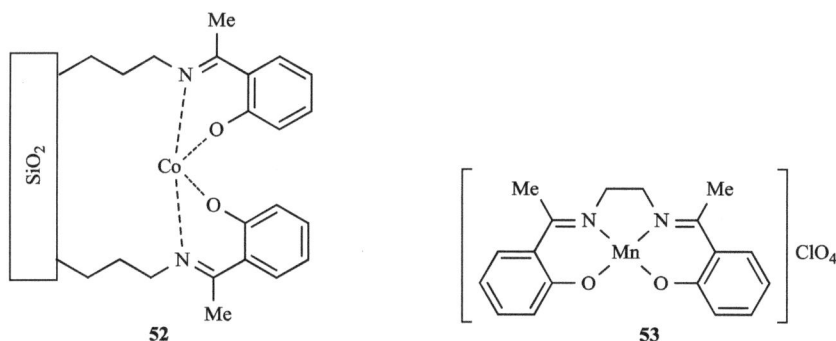

Fig. (2). Complexes **52** and **53** were used for the oxidation of acenaphthene (**2**) into ketone **45** and **49**, respectively.

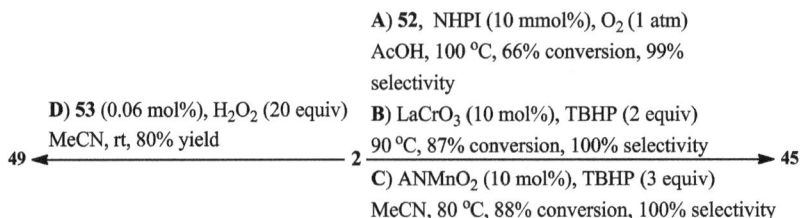

A) **52**, NHPI (10 mmol%), O$_2$ (1 atm)
AcOH, 100 °C, 66% conversion, 99%
selectivity

B) LaCrO$_3$ (10 mol%), TBHP (2 equiv)
90 °C, 87% conversion, 100% selectivity

C) ANMnO$_2$ (10 mol%), TBHP (3 equiv)
MeCN, 80 °C, 88% conversion, 100% selectivity

D) 53 (0.06 mol%), H$_2$O$_2$ (20 equiv)
MeCN, rt, 80% yield

49 ← **2** → **45**

Scheme 25. Oxidation of acenaphthene (**2**) into ketones **45** and **49** using various metal-catalyst.

Organocatalysis constitutes an alternative to metal catalysis for the oxygen-transfer to acenaphthene (**2**) from an oxidation reagent. For instance, Wójtowicz reported the synthesis of silica-supported benzisoselenazol-3(2*H*)-one **54**, which acted as a catalyst for the oxidation alkylaromatic substrates with TBHP into alkyl aryl ketones, aryl aldehydes and aryl carboxylic acids [70]. The authors described that catalyst **54** could easily be removed from the reaction mixture and reused in new catalytic reactions. Organocatalyst **54** exhibited astonishing selectivity for the oxidation of acenaphthene (**2**), since 1-acenaphthenone (**45**) was obtained as the sole product in 88% yield (Scheme **26**, Conditions A).

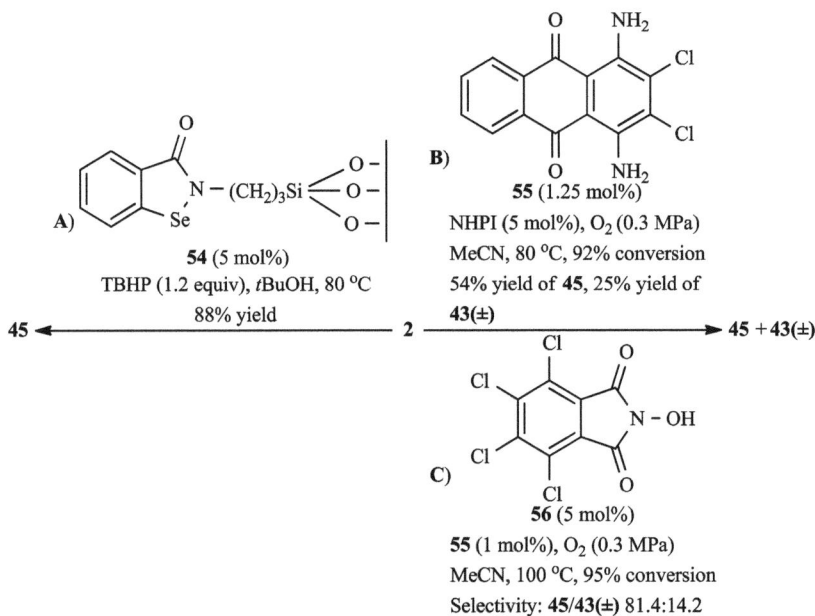

A)
54 (5 mol%)
TBHP (1.2 equiv), *t*BuOH, 80 °C
88% yield

B)
55 (1.25 mol%)
NHPI (5 mol%), O$_2$ (0.3 MPa)
MeCN, 80 °C, 92% conversion
54% yield of **45**, 25% yield of
43(±)

C)
56 (5 mol%)
55 (1 mol%), O$_2$ (0.3 MPa)
MeCN, 100 °C, 95% conversion
Selectivity: **45**/**43**(±) 81.4:14.2

45 ← **2** → **45** + **43**(±)

Scheme 26. Organocatalytic oxidation of acenaphthene (**2**).

Organocatalytic systems for aerobic oxidation of alkylaromatics to the corresponding ketones were reported by Xu and co-workers [71, 72]. These systems consists of two components: 1) a *N*-hydroxyphthalimide that constituted a redox center and 2) a anthraquinone, which acts as redox-active co-factor. The combination of these two components under a dioxygen atmosphere made up the redox catalytic cycle for the oxidation of alkylaromatics [73]. The use of the organocatalytic system NHPI/anthraquinone **55** for the oxidation of acenaphthene (**2**) afforded 1-acenaphthenone (**45**) and (±)-1-acenaphthenol (**43(±)**) in 54 and 25% yield, respectively (Scheme **26**, Conditions B) [71]. Although at a slightly higher temperature, a higher preference of ketone **45** over alcohol **43(±)** (**45/43(±)** 81.4:14.2) was obtained when NHPI was displaced for the tetrachlorinated counterpart **56** (Scheme **26**, Conditions C) [73]. However, it should be noted that the oxidations described in Scheme **26**, Conditions B-C, were performed at slightly different loadings of anthraquinone **55**.

Metabolites of Fluorene
Fluorenols

A frequently used method for obtaining 9-fluorenol (**57**) is reduction of 9-fluorenone (**58**). Various borohydride reagents such as sodium borohydride [74 - 79], sodium cyanoborohydride (NaBH$_3$CN) [80], and zinc borohydride (Zn(BH$_4$)$_2$) [81 - 83] have frequently been applied as reductants under various conditions. A few arbitrary selected examples where borohydride reagents have been utilized for the reduction of 9-fluorenone (**58**) into 9-fluorenol (**57**) are outlined in Scheme **27**. Sodium borohydride is usually employed for reduction of aldehydes and ketones in protic solvents due to its good solubility in these solvents. On the other hand, sodium borohydride displays very poor solubility in acetonitrile, but in the presence of Cu(dmg)$_2$ (dmg = dimethylglyoxime) the solubility is profoundly increased. The combined system sodium borohydride and Cu(dmg)$_2$ in acetonitrile was used to reduce 9-fluorenone (**58**) into 9-fluorenol (**57**) in 97% yield (Scheme **27**, Conditions A) [79]. Both sodium borohydride [76] and sodium cyanoborohydride [80] have been pointed out as efficient reducing agents of carbonyl compounds such as aldehydes, ketones, and α,β-unsaturated enals under solvent free conditions in the presence of wet SiO$_2$, which was obtained by mixing SiO$_2$ with small amounts of water. Under such conditions, sodium borohydride and sodium cyanoborohydride efficiently reduced 9-fluorenone (**58**) into 9-fluorenol (**57**) (Scheme **27**, Conditions B-C). The combination reducing system $Zn(BH_4)_2$/2 NaCl, which was formed when $ZnCl_2$ reacted with sodium borohydride turned out to reduce a variety of carbonyl compounds including 9-fluorenone (**58**) (Scheme **27**, Conditions D) into the corresponding alcohols in high yields [82].

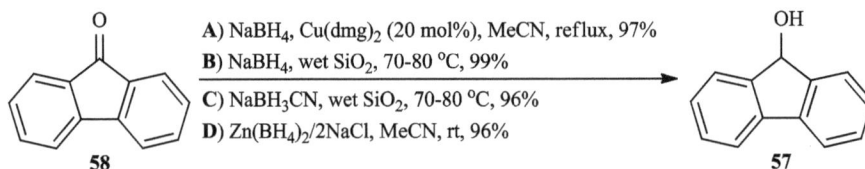

A) NaBH₄, Cu(dmg)₂ (20 mol%), MeCN, reflux, 97%
B) NaBH₄, wet SiO₂, 70-80 °C, 99%
C) NaBH₃CN, wet SiO₂, 70-80 °C, 96%
D) Zn(BH₄)₂/2NaCl, MeCN, rt, 96%

Scheme 27. Reduction of 9-fluorenone (**58**) into 9-fluorenol (**57**) using various borohydride reducing agents.

Interestingly, the heterogeneous Pd/C(en) (en = ethylenediamine) catalyst has been used for chemoselective hydrogenation of aromatic aldehydes and ketones into the corresponding benzyl alcohols [84, 85]. As a matter of fact, when Pd/C(en) was replaced with commercially available Pd/C the carbonyl groups of the investigated aromatic aldehydes and ketones were fully hydrogenolyzed to methylene groups. Thus, it was concluded that the coordinated ethylenediamine in Pd/C(en) acts as a catalyst poison *i.e.* prevents the hydrogenolysis of the carbonyl group. Amongst many substrates investigated, Pd/C(en) turned out to hydrogenolyse 9-fluorenone (**58**) to 9-fluorenol (**57**) in very high chemoselectivity. Another methodology to obtain 9-fluorenol (**57**) has been to employ ruthenium(II) complexes as catalysts for transfer hydrogenation of 9-fluorenone (**58**), in which isopropanol behaves as the hydrogen donor [86 - 90]. In many cases however, this ruthenium(II) catalytic transfer hydrogenation strategy requires synthesis of the catalytic ruthenium(II) complexes since they are not commercially available.

The research group of McNab have shown that flash vacuum pyrolysis (FVP) of ethers **59** or **60** or the oxalate **61** provides the free radicals **62** (from **59** and **60**) and **63** (from **61**), which further led to the fluorene metabolite 1-fluorenol (**64**) and phenol **66** as the major products (from ethers **59** and **60**) along with a small yield of xanthene (**65**) (Scheme **28**) [91, 92]. The fact that phenol **66** was obtained upon FVP of oxalate **61** is evidence for interconversion between benzyl **63** and phenoxyl **62** radicals.

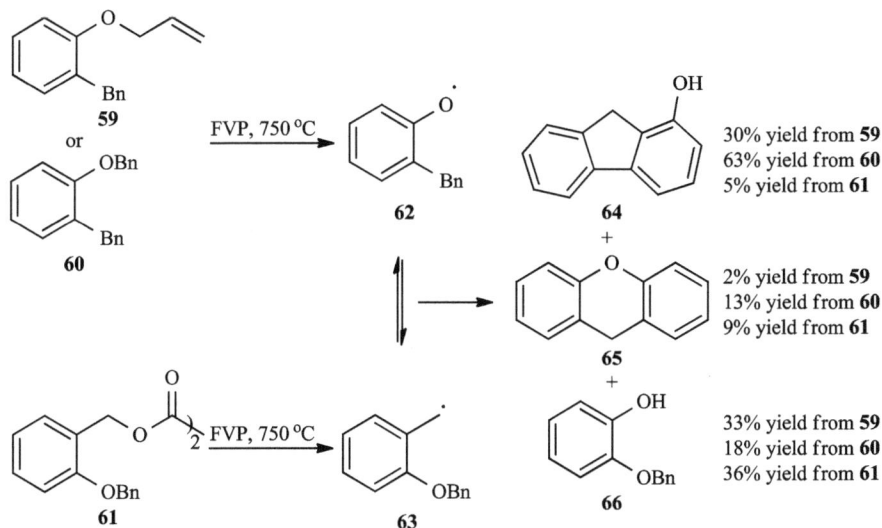

Scheme 28. Formation of the free radicals **62** and **63** by FVP of compounds **59-61** provided products **64-66**.

Photochemistry constitutes a methodology that has been used in order to obtain fluorenols, Kumar and co-workers photolysed benzophenone oxime (**67**) in methanol, using a quartz filter. Substrate **67** underwent a hydroxylation, cyclization of two neighboring phenylgroups and a Wolff-Kishner reduction [93, 94] to provide 1-fluorenol (**64**) in 30-35% yield (Scheme **30a**, Conditions A) [95]. Interestingly, when a solution of benzophenone oxime (**67**) and benzophenone (**68**) in methanol was subjected to photolysis using a Pyrex filter in place of a quartz filter (as in Scheme **29a**, Conditions A) the yield of 1-fluorenol (**64**) increased to 64% (Scheme **29a**, Conditions B) [96, 97]. Noteworthy, neither oxime **67** nor ketone **68** alone provided 1-fluorenol (**64**) when irradiated by using a Pyrex filter. Shi and Wan reported work that dealt with solvolysis and ring closure of quinone methides, which were photogenerated from biaryls systems in which one benzene moiety was decorated with a hydroxyl group and the other benzene moiety included a hydroxymethyl group [98]. In one example, biaryl **69** was irradiated and provided 3-fluorenol (**70**) in 80% yield along with a small yield (7%) of 1-fluorenol (**64**) (Scheme **29b**). The authors included a mechanistic suggestion for the formation of fluorenols **70** and **64** from biaryl **69**.

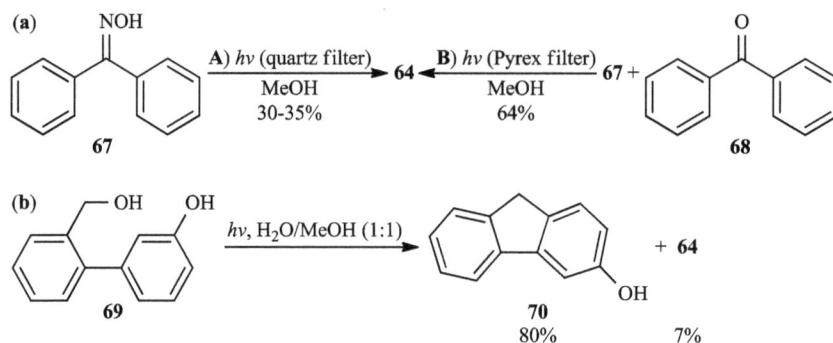

Scheme 29. Photochemical synthesis of 1- (**64**) and 3-fluorenol (**70**).

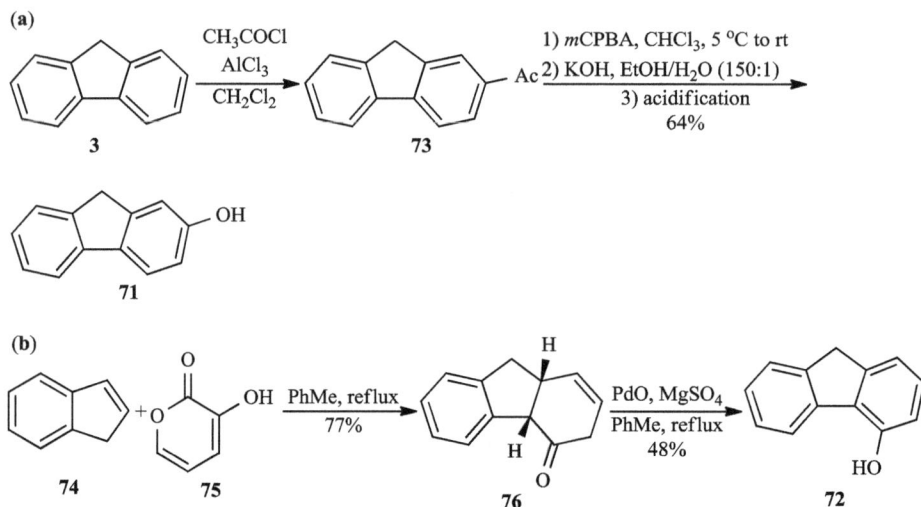

Scheme 30. Conventional methods for synthesis of 2- (**71**) and 4-fluorenol (**72**).

Scheme **30** gives examples of conventional methods that have been utilized in order to obtain 2- (**71**), and 4-fluorenol (**72**). Jones and co-workers devised a three step synthesis in order to obtain 2-fluorenol (**71**) (Scheme **30a**) [99]. The sequence was initiated from fluorene (**3**), which underwent a Friedel–Crafts acylation [100, 101] to give aryl methyl ketone **73**. Baeyer-Villiger oxidation [102] of ketone **73** followed by hydrolysis gave 2-fluorenol (**71**) in 64% yield. 4-Fluorenol (**72**) was obtained *via* a cycloaddition dehydration sequence starting from indene (**74**) (Scheme **30b**) [103], which behaved as a dienophile and underwent a cycloaddition reaction with diene **75** to provide ketone **76** upon loss of carbon dioxide. In the following step ketone **76** was dehydrogenated employing palladium-catalysis in order to give the fluorenol **72** in modest yield (48%).

9-Fluorenone

A common method to prepare 9-fluorenone (**58**) is by oxidation of the corresponding alcohol, namely 9-fluorenol (**57**). For instance, a palladium/phosphine [104, 105] and Pd/C catalyzed [106] activation of phenyl chloride (PhCl) constitutes an efficient method for the selective oxidation of alcohols to carbonyl compounds such as the conversion of fluorenol **57** to ketone **58** (Scheme **31**, Conditions A). During this oxidation process, PhCl was proposed to undergo hydrodechlorination (*i.e.* reduction), and thus generate benzene. Muzart and co-workers prepared 9-fluorenone (**58**) from alcohol **57** in refluxing DCE in the presence of catalytic amounts of $PdCl_2$ and Adogen 464 (Scheme **31**, Conditions B) [107]. It was shown that DCE regenerated (upon formation of ethylene) the catalytic active palladium(II) species upon reaction with a palladium(0) species, which was formed in the catalytic cycle. Adogen 464 on the other hand was suspected to solubilize the palladium(II) species. In order to obtain ketones from the corresponding alcohols using acetone as an oxidizing agent in an Oppenauer-type oxidation, Ajjou developed the first water-soluble transition metal catalyst [108, 109]. When the catalytic system consisting of $(Ir(COD)Cl)_2$, 2,2'-biquinoline-4,4'-dicarboxylic acid dipotassium salt (BQC), and sodium carbonate in acetone/H_2O (1:2) was tested on 9-fluorenol (**57**) the corresponding ketone **58** was obtained in almost quantitative yield (Scheme **31**, Conditions C).

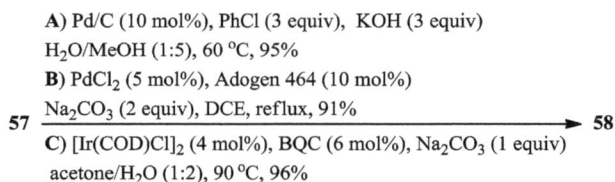

A) Pd/C (10 mol%), PhCl (3 equiv), KOH (3 equiv)
H_2O/MeOH (1:5), 60 °C, 95%

B) $PdCl_2$ (5 mol%), Adogen 464 (10 mol%)
Na_2CO_3 (2 equiv), DCE, reflux, 91%

57 ———————————————————————→ **58**

C) $[Ir(COD)Cl]_2$ (4 mol%), BQC (6 mol%), Na_2CO_3 (1 equiv)
acetone/H_2O (1:2), 90 °C, 96%

Scheme 31. Example of reaction conditions employed in order to oxidize fluorenol **57** into ketone **58**.

Apart from the methods described above for the oxidation of 9-fluorenol (**57**) into the corresponding ketone **58**, dioxygen has been utilized as a stoichiometric oxidant using metal containing catalytic systems such as Pd/C in the presence of sodium borohydride [110], CAN/TEMPO [111], CAN/NHPI [112], and

DDQ/NaNO$_2$ [113] as well as metal free catalytic systems [114, 115].

Perhaps more interesting, fluorene (**3**) can be oxygenated to 9-fluorenone (**58**) using various catalytic systems. In this type of approach, NHPI has been employed in organocatalytic systems in combination with oximes [116] and DDQ [117]. For instance, Yang and co-workers developed an organocatalytic system that included an oxime, particularly dimethylglyoxime (DMG), along with NHPI that oxygenates the benzylic position of hydrocarbons forming the corresponding oxygen containing compounds (*i.e.* alcohol, ketone, and peroxide) with dioxygen. Under the optimized conditions, NHPI (10 mol%)/DMG (10 mol%) under dioxygen (0.5 MPa) in acetonitrile, gave 9-fluorenol (**57**) and 9-fluorenone (**58**) in a 11:89 ratio at 67% conversion (Scheme **32**, Conditions A) [116]. Zhang and co-workers pursued an environmentally friendly oxygenation protocol for converting fluorene (**3**) into 9-fluorenone (**58**). Thus, 9-fluorenone (**58**) was obtained in 92% yield when fluorene (**3**) in THF in the presence of potassium hydroxide was exposed to air at ambient temperature (Scheme **32**, Conditions B) [118].

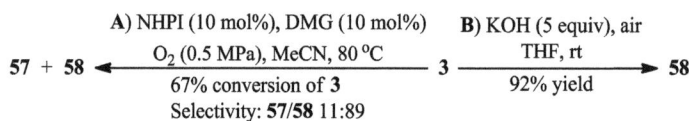

$$
\textbf{57} + \textbf{58} \xleftarrow{\begin{array}{c} \textbf{A)}\ \text{NHPI (10 mol\%), DMG (10 mol\%)} \\ \text{O}_2\ (0.5\ \text{MPa), MeCN, 80\,°C} \\ \hline 67\%\ \text{conversion of } \textbf{3} \\ \text{Selectivity: } \textbf{57/58}\ 11{:}89 \end{array}} \textbf{3} \xrightarrow{\begin{array}{c} \textbf{B)}\ \text{KOH (5 equiv), air} \\ \text{THF, rt} \\ \hline 92\%\ \text{yield} \end{array}} \textbf{58}
$$

Scheme 32. Oxygenation of fluorene (**1**) forming ketone **58** under conditions including dioxygen (Conditions A) and air (Conditions B).

Other protocols for oxidation of fluorene (**3**) into 9-fluorenone (**58**) utilize oxidizing reagents such as sodium hypochlorite (NaOCl) in the presence of *tert*-butyl hydroperoxide [119], sodium chlorite in combination with either *tert*-butyl hydroperoxide or NHPI [120], and *tert*-butyl hydroperoxide under microwave irradiation [121].

Metabolites of Phenanthrene

Phenanthrols

Moriconi *et al.* reported a two-step synthetic protocol in order to obtain 9-phenanthrol (**77**) from phenanthrene (**4**) [122]. Phenanthrene (**4**) was dihydroxylated upon treatment with stoichiometric amounts of osmium tetroxide in the presence of pyridine thus providing *cis*-phenanthrene-diol (**78**) (syntheses of dihydroxy-dihydrophenanthrenes will be specified in the next section). Diol **78** was then subjected to a acid promoted dehydration forming 9-phenanthrol (**77**) in 66% yield over the two steps (Scheme **33a**). In an article published in 1988, Fujishiro and Mitamura show that various polycyclic aromatic aldehydes react with *m*CPBA in a Bayer-Villiger oxidation to give the corresponding phenols *via* an ester intermediate [123]. One of the examples included in their work was the Bayer-Villiger oxidation of 9-formylphenanthrene (**79**) to phenanthrol **77** *via* the phenanthryl formate ester intermediate **80** (Scheme **33b**). The Snieckus research group developed a general two-step protocol for the synthesis of 9-phenanthrols from *o*-(*N,N*-dialkylcarboxamido)phenylboronic acids (in this case, alkyl = Et or *i*Pr) and *o*-bromotoluenes [124, 125]. This protocol was employed in the synthesis of 9-phenanthrol (**77**) and was initiated by a Suzuki-Miyuara cross-coupling [126] between coupling partners **81** and **82** providing biaryl **83**. In the subsequent step, LDA-mediated deprotonation of the 2'-methyl hydrogen in compound **83** triggered the cyclization into phenanthrol **77** (Scheme **33c**).

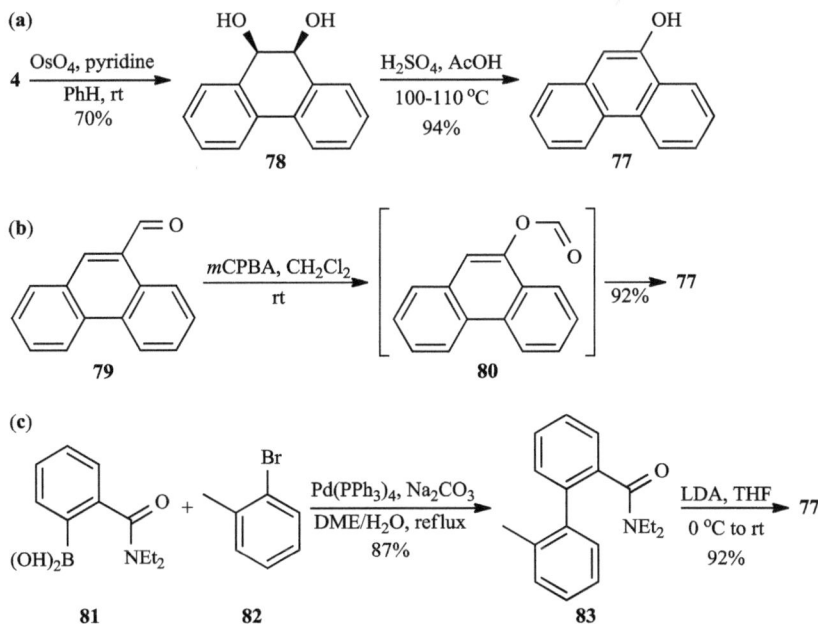

Scheme 33. Syntheses of 9-phenanthrol (**77**) from various starting materials.

Mosettig and Duvall have investigated the dehydrogenation of ketotetrahydrophenanthrenes with regard to solvent, temperature, and catalyst in order to obtain the corresponding phenanthrols [127]. Their work concluded that palladium black catalyzed the dehydrogenation of 1-keto-1,2,3-4-tetrahydrophenanthrene (**85**) into 1-phenanthrol (**84**) in refluxing naphthalene (Scheme **34a**). Using the same catalyst, 4-phenanthrol (**92**) was obtained in 56-63% yield from 4-keto-1,2,3,4-tetrahydrophenanthrene in refluxing xylene. Thermal reactions between nitronaphthalene and butadienes have been investigated for the preparation of phenanthrols [128, 129]. Under optimized conditions, it was discovered that a domino reaction occurred: (1) polar Diels-Alder reaction between nitronaphthalene, which behaved as an electrophile, and a nucleophilic diene carrying a siloxy group to generate a Diels-Alder adduct and (2) thermal elimination of the nitro group as nitrous acid to provide the phenanthrol. Hence, when Danishefsky's diene (**86**) and 1-nitronapthalene (**87**) dissolved in benzene was heated at 150 °C in a sealed tube 2-phenanthrol (**88**) was regioselectively obtained (Scheme **34b**, Conditions A). It should be mentioned that the hydroxyl group was derived from the siloxy group and that the methoxy group was thermally eliminated as methanol. When 1-nitronapthalene (**87**) was replaced with 2-nitronapthalene (**89**) 3-phenanthrol (**90**) was formed (Scheme **34b**, Conditions B). Switching from Danishefsky's diene (**86**) (Scheme **34b**, Conditions A) to (*E*)-1-(trimethylsilyloxy)-1,3-butadiene (**91**) (Scheme **34c**), where both dienes possess essentially the same nucleophilicity provided 4-phenanthrol (**92**).

In a two-step protocol for the preparation of phenanthrols along with other polynuclear aromatic hydroxyl compounds, gold(III) catalysis has been utilized in an intramolecular Diels-Alder reaction with various 1-(2-furyl)-hex-1-e--5-yn-3-ol derivatives. This protocol was useful for the synthesis of 2-phenanthrol (**88**) in which 2-(2-furyl)benzaldehyde (**93**) underwent indium mediated propargylation to afford propargyl alcohol **94** [130]. The obtained compound **94** behaved as a substrate for the intramolecular Diels-Alder reaction where the furan moiety constituted the diene and the alkyne group the dienophile upon treatment with catalytic amounts of gold(III) chloride in acetonitrile, thus forming 2-phenanthrol (**88**) (Scheme **34d**). Wang and co-workers have identified

(Rh$_2$(OAc)$_4$) as a useful catalyst for the intramolecular cyclization of 2,2'-diformylbiphenyls into C=C bonds [131]. The method required that 2,2'-diformylbiphenyls was converted to the corresponding bis(*N*-tosylhydrazone)s prior to treatment with *p*-toluenesulfonyl hydrazide followed by subjecting the reaction mixture to a rhodium(II) catalyzed cyclization. This method was efficient for the synthesis of 2-phenanthrol (**88**), which was formed when bis(*N*-tosylhydrazone) (**95**) underwent cyclization in the presence of [Rh$_2$(OAc)$_4$] (Scheme **34e**).

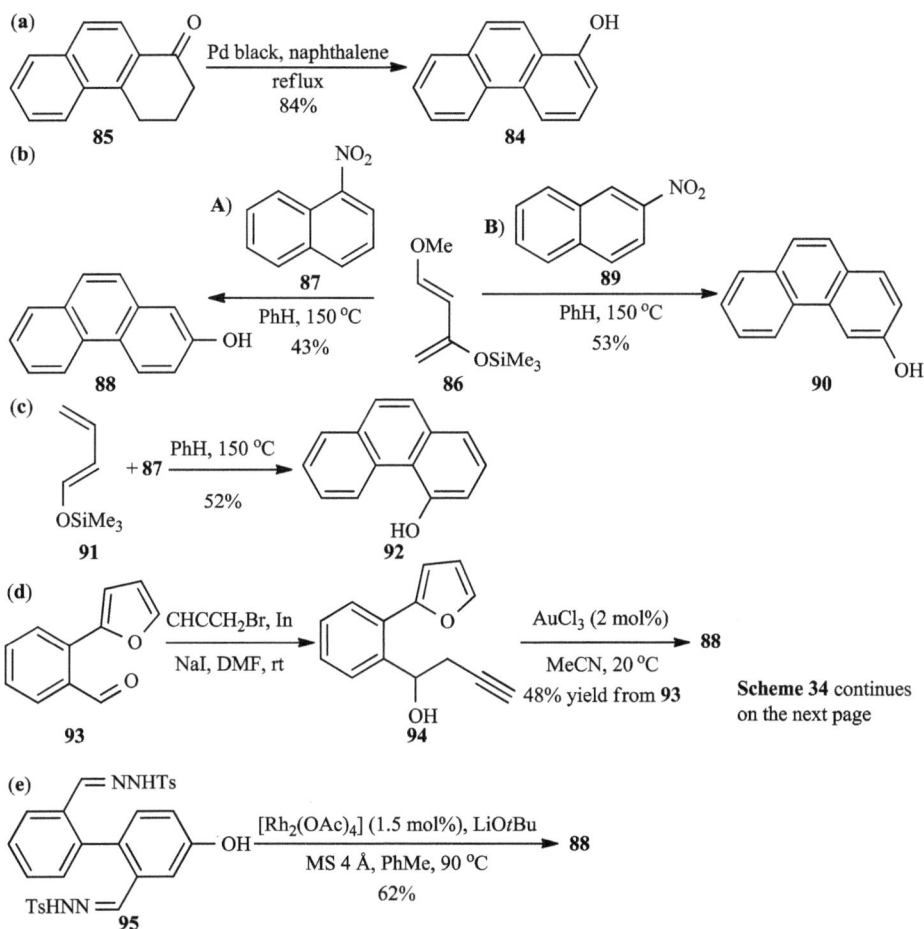

Scheme 34 continues on the next page

Scheme 34. Various syntheses of phenanthrols.

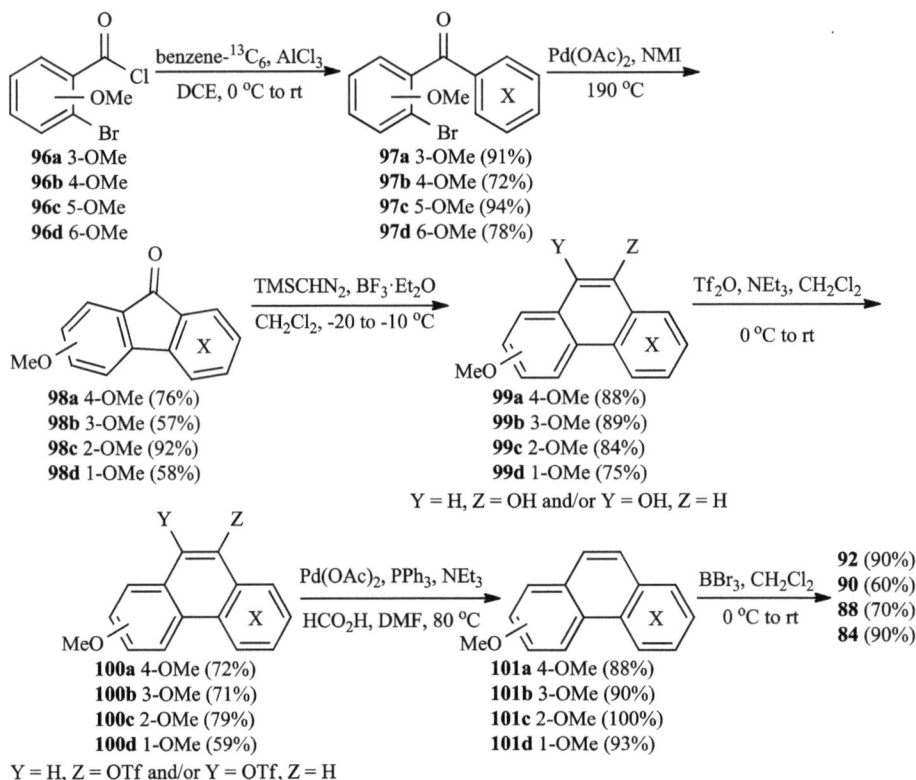

Scheme 35. Synthesis of $^{13}C_6$ labeled 1- (**87**), 2- (**91**), 3- (**93**), and 4-phenanthrol (**95**) (all carbons in the ring marked X are labeled with ^{13}C).

Reported in the literature there is a few examples where phananthrols have been ^{13}C labeled in one [132, 133] or several [134] positions. For instance, $^{13}C_6$ labeled phenanthrols **84**, **88**, **90**, and **92** have been synthesized in six steps where benzene-$^{13}C_6$ constituted the ^{13}C source (Scheme **35**) [134]. Thus, benzene-$^{13}C_6$ underwent Friedel-Crafts acylation with acyl chloride **96** to provide ^{13}C labeled bromobenzophenone **97**. In the following step, compound **97** underwent intramolecular palladium catalyzed cross-coupling in *N*-methylimidazole (NMI) to form fluorenone **98**. Compound **98** ring-expanded by one carbon atom to give hydroxy-methoxyphenanthrenes **99** when treated with trimethylsilyldiazomethane in the presence of boron trifluoride diethyl etherate. The hydroxyl group within compounds **99** was removed in a two-step sequence including: (1) triflation into aryl triflates **100** and (2) palladium catalyzed reductive removal of the triflate group to give methoxyphenanthrenes **101**. Finally a BBr₃ promoted de-*O*-

methylation gave the $^{13}C_6$ labeled phenanthrols **84**, **88**, **90**, and **92**.

Phenanthrene-Diols

In the section discussing the synthesis of naphthalene diols we described that Platt and Oeasch found that 1,2-naphthoquinone (**19**) was efficiently reduced to *trans*-(±)-dihydrodiol **17(±)** with sodium borohydride under an atmosphere of air (Scheme **10**). The authors also showed that the same protocol can be used for the reduction of 9,10-phenanthrenedione (**102**) into *trans*-9,10-dihydrophenanthre-e-9,10-diol (**103**) [135]. Likewise, using the same conditions 1,2- (**104**) and 3,4-phenanthrenedione (**105**) were reduced to *trans*-1,2-dihydroxy-1,2- (**106**) and *trans*-3,4-dihydrophenanthrene-3,4-diol (**107**), respectively [33].

Intramolecular pinacol coupling [136] constitutes a type of reaction that has been quite frequently used in the synthesis of *trans*-9,10-dihydroxy-9-10-dihydrophenanthrene (**106**). For example, Li and co-workers found that manganese under aqueous conditions in the presence of either ammonium chloride or acetic acid behaves as a reducing agent for the pinacol coupling of aromatic aldehydes [137]. To be more precise, when (1,1'-biphenyl)-2,2'-dicarbaldehyde (**108**) was employed as substrate, an intramolecular pinacol coupling took place to form *trans*-9,10-dihydrophenanthrene-9,10-diol (**103**) as the sole diastereomer (Scheme **36**, Conditions A). In another study, when samarium(II) iodide in THF was used as reducing agent, dialdehyde **108** underwent the same type of intramolecular pinacol coupling to give *trans*-diol **103** as the sole diastereomer (Scheme **36**, Conditions B) [138]. In the presence of zinc as reducing agent, Itoh and co-workers have identified Cp$_2$Ti(Ph)Cl as an efficient catalyst for the intramolecular pinacol reaction of both aromatic dialdehydes and aliphatic dialdehydes to provide the corresponding 1,2-diols in high *trans*-selectivity [139, 140]. When dialdehyde **104** was investigated as the substrate, *trans*-diol **103** was formed as the major diastereomer (*trans*/*cis* 91:9) (Scheme **36**, Conditions C). Another catalyst that has been used in the intramolecular pinacol reaction of dialdehyde **108** is indium(III) chloride, which generated substrate **103** as the only diastereomer in the presence of magnesium as reducing reagent (Scheme **36**, Conditions D) [141].

A) Mn (3 equiv), NH$_4$Cl (3 equiv)
H$_2$O/THF (4:1), rt, 87%

B) SmI$_2$ (2.1 equiv), THF, 0 °C, 99%

C) Cp$_2$Ti(Ph)Cl (3 mol%), Zn (1 equiv)
Me$_3$SiCl (1.5 equiv), THF, rt, 70% (*trans*/*cis* 91:9)

D) InCl$_3$ (10 mol%), Mg (2 equiv)
TMSCl (4 equiv), THF, rt, 76%

108

103

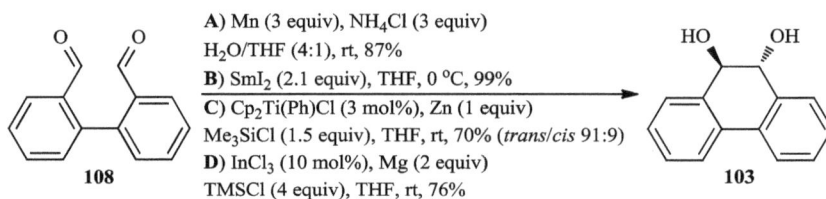

Scheme 36. Intramolecular pinacol reaction of (1,1'-biphenyl)-2,2'-dicarbaldehyde (**108**) under various conditions to form *trans*-diol **103**.

Lin and You have developed a one-pot protocol for the synthesis of *trans*-9,1--dihydrophenanthrene-9,10-diol (**103**) [142]. The one-pot protocol consisted of two consecutive steps: (1) (Ph$_3$P)$_2$NiCl$_2$ catalyzed homocoupling between *o*-halobenzaldehydes in the presence of PPh$_3$ and Zn to form biphenyl-2,2--dialdehydes along with Zn(II) halides and (2) a second Pinacol reaction to form *trans*-9,10-dihydroxy-9,10-dihydrophenanthrenes (**103**), in which the Zn(II) halides formed during the initial homocoupling served as a promoter (Scheme **37**).

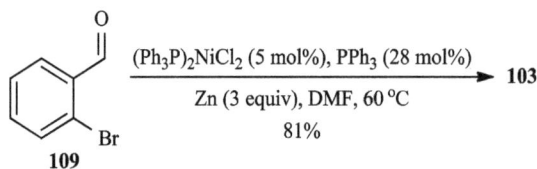

$$\frac{(Ph_3P)_2NiCl_2 \ (5 \ mol\%), \ PPh_3 \ (28 \ mol\%)}{Zn \ (3 \ equiv), \ DMF, \ 60 \ ^\circ C} \longrightarrow \mathbf{103}$$

81%

109

Scheme 37. One-pot procedure for the synthesis of *trans*-9,10-dihydrophenanthrene-9,10-diol (**103**).

Ruthenium catalyzed *cis*-dihydroxylation of phenanthrene (**4**) has emerged as a methodology to generate *cis*-9,10-dihydrophenanthrene-9,10-diol (**78**). The Shing research group verified that RuCl$_3$•(H$_2$O)$_3$ was able to catalyze the *cis*-dihydroxylation of a wide range of alkenes using sodium metaperiodinate as oxidizing agent [143]. The authors pointed out that long reaction time lowered the yield of desired products presumably due to oxidative 1,2-glycole cleavage caused by the sodium metaperiodate present in the reaction mixtures. Along with a wide range of alkene substrates, the protocol is suitable for phenanthrene (**4**) as it generated *cis*-9,10-dihydroxy-9,10-dihydrophenanthrene (**78**) (Scheme **38**, Conditions A). Ruthenium nanoparticles grafted onto hydroxyapatite (nano-RuHAP) constitutes another catalyst for facilitating the *cis*-dihydroxylation of alkenes, including phenanthrene (**4**) to *cis*-diol **78** (Scheme **38**, Conditions B)

[144]. The nano-RuHAP catalyst could be recycled up to four times without any apparent loss of activity.

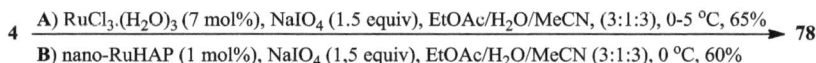

4 $\xrightarrow{\text{A) RuCl}_3.(\text{H}_2\text{O})_3 \text{ (7 mol\%), NaIO}_4 \text{ (1.5 equiv), EtOAc/H}_2\text{O/MeCN, (3:1:3), 0-5 °C, 65\%}}$ **78**
$\phantom{\text{4}}$ B) nano-RuHAP (1 mol%), NaIO$_4$ (1,5 equiv), EtOAc/H$_2$O/MeCN (3:1:3), 0 °C, 60%

Scheme 38. Ruthenium catalyzed *cis*-dihydroxylation of phenanthrene (**4**)

Scherf and co-workers have reported a protocol that leads to the formation of *cis*-9,10-dihydroxy-9,10-dihydrophenanthrenes from *o*-carbonyl-substituted bromo-benzenes [145]. The one-pot protocol consists of two sequential steps (both catalyzed by Ni(COD)$_2$): homocoupling between *o*-carbonyl-substituted bromobenzenes to form 2,2'-keto- or aldehyde substituted biphenyls, which underwent an intramolecular Pinacol reaction forming *cis*-9,10-dihydro-phenanthrene-diol (**78**). The one-pot protocol where both steps was facilitated by Ni(COD)$_2$ was useful for the synthesis of *cis*-phenanthrene-diol (**78**) in high stereoseletivity, when *o*-bromobenzaldehyde (**109**) was employed as the starting material (Scheme **39**).

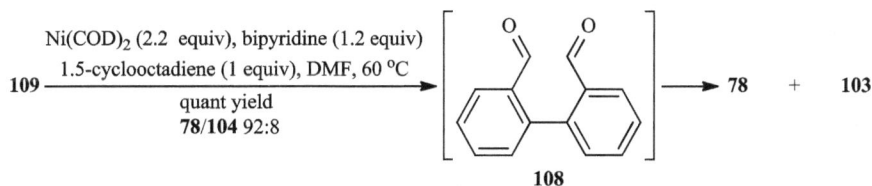

Scheme 39. Ni(COD)$_2$ catalyzed one-pot protocol for the synthesis of *cis*-9,10-dihydroxy-9-10-dihydrophenanthrene (**78**).

Recently, Sydnes and co-workers pursued an enantioselective total synthesis of (-)-(1*R*,2*R*)-1,2-dihydroxy-1,2-dihydroxyphenanthrene (**109(-)**) [146] Scheme **40**. A Shi asymmetric epoxidation reaction [147] constituted the key step in order to install the *trans*-1*R*,2*R*-glycol moiety within compound **110(-)**. The synthesis commenced with a Negishi cross-coupling [148] reaction between 1-bromonaphthlalen (**10**) and the organozinc reagent of ethyl-4-bromobutyrate (**111**) to give ester **112**, which cyclized into ketone **113** upon treatment with Eaton's reagent (7.7 wt% P$_2$O$_5$ in MsOH). Ketone **113** was deprotonated with KHMDS (HMDS = hexamethyldisilazane) followed by quenching with TBDPSCl (TBDPS

= *tert*-butyldiphenylsilyl) to provide silyl enol ether **114**. The protected enol **114** constituted the substrate in the key step in which it underwent a Shi asymmetric epoxidation when treated with the (-)-Shi catalyst in the presence of oxone resulting in the formation of silyloxy epoxide **115**. Epoxide **115** underwent stereoselective reduction with hydroborane to provide the *trans*-diol monosilyl ether **116**. In the following two steps, the TBDPS protective group was removed using *tetra-N*-butylammonium fluoride (TBAF) and then the hydroxyl groups were acetylated with acetic anhydride in the presence of pyridine to give compound **117**. In order to install a double-bond, compound **117** was subjected to bromination with *N*-bromosuccinimide (NBS) followed by base promoted elimination of hydrogen bromide using 1,5-diazabicyclo(4.3.0)non-5-ene (DBN) to give compound **118**. Final alcoholysis of **118** provided target compound **110(-)**.

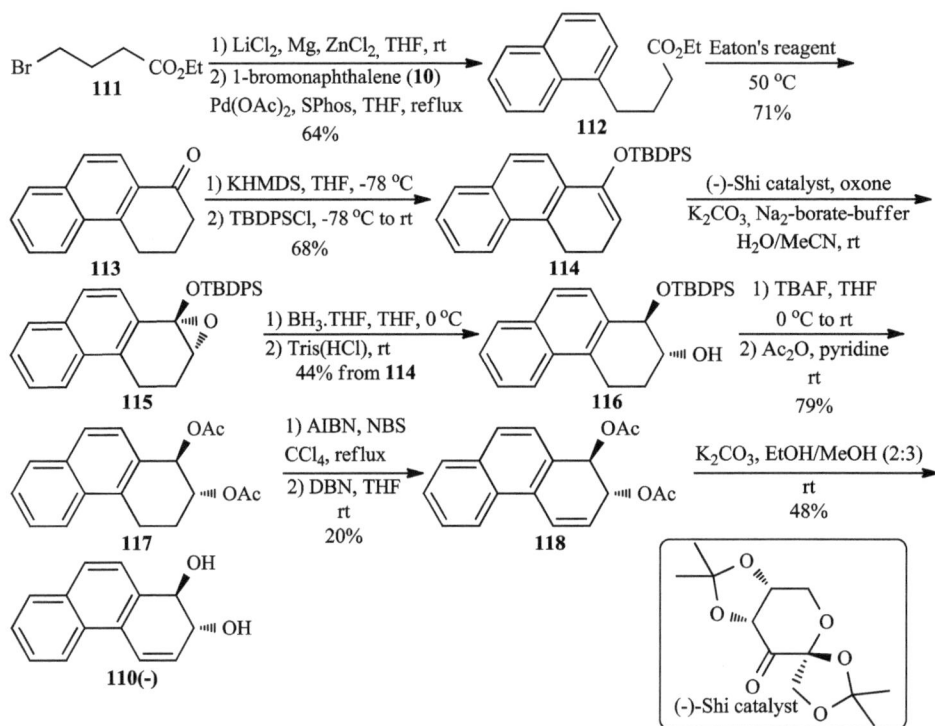

Scheme 40. Enantioselective synthesis of (-)-(1*R*,2*R*)-1,2-dihydrophenanthrene-1,2-diol (**110(-)**).

Both *cis*-9,10-dihydroxy-9,10-dihydrophenanthrene (**78**) and its *trans*-counterpart **103** have been used as substrates in the synthesis of 9,10-epoxy-9,-

0-dihydrophenanthrene (**119**). For example, Dansette and Jerina transformed *cis*-9,10-dihydroxy-9,10-dihydrophenanthrene (**78**) to epoxide **119** over three steps (Scheme **41a**) [149]. In the first step, orthoester **120** was formed when *cis*-diol **78** was reacted with trimethyl orthoformate in the presence of benzoic acid. In the subsequent step, orthoester **120** was treated with trimethylsilyl chloride to provide the acetylated chlorohydrin **121**, which underwent deacetylation and cyclization into epoxide **119** upon treatment with sodium methoxide. Cortez and Harvey on the other hand have shown that epoxide **119** was accessible in one step from *trans*-9,10-dihydroxy-9,10-dihydrophenanthrene   (**103**)   when   it   was cyclodehydrated with *N,N*-dimethylformamide dimethyl acetal (Scheme **41b**) [150]. Another useful cyclodehydrating reagent for *trans*-9,10-dihydroxy-9-10-dihydrophenanthrene (**103**) is diethoxytriphenylphosphorane, $Ph_3P(OEt)_2$, which provided epoxide **119** in 99% yield [151].

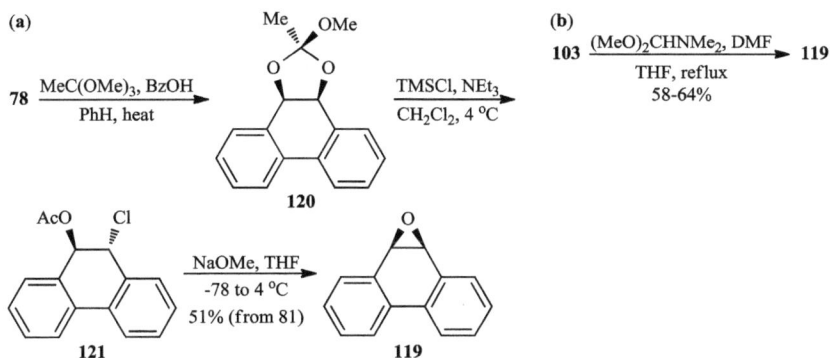

Scheme 41. Syntheses of 9,10-epoxy-9,10-dihydrophenanthrene (**119**) from 9,10-dihydrophenanthrene-9, 10-diols.

In order to epoxidize the 9,10-position of phenanthrene (**4**) in order to provide 9,10-epoxy-9,10-dihydrophenanthrene (**119**), the literature offers a few options (Scheme **42**). For example, Hamilton and co-workers have found that carbamimidoperoxoic amide **121**, which formed *in situ* from *N,N'*-diisopropylcarbodiimide (DIC) and hydrogen peroxide epoxidized phenanthrene (**4**) forming epoxide **119** (Scheme **42**, Conditions A) [152]. The conditions previously described using sodium hypochlorite [40] or methyl(tri-fluoromethyl)dioxirane [42] as oxidation reagent for the epoxidation of naphthalene (**1**) were both productive for the epoxidation of phenanthrene (**4**) to epoxide **119** (Scheme **42**, Conditions B-C)

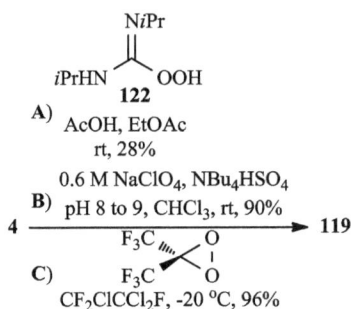

Scheme 42. Epoxidation of phananthrene (**4**) to epoxide **119**.

Chrysene Metabolites

Chrysenols

Chrysenols are formed as metabolites of chrysene (**5**) *in vivo*. Jørgensen and Joensen reported a three step synthesis to obtain 1-, 2-, 3-, and 4-chrysenoles (**122a-125a**) (Scheme **43**) [153]. The syntheses commenced from the methoxybenzyl chlorides **122b-124b**, which were treated with triphenylphosphine to furnish the corresponding phosphonium salts. Subsequent Wittig reaction [154] with 1-naphthaldehyde gave stilbenes **122c-124c** in a 1:1 *E*/*Z*-ratio in all three cases. In order to obtain the chrysene core structure the photochemical oxidative cyclization, namely the slightly modified Mallory reaction [155] was applied using iodine as a terminal oxidant and 1,2-epoxubutane as a scavenger to capture the *in situ* generated hydrogen iodide. Thus, cyclization of **122c** and **124c** gave **122d** and **124d**, respectively. Stilbene **123c** on the other hand contains two reactive positions, namely position 1 and 5. Subjecting compound **123c** to the slightly modified Mallory conditions gave **123d** and **125d** in a 1:1 ratio in a 80% combined yield. Two different protocols were employed for the final de-*O*-methylation of **122d-125d**: 1) potassium in THF followed by quenching of the formed radicals with EtOH, and 2) BBr$_3$ in methylene chloride. The former protocol furnished **122a-125a** in 60-73% yield together with unreacted starting material. The latter protocol worked significantly better in three out of the four examples resulting in the formation of the desired products **122a-125a** in 58-96% yield.

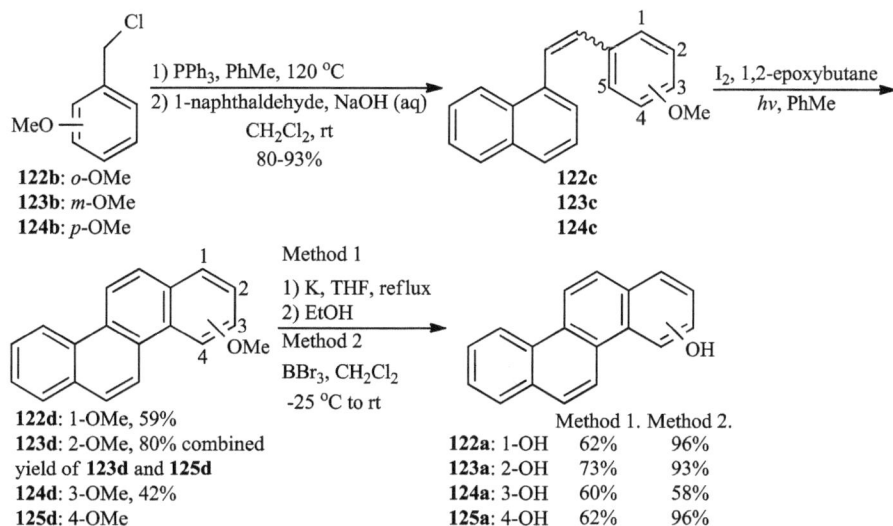

122b: *o*-OMe
123b: *m*-OMe
124b: *p*-OMe

122c
123c
124c

122d: 1-OMe, 59%
123d: 2-OMe, 80% combined
yield of **123d** and **125d**
124d: 3-OMe, 42%
125d: 4-OMe

	Method 1.	Method 2.
122a: 1-OH	62%	96%
123a: 2-OH	73%	93%
124a: 3-OH	60%	58%
125a: 4-OH	62%	96%

Scheme 43. Synthesis of chrysenols **122a-125a** utilizing a slightly modified Mallory reaction.

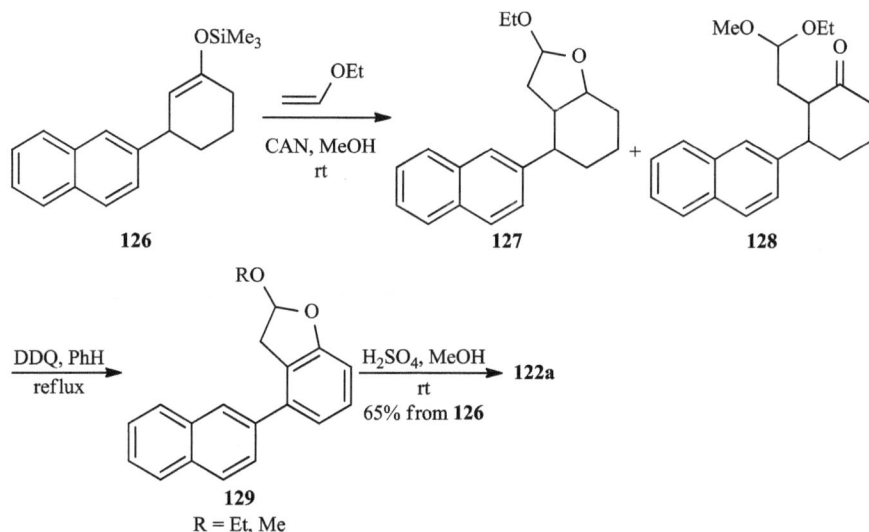

126

127

128

129
R = Et, Me

Scheme 44. Synthesis of 1-chrysenol (**122a**) *via* oxidative addition of silyl enol ether **126** to ethyl vinyl ether.

Ruzziconi and co-workers introduced another methodology to prepare 1-chrysenol (**122a**) (Scheme **44**) [156]. The methodology included CAN-promoted oxidative addition of silyl enol ethere **126** to ethyl vinyl ether, thus forming a mixture of cyclic **127** and acyclic **128** acetals. Subsequent oxidation of compounds **127** and **128** with DDQ followed by acid catalyzed cyclization of substrate **129** gave 1-

chrysenol (**122a**) in 65% yield from starting material **126**.

Johnson *et al.* also reported a procedure for the synthesis of 1-chrysenol (**122a**) (Scheme **45**) [157]. The synthesis was initiated by treating phenylacetylene (**130**) with potassium forming (phenylethynyl)potassium, which underwent a 1,2-addition to decalin-1,5-dione **131** (mixture of *cis* and *trans* products) to give α-acetylenic carbinol **132**. Palladium catalyzed hydrogenation of **132** gave the saturated counterpart **133**. Subsequent acid promoted cyclization of substrate **133** afforded the ABCD ring system of compound **134** of 1-chrysenol (**122a**). Finally, ketone **134** was subjected to palladium-catalyzed dehydrogenation using a Heymann type of apparatus to furnish 1-chrysenol (**122a**).

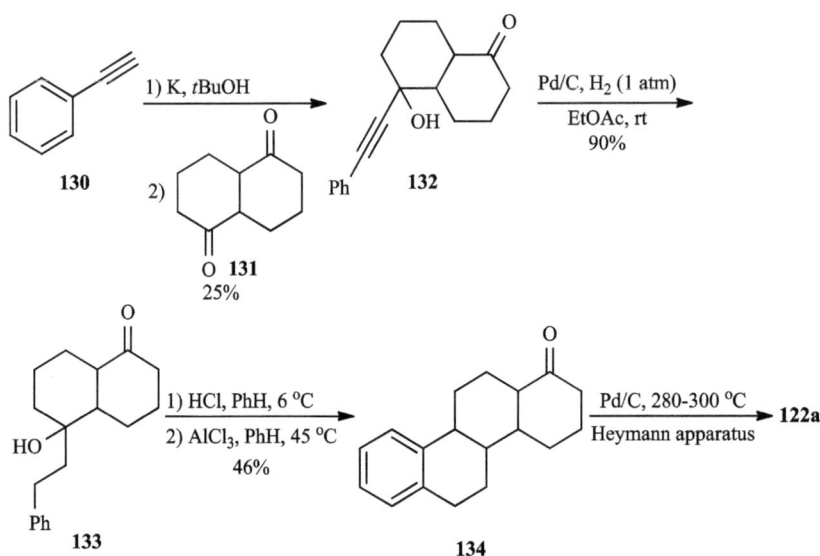

Scheme 45. Synthesis of 1-chrysenol (**122a**) from phenylacetylene (**130**).

Kumar pursued two synthetic pathways for the preparation of 2-chrysenol (**123a**) (Scheme **46**). Following the successful preparation of 2-chrysenol (**123a**), the same synthetic strategy was applied for the synthesis of 11-methyl-2-chrysenol (**135**) (**Scheme 47**) [158]. The synthesis commenced with a Suzuki-Miyaura cross-coupling reaction in the syntheses of all three products. Hence, nucleophile **13** was coupled with electrophiles **137** and **139** to give coupling products **140** and **141**, respectively, whereas boronic acid **136** was coupled with aryl bromide **138** to give compound **142**. The two pathways leading to 2-chrysenol (**123a**) utilized the common intermediate 2-methoxychrysene (**144**). In the first methodology leading to 2-chrysenol (**123a**), the aldehyde functionality within substrate **140** was extended by one methylene group to furnish epoxide **143** when aldehyde **140** was treated with trimethylsulfonium iodide in refluxing methylene chloride in the presence of aqueous sodium hydroxide using TBAI as a phase transfer catalyst. The obtained epoxide **143** was subjected to acid catalyzed cyclodehydration providing 2-methoxychrysene (**144**). It is noteworthy that the cyclodehydration proceeded with complete regioselectivity, *i.e.* cyclodehydration only took place at the C-1 position of the naphthalene moiety of epoxide **143** and not at the C-3 position, which would have furnished the benzo[α]anthracene analogue of 2-methoxychrysene (**144**).

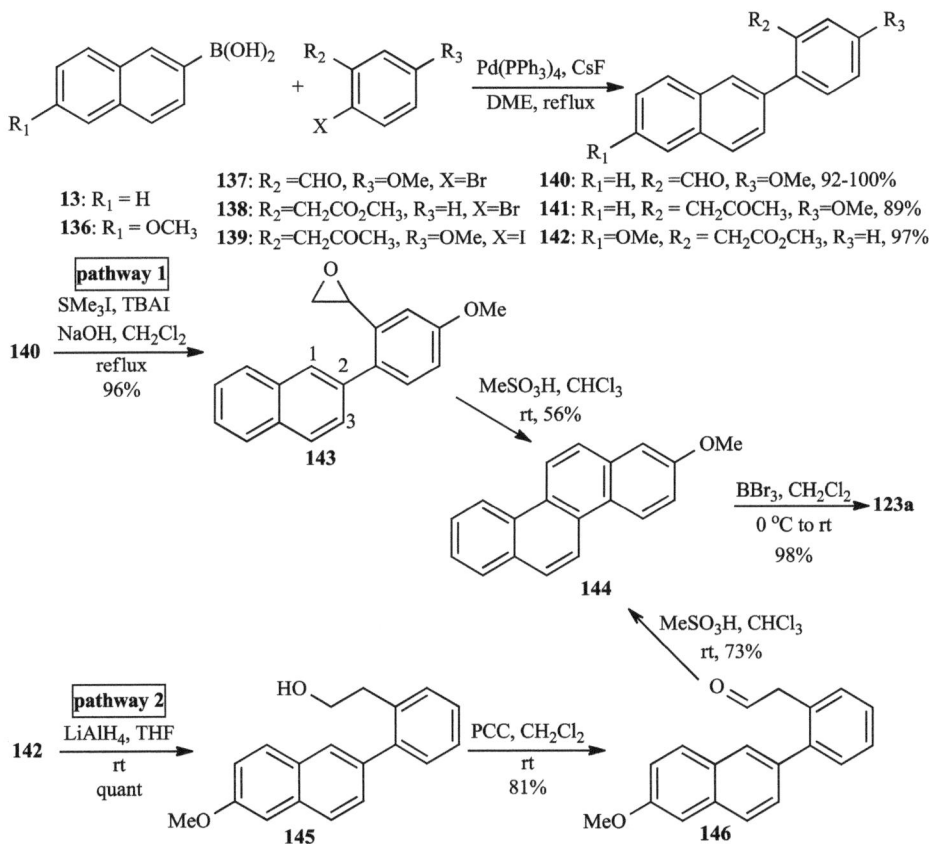

13: R_1 = H
136: R_1 = OCH$_3$

137: R_2 =CHO, R_3=OMe, X=Br
138: R_2=CH$_2$CO$_2$CH$_3$, R_3=H, X=Br
139: R_2=CH$_2$COCH$_3$, R_3=OMe, X=I

140: R_1=H, R_2 =CHO, R_3=OMe, 92-100%
141: R_1=H, R_2 = CH$_2$COCH$_3$, R_3=OMe, 89%
142: R_1=OMe, R_2 = CH$_2$CO$_2$CH$_3$, R_3=H, 97%

Scheme 46. Two synthetic pathways for the preparation of 2-chrysenol (**123a**).

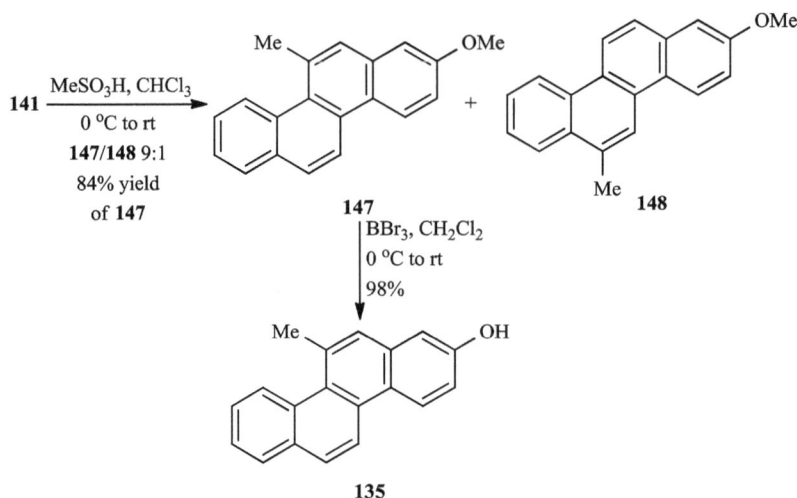

Scheme 47. Synthesis of compound **135** *via* acid catalyzed cyclodehydration of ketone **141**.

The second methodology leading to 2-chrysenol (**123a**) included a reduction/oxidation sequence to furnish aldehyde **146** *via* alcohol **145** from ester **142**. Followed by an acid catalyzed cyclodehydration of aldehyde **146** to form 2-methoxychrysene (**144**) in complete regioselectivity. Final boron tribromide mediated de-*O*-methylation of compound **144** provided 2-chrysenol (**123a**). Acid catalyst cyclodehydration was also useful in the synthesis of 11-methyl-2-chrysenol (**135**) (Scheme **47**), as ketone **141** underwent cyclodehydration to 2-methoxy-11-methylchrysene (**147**) in 84% yield, after recrystallization. However, the reaction was not completely regioselective as crude ^{1}H NMR showed formation of both compounds **147** and **148** in a 9 to 1 ratio. 11-Methyl-2-chrysenol (**135**) was obtained when compound **147** was subjected to de-*O*-methylation using BBr$_3$ in DCM.

In the section describing the synthetic strategy for the formation of phenanthrols the work of Sarkar and co-workers was discussed where propargyl alcohol **94** underwent an intramolecular Diels-Alder reaction to provide 2-phenanthrol (**88**) (Scheme **34d**) using gold(III) chloride as catalyst [130]. This intramolecular Diels-Alder methodology also turned out to be useful for the synthesis of 2-chrysenol (**123a**) (Scheme **48**). Hence, the precursor for the intramolecular Diels-Alder reaction was prepared from α,β-unsaturated aldehyde **149** by an indium mediated propargylation to furnish an inseparable mixture of the desired propargyl alcohol **150** along with the undesired allene **151** (**150/151** 1.7:1). The crude mixture was then treated with a catalytic amount of gold(III) chloride to give 2-chrysenol (**123a**), which could be separated from unreacted allene **151**.

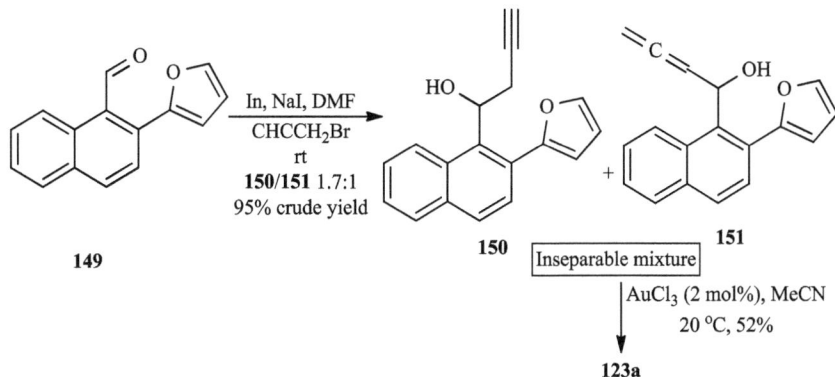

Scheme 48. Synthesis of 2-chrysenol (**123a**) *via* an intramolecular Diels-Alder reaction in propargyl alcohol **150**.

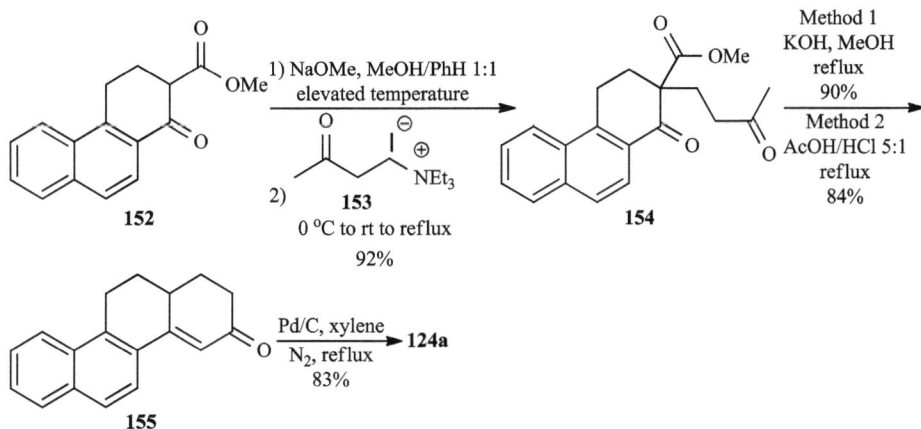

Scheme 49. Synthesis of 3-chrysenol (**124a**).

In 1943 Wilds and Shunk published a three step synthesis of 3-chrysenol (**124a**) starting from β-keto ester **152** (Scheme **49**) [159]. Compound **152** was subjected to sodium methoxide to form the corresponding sodium enolate, which was quenched upon addition of the methiodide **153** to give diketone **154**. To promote intramolecular Aldol condensation, hydrolysis and decarboxylation of compound

154 in order to form α,β-unsaturated ketone **155** two different reaction conditions were successfully employed: 1) methanolic potassium hydroxide at reflux, and 2) an acidic mixture consisting of hydrogen chloride and acetic acid (1:5) at reflux. Desired 3-chrysenol (**124a**) was finally formed when **155** was subjected to palladium catalyzed dehydrogenation under a N$_2$ atmosphere.

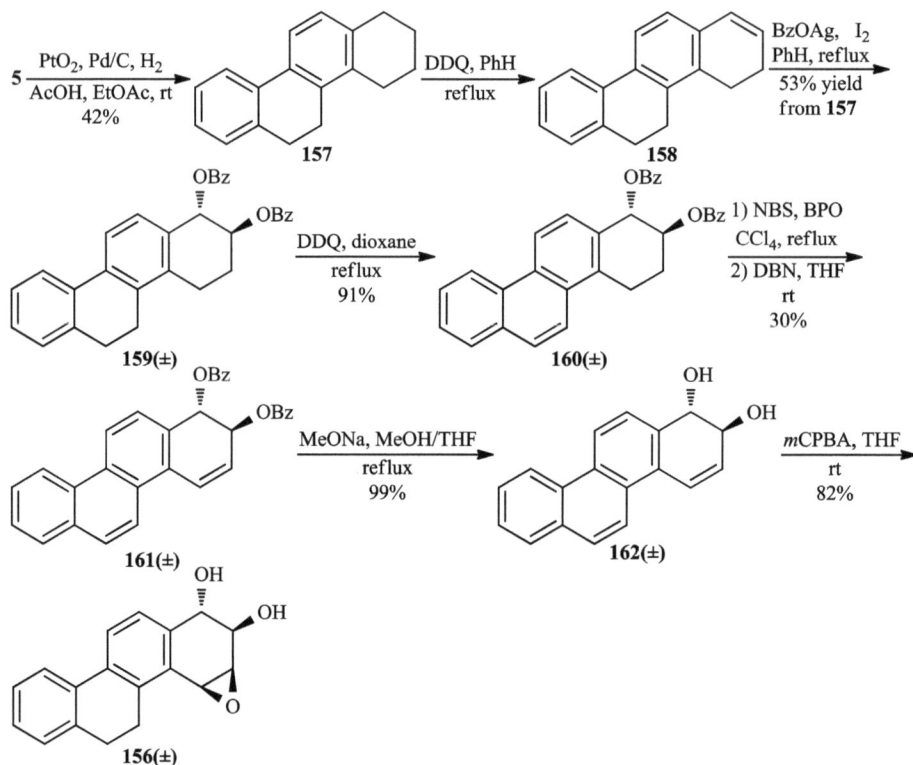

Scheme 50. Synthesis of the chrysene bay region *anti*-(±)-diol epoxide **156**(±).

Syntheses of Dihydro Diols and Diol Epoxides of Chrysene

In the late 1970s Harvey and co-workers settled the synthesis of the chrysene bay region *anti*-(±)-diol epoxide **156**(±) from chrysene (**5**) (Scheme **50**) [160, 161]. The synthesis was initiated by a regioselective hydrogenation of chrysene (**5**) employing a mixed Pd/C-PtO$_2$ catalyst to furnish the partially saturated chrysene analogue **157**. In the following step, compound **157** was subjected to partial dehydrogenation using DDQ as oxidation reagent to generate intermediate **158**, which underwent a Prévost reaction [162] to furnish *trans*-(±)-1,2bis(benzoyloxy)-1,2,3,4,5,6-hexahydrochrysene **(159(±))**. DDQ was then once again used in order to regioselectively dehydrogenate compound **159(±)** to the corresponding 1,2,3,4-tetrahydro analogue **160(±)** in excellent yield (91%). Subsequent dehydrogenation of substrate **160(±)** to generate *trans*-(±)-1,2bis(benzoylox-)-1,2-dihydrochrysene **(161(±))** was performed by employing a bromination/dehydrobromination sequence using NBS and benzoyl peroxide (BPO) in refluxing tetrachloromethan followed by elimination with DBN. *trans*-(±)-1,2-Dihydroxy-1,2-dihydrochrysene **(162(±))** was obtained when compound **161(±)** was de-*O*-benzoylated with sodium methoxide, which then was subjected to Prilezhaev reaction [163] to stereospecifically give the desired product **156(±)**.

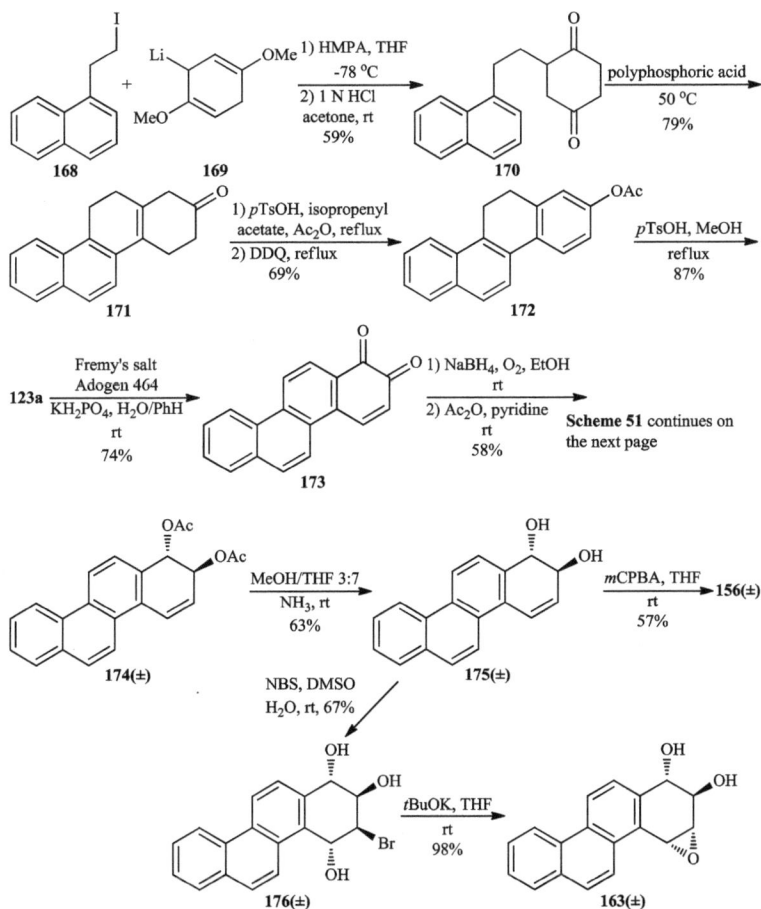

Scheme 51. Synthesis of the chrysene bay region *anti*-**156(±)** and *syn*-(±)-diol epoxide **163(±)**.

Later on, Harvey's research group reported another synthetic pathway in order to make *anti-*(±)-diol epoxide **156**(±), which was reported to be more efficient than their previous synthesis outlined in Scheme **50**. The work also included the synthesis of the diastereomeric chrysene bay region *syn-*(±)-diol epoxide **163**(±) (Scheme **51**). Moreover, the same report included the synthesis of *anti-* and *syn-* (±)-1,2-diol-3,4-epoxide derivatives of 5-methylchrysene, namely compounds **164**(±) and **165**(±) (the synthesis of **164**(±) had previously been reported in a preliminary report [164]) (Scheme **52**). In addition, the same chemistry was used in order to prepare *anti-***166**(±) and *syn-*(±)-1,2-diol-3,4-epoxide **167**(±) derivatives of 11-methylchrysene (Scheme **53**) [165].

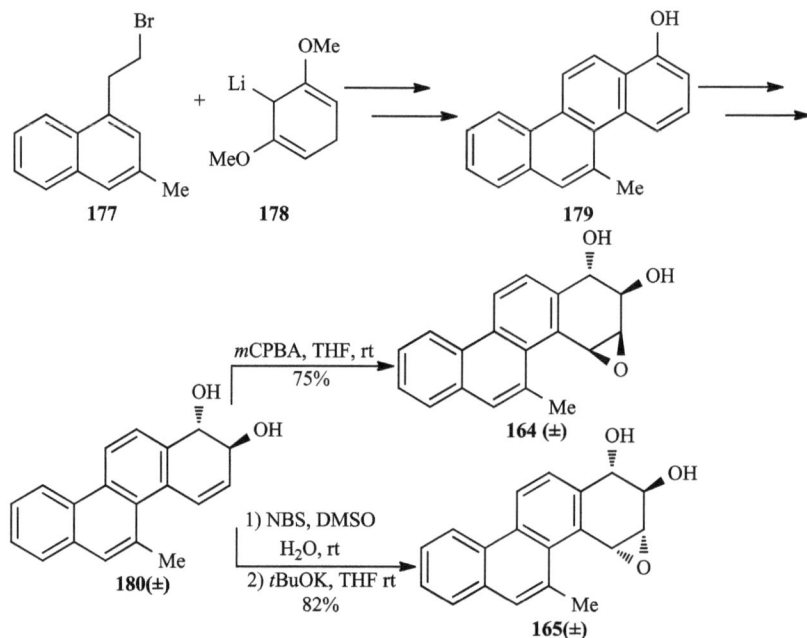

Scheme 52. Synthesis of the *anti-* and *syn-*(±)-1,2-diol-3,4-epoxide derivatives **164**(±) and **165**(±), respectively, of 5-methylchrysene. The synthetic pathway followed the same features as the one leading to the *anti-* and *syn-*(±)-diol epoxides **156**(±) and **163**(±) in Scheme **51**.

The synthesis leading to both compound **156**(±) and **163**(±) was initiated by alkylation of the lithium salt **169** with 1-(2-iodoethyl)naphthalene (**168**) to give diketone **170** after HCl mediated hydrolysis of the enol ether groups. In the following step, diketone **170** underwent acid promoted cyclization to form the ABCD ring system of ketone **171**. The enol tautomer was acetylated with isopropenyl acetate and dehydrogenated with DDQ to give ester **172**. Subsequent methanolysis of compound **172** in the presence of *p*-toluenesulfonic acid liberated the acetyl group to give 2-chrysenol (**123a**). Oxidation of 2-chrysenol (**123a**) with Fremy's salt ($Na_2NO(SO_3)_2$) gave 1,2-chrysenedione (**173**), which was reduced with sodium borohydride while oxygen was bubbled through the reaction mixture in order to reoxidize undesired by-products back to compound **173**. The crude mixture obtained after reduction of substrate **173** was immediately acetylated with acetic anhydride and then purified to furnish *trans*-(±)-diacetate **174(±)**, prior to removal of the acetyl groups by aminolysis resulting in the formation of unprotected *trans*-(±)-1,2-dihydrodiol **175(±)** of chrysene. Compound **175(±)** was then converted to target compound **156(±)** by Prilezhaev reaction. The diastereomeric counterpart **163(±)** was also achieved stereospecifically *via* bromination of **175(±)** with NBS to furnish bromohydrin **176(±)**, which cyclized into the *syn*-(±)-diol epoxide **163(±)** using *t*BuOK as catalyst.

The synthetic route leading to the *anti*- and *syn*-(±)-1,2-diol-3,4-epoxide derivatives **164(±)** and **165(±)** (Scheme **52**), respectively, of 5-methylchrysene was essentially the same as the synthetic pathway leading to the *anti*- and *syn*-(±)-diol epoxides **156(±)** and **63(±)** in Scheme **51**. The first step in the synthesis of compounds **164** and **165** constituted, however, an alkylation of 6-lithio-1-5-dimethoxycyclohexa-1,4-diene (**178**) with alkyl bromide **177** in contrast to alkylation of 3-lithio-1,4-dimethoxycyclohexane-1,4-diene (**169**). The remaining synthesis toward the desired products **164(±)** and **165(±)** proceeded consecutively over intermediates 1-hydroxy-5-methylchrysene (**179**) and *trans*-(±)-5-metyl-1,2-dihydroxy-1,2-dihydrochrysene (**180(±)**).

The prepared diastereomeric diol epoxide derivatives **166(±)** and **167(±)** of 11-methylchrysene were synthesized (Scheme **53**) following a modified procedure compared to the preparation of *anti*- and *syn*-(±)-diol epoxides **156(±)** and **163(±)** outlined in Scheme **51**. Hence, the lithium salt **181** was alkylated with electrophile **182** in THF and then converted to diketone **183**, which cyclized to ketone **184** upon treatment with polyphosphoric acid at 115 °C. Oxidation of the obtained ketone **184** into the corresponding partially dehydrogenated ketone **185** took place in the presence of trityl trifluoroacetate, which was generated *in situ* from trityl alcohol (TrOH) and trifluoroacetic acid at reflux. 11-Methyl-3,4-dihydrochrysene

(**186**) was formed when ketone **185** was reduced with sodium borohydride to the corresponding alcohol, which was dehydrated under acidic conditions at reflux to give compound **186** in 92% yield. Subsequently, the vinyl group within dihydrochrysene **186** was *trans*-dibenzoylated with silver benzoate employing iodine as a promoter to afford the *trans*-(±)-dibenzoate compound **187(±)**, which underwent an acid catalyzed elimination of benzoic acid to yield benzoyl enol ether **188**. Complete aromatization of benzoyl enol ether **188** to 11-methyl-2-chrysenol (**135**) was achieved with sequential dehydrogenation and ester hydrolysis using DDQ and hydrochloric acid, respectively. In the following step, Fremy's reagent oxidized chrysenol **135** into quinone **189**, which was reduced with sodium borohydride to *trans*-(±)-diol **190(±)**. Epoxidation of compound **190(±)** with *m*CPBA gave the desired *anti*-(±)-3,4-epoxy-*trans*-1,2-dihydroxy-11-methyl-1,2,3,4-tetrahydrochrysene (**166(±)**). The diastereomeric counterpart *syn*-(±)-3,4-epoxy-*trans*-1,2-dihydroxy-11-methyl-1,2,3,4-tetrahydrochrysene (**167(±)**) was formed when compound **190(±)** was reacted with NBS to generate a bromohydrin intermediate, which underwent an intramolecular substitution reaction in the presence of *t*BuOK.

Harvey and co-workers has also reported the synthesis of both the *anti*- and *syn*-isomers (**191(±)**) and (**192(±)**), respectively, of (±)-3,4-epoxy-*trans*-1,2-dihydroxy-1,2,3,4-tetrahydro-6-methylchrysene (Scheme **54**) [166]. The synthesis of both isomers commenced by treating 1,4-dimethylnaphthalene (**193**) with 1 equiv. of NBS. The naphthylmethyl bromide intermediate was reacted with triphenylphosphine in refluxing benzene to give the phosphonium salt **194**. The resulting Wittig salt **194** was then used in a Wittig reaction with *m*-methoxybenzaldehyde to generate stilbene **195** as a mixture of geometrical isomers, which was subjected to a photo-induced cyclization to afford chrysene derivative **196**. Subsequent BBr$_3$-mediated de-*O*-methylation gave diketone **198**, after oxidation of 2-chrysenol **197** with Fremy's salt. Subsequent reduction of diketone **198** with sodium borohydride gave the *trans*-(±)-diol **199(±)** that constituted the last common intermediate leading to the target compounds **191(±)** and **192(±)**. Thus, *anti*-(±)-isomer **191(±)** was obtained when the alkene moiety within compound **199(±)** was epoxidized with *m*CPBA. *syn*-(±)-Isomer **192(±)** was generated by utilizing a two-step sequence from precursor **199(±)**: 1) treatment with NBA to form the bromohydrin intermediate (the exact structure of this intermediate was not reported), and 2) treatment of the bromohydrin intermediate with base in order to promote an intramolecular cyclization that finally gave the target compound **192(±)**.

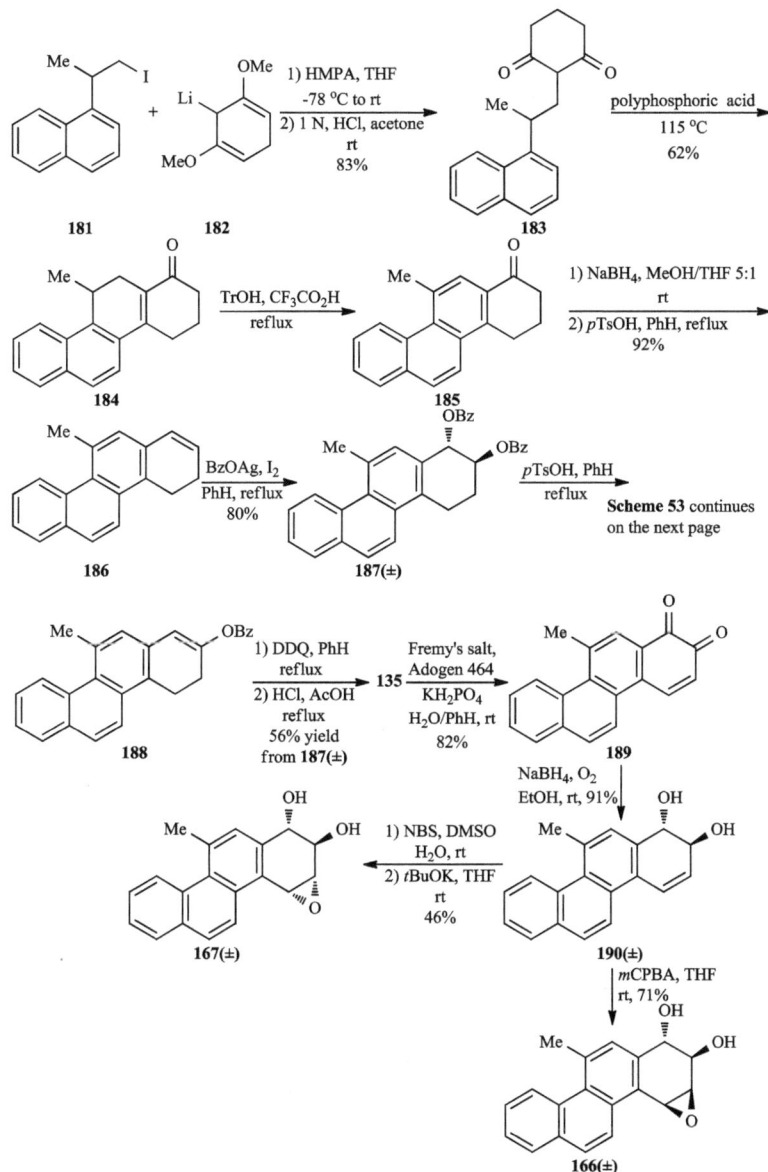

Scheme 53. Synthesis of diastereomeric *anti-* (**166(±)**) and *syn-*(±)-3,4-epoxy-*trans*-1,2-dihydroxy-11-methyl-1,2,3,4-tetrahydrochrysene (**167(±)**).

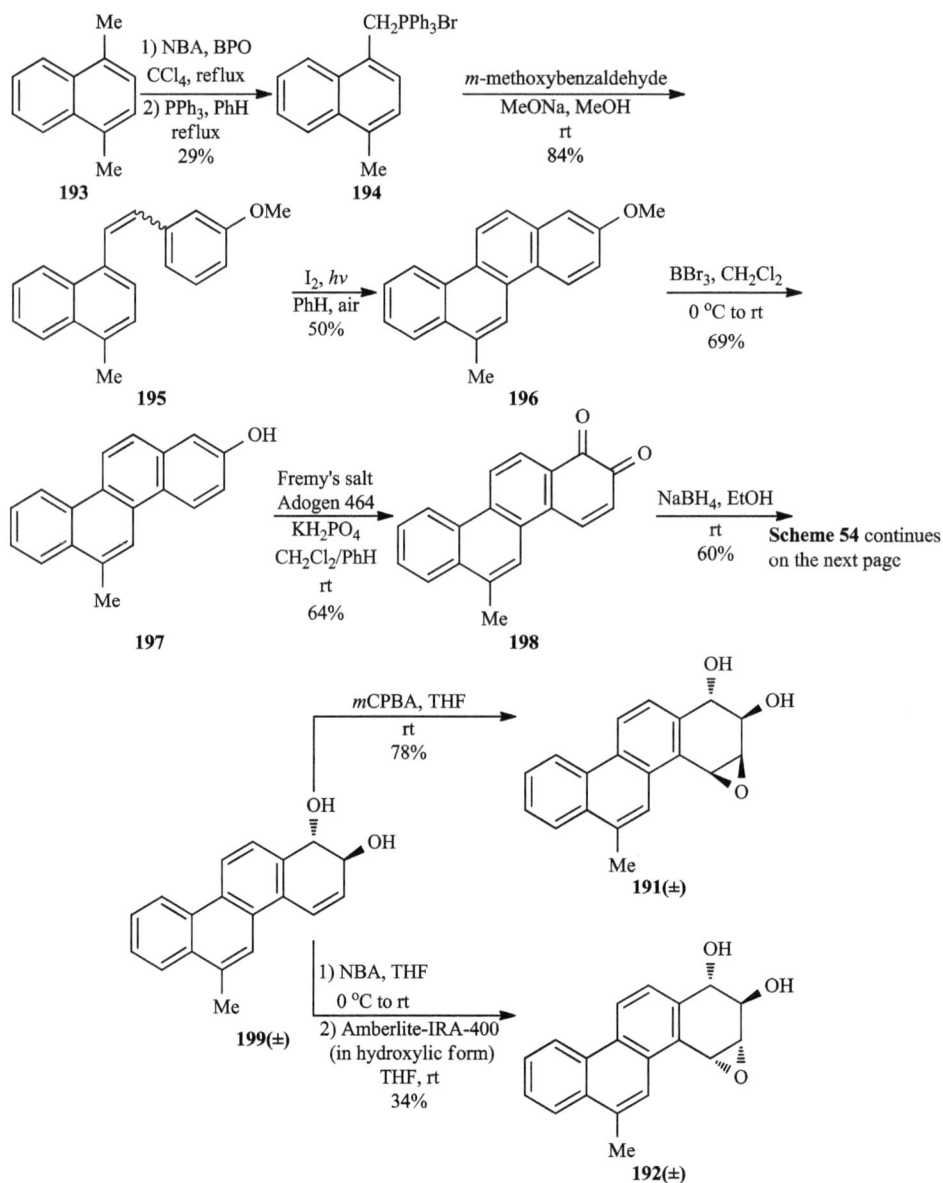

Scheme 54. Synthesis of *anti-* and *syn*-(±)-3,4-epoxy-*trans*-1,2-dihydroxy-3,-6-methyl-1-2,3,4-tetrahydrochrysene **191(±)** and **192(±)**, respectively.

Hecht and his colleagues reported the synthesis of the bay region *anti*-(±)-3-4-epoxy-*trans*-1,2-dihydroxy-1,2,3,4-tetrahydro-chrysenes **200a(±)** and **200b(±)**

containing a ethyl- and propyl-group, respectively, in 5-position [167] Scheme **55**. 5-Formyl-2-methoxychrysene (**201**) constituted the starting material for both target compounds **200a(±)** and **200b(±)**. Thus, aldehyde **201** was reacted with methyl and ethyl Grignard reagent to furnish the corresponding alcohols, which without purification were oxidized with pyridinium chlorochromate (PCC) to alkyl-aryl ketones **202a** and **202b**. The following step constituted a Clemmensen reduction [168] of the oxo-substituent within **202a** and **202b** to generate the ethyl- and propyl-arenes **203a** and **203b**, respectively. Substrates **203a** and **203b** was then subjected to a de-*O*-methylation with BBr$_3$ in methylene chloride to yield chrysenols **204a** and **204b**. Both chrysenols were then oxidized with Fremy's salt to give the chrysene-1,2-quinones **205a** and **205b**. Subsequently, compounds **205a** and **205b** were reduced to the corresponding *trans*-1,2-diols **206a(±)** and **206b(±)**, respectively, which was subjected to Prilezhaev reaction to give the target chrysene analogue **200a(±)** and the corresponding 5-propyl analogue **200b(±)**.

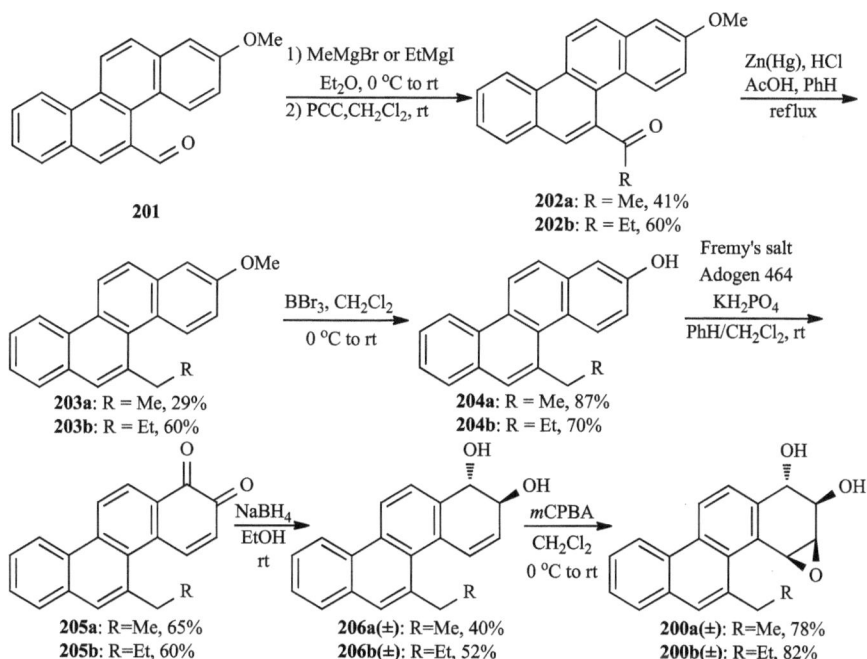

Scheme 55. Synthesis of diol epoxides **200a(±)** and **200b(±)**.

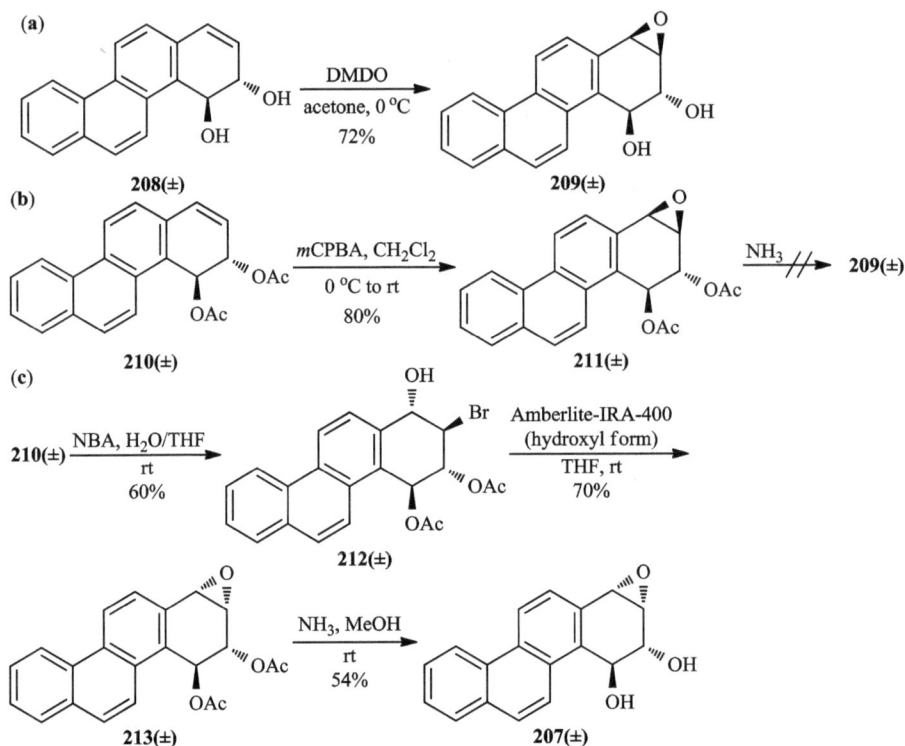

Scheme 56. Synthesis of diol epoxides **207(±)** and **209(±)**.

Seidel and co-workers developed a stereospecific synthetic protocol leading to chrysene analogue **207(±)** [169]. In this study, dimethyldioxirane (DMDO) was found useful for the stereospecific epoxidation of (±)-*trans*-3,4-dihydrochrysen-e-3,4-diol (**208(±)**) to the corresponding *syn*-(±)-diol epoxide **209(±)** (Scheme **56a**). Likewise, *m*CBPA also gave a similar *syn*-selectivity for the epoxidation of *trans*-diacetate **210(±)** to *syn*-(±)-bis(acetoxy) epoxide **211(±)** (Scheme **56b**). Unfortunately, the acetate groups of substrate **211(±)** could not be efficiently removed by aminolysis due to decomposition of the oxirane ring. The authors settled that NBA constituted a key reagent in order to achieve the *anti*-(±)-diol epoxide **207(±)** (Scheme **56c**). Thus, (±)-*trans*-diacetate **210(±)** underwent bromination with NBA to give one single stereoisomer **212(±)** that possessed all its neighboring substituents in a *trans*-relationship. The *anti*-(±)-*trans*-bis(acetoxy) epoxide **213(±)** was then obtained upon treatment of **212(±)** with the hydroxylic form of Amberlite-IRA-400. Finally, the desired *anti*-(±)-*trans*-diol epoxide **207(±)** was generated when the ester groups of compound **213(±)** was removed by aminolysis.

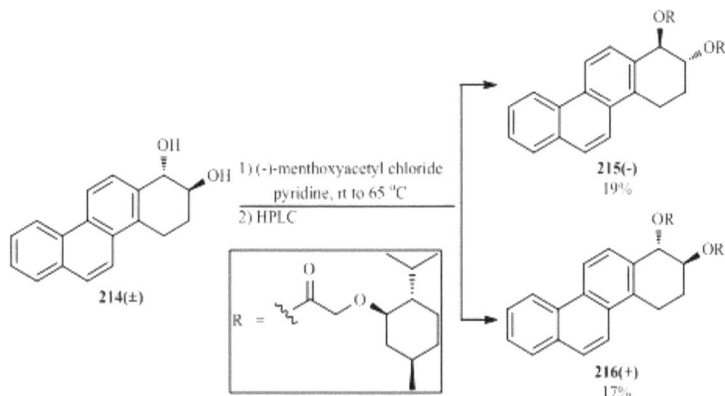

Scheme 57. Formation of diastereomers **215(-)** and **216(+)** upon treatment of (±)-*trans*-diol **214(±)** with (-)-menthoxyacetyl chloride.

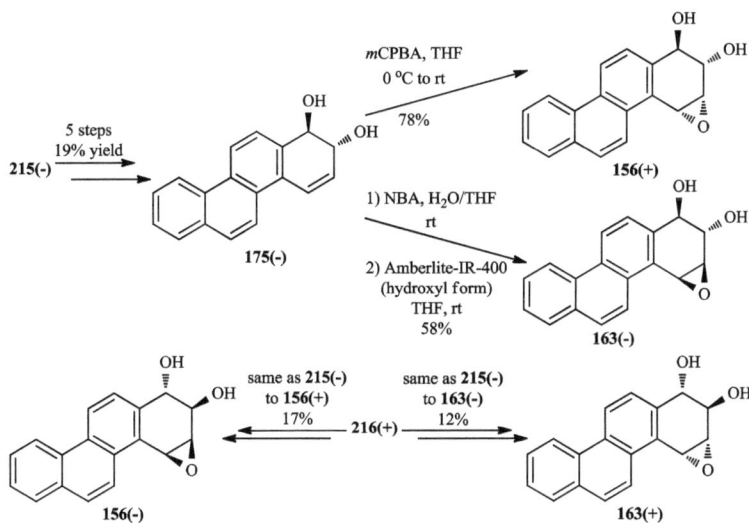

Scheme 58. Synthesis of enantiomerically pure diol epoxides **156(+)**, **156(-)**, **163(+)**, and **163(-)**.

All synthetic routes, presented for the formation of the chrysene bay region diol epoxides **156(±)** and **163(±)** have resulted in formation of racemic mixtures. In order to obtain the diol epoxides **156(+)**, **156(-)**, **163(+)**, and **163(-)** as pure enantiomers [170], a racemic mixture (±)-*trans*-diol **214(±)** was acylated with (−)-menthoxyacetyl chloride to form a set of two diastereomers **215(-)** and **216(+)**, which could be separated by high pressure liquid chromatography (HPLC) (Scheme **57**).

The obtained ester **215(-)** was then used as an enantiomerically pure building block in order to prepare *anti*-(+)-diol epoxide **156(+)** and *syn*-(±)-diol epoxide **163(-)** in six and seven steps, respectively. By similar means *anti*-(-)-diol epoxide **156(-)** and *syn*-(+)-diol epoxide **163(+)** were obtained from ester **216(+)** (Scheme **58**).

As outlined above, turning a racemic mixture into a set of two diastereomers followed by chromatographic separation constitutes a methodology to obtain enantiomerically pure chrysene metabolites. A more conventional alternative to obtain enantiomerically pure metabolites of chrysene was settled by Jørgensen and co-workers when they reported the enantioselective synthesis of (1R,2R)-(-)-1,2-dihydrochrysene-1,2-diol (**175(-)**) (Scheme **59**) [171]. The synthesis commenced from *N,N*-diethylbenzamide (**217**), which underwent *ortho*-lithiation with *s*BuLi and then transmetalation with zinc(II)chloride before it was used as a coupling partner with ethyl-4-bromobutyrate in the nickel catalyzed Negishi cross-coupling reaction giving benzamide **218**. Subsequent alkaline hydrolysis of the ester moiety within compound **218** gave the corresponding benzamide **219**, which underwent cyclization forming ketone **220** upon treatment with P_2O_5 in methanesulfonic acid. Ketone **220** was then treated with potassium KHMDS followed by trapping of the potassium enolate with TBDPSCl to generate enol ether **221**. The formed potassium enolate was then captured with a TBDPS-protective group to generate silyl enol ether **221**. The following two steps in order to generate *trans*-diol monosilyl ether **223(-)** in 83% *ee* was pointed out to be the key steps in order to reach the target compound **156(-)**. The key steps constituted: 1) Shi-epoxidation of alkene **221** to furnish epoxide **222** and 2) regio- and stereoselective addition of hydride to epoxide **222** gave *trans*-diol monosilyl ether **223(-)**. In the subsequent step, a TBDPS-protective group was installed on the free hydroxyl group of *trans*-diol monosilyl ether **223(-)** to furnish the corresponding *trans*-diol disilyl ether **220(-)**, which underwent bromination and then elimination of hydrogen bromide to give compound **225(-)**. Amide **225(-)** was subjected to chemoselective reduction with Schwartz's reagent to furnish benzaldehyde **226**, which was used in a Wittig reaction together with benzyltriphenylphosphonium chloride to give stilbene **227** in a *E/Z*-ratio of 3/2. The ABCD ring system of compound **228** was generated photochemically when stilbene **227** was exposed to light in the presence of catalytic amount of iodine. Target compound **175(-)** was finally obtained in 87% *ee* when the TBDPS-protective groups of substrate **228** was removed using *tetra-n*-butylammonium fluoride (TBAF) in THF.

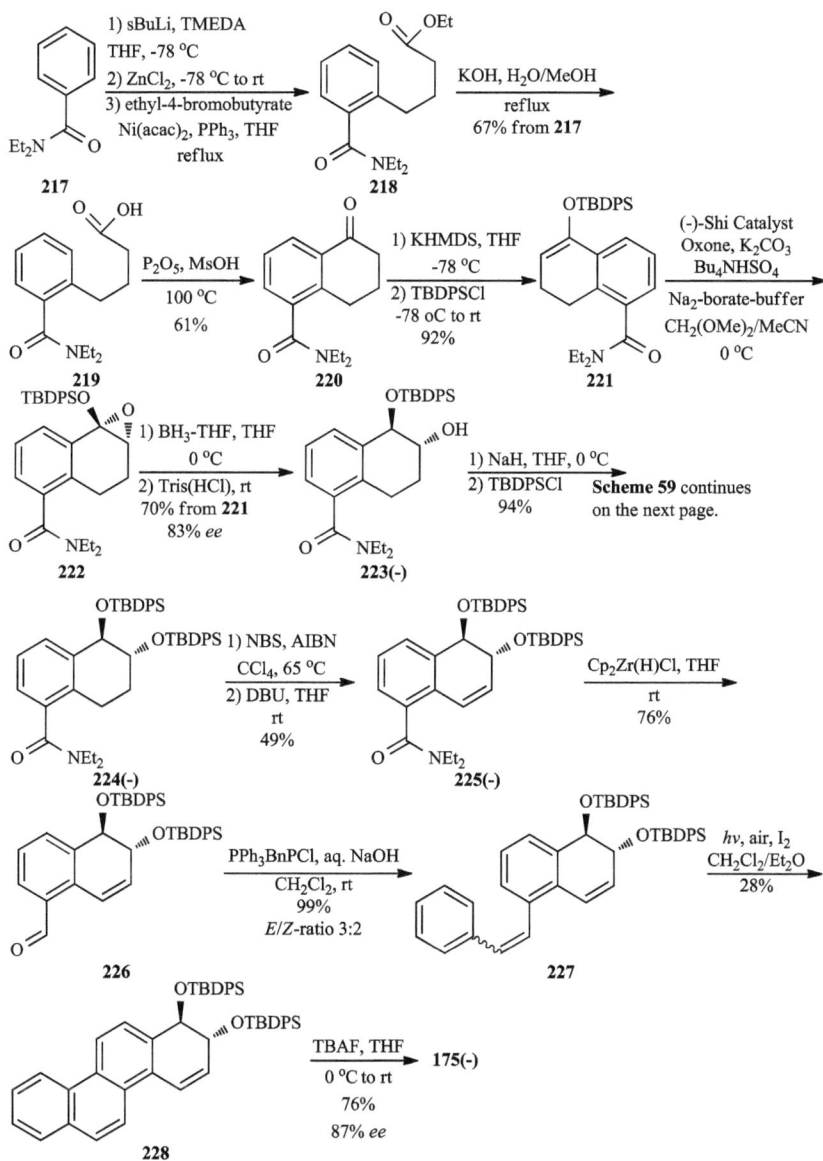

Scheme 59. Synthesis of (1*R*,2*R*)-(-)-1,2-dihydrochrysene-1,2-diol (**175(-)**).

CONCLUDING REMARKS

Most of the prepared PAH metabolites have been prepared as racemic mixtures. This is not surprising since most of the work in this field has been done prior to the development of efficient asymmetric catalysts. In the future the synthesis of these metabolites will be dominated by asymmetric strategies preparing only the desired enantiomer in high enatiomericall purity.

CONFLICT OF INTEREST

The authors confirm that they have no conflict of interest to declare for this publication.

ACKNOWLEDGEMENTS

The University of Stavanger (UiS), the Research Council of Norway (PETROMAKS 2, project # 229153), and the research program Bioactive is gratefully acknowledged for supporting the work conducted at UiS.

REFERENCES

[1]   Jacob, J. The significance of polycyclic aromatic hydrocarbons as environmental carcinogens. 35 Years research on PAH – A retrospective. *Polycycl. Aromat. Compd.,* **2008**, *28*, 242-272.
        [http://dx.doi.org/10.1080/10406630802373772]

[2]   Pampanin, D.M.; Sydnes, M.O. Polycyclic aromatic hydrocarbons a constituent of petroleum: presence and influence in the aquatic environment. In: *Hydrocarbons*; Kutcherov, V.; Kolesnikov, A., Eds.; InTech: Rijeka, **2013**; pp. 83-118.

[3]   Keith, L.H.; Telliard, W.A. Priority pollutants: I. A perspective view. *Environ. Sci. Technol.,* **1979**, *13*, 416-423.
        [http://dx.doi.org/10.1021/es60152a601]

[4]   Zhao, D.; Wu, N.; Zhang, S.; Xi, P.; Jingbo, L.; You, J. Synthesis of phenol, aromatic ether, and benzofuran by copper-catalyzed hydroxylation of aryl halides. *Angew. Chem. Int. Ed.,* **2009**, *48*, 8729-8732.
        [http://dx.doi.org/10.1002/anie.200903923]

[5]   Hashimoto, T.; Maruoka, K. The basic principle of phase-transfer catalysis and some mechanistic aspects. In: *Asymmetric Phase Transfer Catalysis*; Maruoka, K., Ed.; Wiley-VCH GmbH & Co: Weinhem, **2008**; pp. 1-8.
        [http://dx.doi.org/10.1002/9783527622627.ch1]

[6]   Wang, D.; Kuang, D.; Zhang, F.; Tang, S.; Jiang, W. Triethanolamine as an inexpensive and efficient ligand for copper-catalyzed hydroxylation of aryl halides in water. *Eur. J. Org. Chem.,* **2014**, 315-318.
        [http://dx.doi.org/10.1002/ejoc.201301370]

[7]   Xiao, Y.; Xu, Y.; Cheon, H.S.; Chae, J. Copper(II)-catalyzed hydroxylation of aryl halides using glycolic acid as a ligand. *J. Org. Chem.,* **2013**, *78*(11), 5804-5809.
[http://dx.doi.org/10.1021/jo400702z] [PMID: 23713792]

[8]   Priyadarshini, S.; Amal Joseph, P.J.; Lakshmi Kantam, M.; Sreedhar, B. Copper MOF: scope and limitation in catalytic hydroxylation and nitration of aryl halides. *Tetrahedron,* **2013**, *69*, 6409-6414.
[http://dx.doi.org/10.1016/j.tet.2013.05.102]

[9]   Yang, K.; Li, Z.; Wang, Z.; Yao, Z.; Jiang, S. Highly efficient synthesis of phenols by copper-catalyzed hydroxylation of aryl iodides, bromides, and chlorides. *Org. Lett.,* **2011**, *13*(16), 4340-4343.
[http://dx.doi.org/10.1021/ol2016737] [PMID: 21786767]

[10]   Chen, J.; Yuan, T.; Hao, W.; Cai, M. Simple and efficient CiI/PEG-400 system for hydroxylation of aryl halides with potassium hydroxide. *Catal. Commun.,* **2011**, *12*, 1463-1465.
[http://dx.doi.org/10.1016/j.catcom.2011.06.002]

[11]   Xu, H.J.; Liang, Y.F.; Cai, Z.Y.; Qi, H.X.; Yang, C.Y.; Feng, Y.S. CuI-nanoparticles-catalyzed selective synthesis of phenols, anilines, and thiophenols from aryl halides in aqueous solution. *J. Org. Chem.,* **2011**, *76*(7), 2296-2300.
[http://dx.doi.org/10.1021/jo102506x] [PMID: 21361386]

[12]   Tlili, A.; Xia, N. Monnier. F.; Taillefer, M. A very simple copper-catalyzed synthesis of phenols employing hydroxide salts. *Angew. Chem. Int. Ed.,* **2009**, *48*, 8725-8728.
[http://dx.doi.org/10.1002/anie.200903639]

[13]   Ren, Y.; Cheng, L.; Tian, X.; Zhao, S.; Wang, J.; Hou, C. Iron-catalyzed conversion of unactivated aryl halides to phenols in water. *Tetrahedron Lett.,* **2010**, *51*, 43-45.
[http://dx.doi.org/10.1016/j.tetlet.2009.10.036]

[14]   Xu, J.; Wang, X.; Shao, C.; Su, D.; Cheng, G.; Hu, Y. Highly efficient synthesis of phenols by copper-catalyzed oxidative hydroxylation of arylboronic acids at room temperature in water. *Org. Lett.,* **2010**, *12*(9), 1964-1967.
[http://dx.doi.org/10.1021/ol1003884] [PMID: 20377271]

[15]   Kaboudin, B.; Abedi, Y.; Yokomatsu, T. Cu^{II}-β-cyclodextrin complex as a nanocatalyst for homo- and cross-coupling of arylboronic acids under ligand- and base-free conditions in air: chemoselective cross-coupling of arylboronic acids in water. *Eur. J. Org. Chem.,* **2011**, 6656-6662.
[http://dx.doi.org/10.1002/ejoc.201100994]

[16]   Sawant, S.D.; Hudwekar, A.D.; Aravinda Kumar, K.A.; Venkateswarlu, V.; Sing, P.P.; Vishwakarma, R.A. Ligand- and base-free synthesis of phenols by rapid oxidation of arylboronic acids using iron(III) oxide. *Tetrahedron Lett.,* **2014**, *55*, 811-814.
[http://dx.doi.org/10.1016/j.tetlet.2013.12.003]

[17]   Yang, D.; An, B.; Wei, W.; Jiang, M.; You, J.; Wang, H. A novel sustainable strategy for the synthesis of phenols by magnetic CuFe$_2$O$_4$-catalyzed oxidative hydroxylation of arylboronic acids under mild conditions in water. *Tetrahedron,* **2014**, *70*, 3630-3634.
[http://dx.doi.org/10.1016/j.tet.2014.03.076]

[18]   Mulakayala, N. Ismail; Kottur, M. K.; Rapolu, R. K.; Kandagatla, B.; Rao, P.; Oruganti S.; Pal, M.

Catalysis by Amberlite IR-120: a rapid and green method for the synthesis of phenols from arylboronic acids under metal, ligand, and base-free conditions. *Tetrahedron Lett.,* **2012**, *53*, 6004-6007. [http://dx.doi.org/10.1016/j.tetlet.2012.08.087]

[19]   Wang, L.; Dai, D-Y.; Chen, G.; He, M-Y. Rapid and green synthesis of phenols catalyzed by a deep eutectic mixture based on fluorinated alcohol in water. *J. Fluor. Chem.,* **2014**, *158*, 44-47. [http://dx.doi.org/10.1016/j.jfluchem.2013.12.006]

[20]   Wang, L.; Dai, D.Y.; Chen, Q.; He, M.Y. Rapid, sustainable, and gram-scale synthesis of phenols catalyzed by a biodegradable deep eutectic mixture in water. *Asian J. Org. Chem.,* **2013**, *2*, 1040-1043. [http://dx.doi.org/10.1002/ajoc.201300192]

[21]   Zhu, C.; Wang, R.; Falck, J.R. Mild and rapid hydroxylation of aryl/heteroaryl boronic acids and boronate esters with *N*-oxides. *Org. Lett.,* **2012**, *14*(13), 3494-3497. [http://dx.doi.org/10.1021/ol301463c] [PMID: 22731862]

[22]   Gogoi, P.; Bezboruah, P.; Gogoi, J.; Boruah, R.C. *ipso*-Hydroxylation of arylboronic acids and boronate esters by using sodium as an oxidant in water. *Eur. J. Org. Chem.,* **2013**, 7291-7294. [http://dx.doi.org/10.1002/ejoc.201301228]

[23]   Gohain, M.; du Plessis, M.; van Tonder, J.H.; Bezuidenhoudt, B.C. Preparation of phenolic compounds through catalyst-free *ipso*-hydroxylation of arylboronic acids. *Tetrahedron Lett.,* **2014**, *55*, 2082-2084. [http://dx.doi.org/10.1016/j.tetlet.2014.02.048]

[24]   Dai, Y.; Feng, X.; Liu, H.; Jiang, H.; Bao, M. Synthesis of 2-naphthols *via* carbonylative Stille coupling reaction of 2-bromobenzyl bromides with tributylallylstannane followed by the Heck reaction. *J. Org. Chem.,* **2011**, *76*(24), 10068-10077. [http://dx.doi.org/10.1021/jo201907t] [PMID: 22073924]

[25]   Mizoroki, T.; Mori, K.; Ozaki, A. Arylation of olefin with aryl iodide catalyzed by palladium. *Bull. Chem. Soc. Jpn.,* **1971**, *44*, 581. [http://dx.doi.org/10.1246/bcsj.44.581]

[26]   Heck, R.F.; Nolley, J.P., Jr Palladium-catalyzed vinylic hydrogen substitution reactions with aryl, benzyl, and styryl halides. *J. Org. Chem.,* **1972**, *37*, 2320-2322. [http://dx.doi.org/10.1021/jo00979a024]

[27]   Peng, F.; Fan, B.; Shao, Z.; Pu, X.; Li, P.; Zhang, H. Cu(OTf)₂-Catalyzed Isomerization of 7-Oxabicyclic Alkenes: A practical Route to the Synthesis of Naphthol Derivatives. *Synthesis,* **2008**, 3043-3046.

[28]   Sawama, Y.; Kawamoto, K.; Satake, H.; Krause, N.; Kita, Y. Regioselective gold-catalyzed allylative ring opening of 1,4-epoxy-1,4-dihydronaphthalenes. *Synlett,* **2010**, 2151-2155. [http://dx.doi.org/10.1055/s-0030-1258528]

[29]   Villeneuve, K.; Tam, W. Ruthenium-catalyzed isomerization of oxa/azabicyclic alkenes: an expedient route for the synthesis of 1,2-naphthalene oxides and imines. *J. Am. Chem. Soc.,* **2006**, *128*(11), 3514-3515. [http://dx.doi.org/10.1021/ja058621l] [PMID: 16536513]

[30] Ballantine, M.; Menard, M.L.; Tam, W. Isomerization of 7-oxabenzonorbornadienes into naphthols catalyzed by [RuCl(2)(CO)(3)](2). *J. Org. Chem.,* **2009**, *74*(19), 7570-7573.
[http://dx.doi.org/10.1021/jo901504n] [PMID: 19725525]

[31] Kundu, N.G. Convenient Method for the Reduction of *ortho*-Quinones to Dihydrodiols. *J. Chem. Soc. Chem. Commun.,* **1979**, 564-565.
[http://dx.doi.org/10.1039/c39790000564]

[32] Kundu, N.G. Reduction of *ortho*-Quinones to Dihydrodiols. *J. Chem. Soc., Perkin Trans. 1,* **1980**, 1920-1923.
[http://dx.doi.org/10.1039/p19800001920]

[33] Platt, K.L.; Oesch, F. Efficient synthesis of non-K-region *trans*-dihydro diols of polycyclic aromatic hydrocarbons from *o*-Quinones and Catechols. *J. Org. Chem.,* **1983**, *48*, 265-268.
[http://dx.doi.org/10.1021/jo00150a027]

[34] Jeffrey, A.M.; Yeh, H.J.; Jerina, D.M. Synthesis of *cis*-1,2-Dihydroxy-1,2-dihydronaphthalene and *cis*-1,4-Dihydroxy-1,4-dihydronaphthalene. *J. Org. Chem.,* **1974**, *39*, 1405-1407.
[http://dx.doi.org/10.1021/jo00926a018]

[35] Rashid, A.; Read, G. Quinone Epoxides. Part III. Stereospecific Reductions with Metal Hydrides. *J. Chem. Soc. C,* **1969**, 2053-2058.
[http://dx.doi.org/10.1039/j39690002053]

[36] Yagi, H.; Thakker, D.R.; Hernandez, O.; Koreeda, M.; Jerina, D.M. Synthesis and reactions of the highly mutagenic 7,8-diol 9,10-epoxides of the carcinogen benzo[*a*]pyrene. *J. Am. Chem. Soc.,* **1977**, *99*(5), 1604-1611.
[http://dx.doi.org/10.1021/ja00447a053] [PMID: 839009]

[37] Becker, A.R.; Janusz, J.M.; Bruice, T.C. Solution Chemistry of the *syn*- and *anti*-Tetrahydrodiol Epoxides, the *syn*- and *anti*-Tetrahydrodimethoxy Epoxides, and the 1,2- and 1,4-Tetrahydro Epoxides of Naphthalene. *J. Am. Chem. Soc.,* **1979**, *101*, 5679-5687.
[http://dx.doi.org/10.1021/ja00513a037]

[38] Schmidt, R.R.; Angerbauer, R. Eine neuer Weg zu Naphthalinoxiden. *Angew. Chem.,* **1979**, *91*, 325-326.
[http://dx.doi.org/10.1002/ange.19790910409]

[39] Tsui, G.C.; Lautens, M. Rhodium(I)-catalyzed domino asymmetric ring opening/enantioselective isomerization of oxabicyclic alkenes with water. *Angew. Chem. Int. Ed. Engl.,* **2012**, *51*(22), 5400-5404.
[http://dx.doi.org/10.1002/anie.201200390] [PMID: 22511539]

[40] Krishnan, S.; Kuhn, D.G. Hamilton, G. A. Direct oxidation in high yield of some polycyclic aromatic compounds to arene oxides using hypochlorite and phase transfer catalysts. *J. Am. Chem. Soc.,* **1977**, *99*, 8121-8123.
[http://dx.doi.org/10.1021/ja00466a093]

[41] Ishikawa, K.; Griffin, G.W. A complementary route to *anti*-1,2:3,4-Naphthalene Dioxide; direct oxidation of arenes with *m*-Chloroperbenzoic acid. *Angew. Chem. Int. Ed. Engl.,* **1977**, *16*, 171-172.
[http://dx.doi.org/10.1002/anie.197701711]

[42] Mello, R.; Ciminale, F.; Fiorentino, M.; Fusco, C.; Prencipe, T.; Curci, R. Oxidations by methyl(trifluoromethyl)dioxirane. 4.[1] Oxyfunctionalization of aromatic hydrocarbons. *Tetrahedron Lett.,* **1990**, *31*, 6097-6100.
[http://dx.doi.org/10.1016/S0040-4039(00)98039-0]

[43] Erenler, R.; Demirtas, I.; Buyukkidan, B.; Cakmak, O. Synthesis of hydroxy, epoxy, nitrato, and methoxy derivatives of tetralins and naphthalenes. *J. Chem. Res.,* **2006**, 753-757.
[http://dx.doi.org/10.3184/030823406780199721]

[44] Boyd, D.R.; Sharma, N.D.; O'Dowd, C.R.; Hempenstall, F. Enantiopure arene dioxides: chemoenzymatic synthesis and application in the production of *trans*-3,4-dihydrodiols. *Chem. Commun. (Camb.),* **2000**, 2151-2152.
[http://dx.doi.org/10.1039/b006837n]

[45] Hu, Y.; Ziffer, H.; Silverton, J.V. Preparation and absolute stereochemistry of (-)-acenaphthenol. *Can. J. Chem.,* **1989**, *67*, 60-62.
[http://dx.doi.org/10.1139/v89-010]

[46] Aribi-Zouioueche, L.; Fiaud, J.C. Kinetic resolution of 1-acenaphthenol and 1-acetoxynaphthene through lipase-catalyzed acylation and hydrolysis. *Tetrahedron Lett.,* **2000**, *41*, 4085-4088.
[http://dx.doi.org/10.1016/S0040-4039(00)00544-X]

[47] Merabet-Khelassi, M.; Houiene, Z.; Aribi-Zouioueche, L.; Riant, O. Green methodology for enzymatic hydrolysis of acetates in non-aqueous media *via* carbonate salts. *Tetrahedron Asymmetry,* **2012**, *23*, 828-833.
[http://dx.doi.org/10.1016/j.tetasy.2012.06.001]

[48] Bouzemi, N.; Debbeche, H.; Aribi-Zouioueche, L.; Fiaud, J.C. On the use of succinic anhydride as acylating agent for practical resolution of aryl-alkyl alcohols through lipase-catalyzed acylation. *Tetrahedron Lett.,* **2004**, *45*, 627-630.
[http://dx.doi.org/10.1016/j.tetlet.2003.10.208]

[49] Debbeche, H.; Toffano, M.; Fiaud, J.C.; Aribi-Zouioueche, L. Multi-substrate screening for lipase-catalyzed resolution of arylalkylethanols with succinic anhydride as acylating agent. *J. Mol. Catal., B Enzym.,* **2010**, *66*, 319-324.
[http://dx.doi.org/10.1016/j.molcatb.2010.05.016]

[50] Mitsunobu, O.; Yamada, M. Preparation of esters of carboxylic and phosphoric acid *via* quaternary phosphonium salts. *Bull. Chem. Soc. Jpn.,* **1967**, *40*, 2380-2382.
[http://dx.doi.org/10.1246/bcsj.40.2380]

[51] Bouzemi, N.; Aribi-Zouioueche, L.; Fiaud, J.C. Combined lipase-catalyzed resolution/Mitsunobu esterification for the production of enantiomerically enriched arylalkyl carbinols. *Tetrahedron Asymmetry,* **2006**, *17*, 797-800.
[http://dx.doi.org/10.1016/j.tetasy.2006.02.016]

[52] Houiene, Z.; Merabet-Khelassi, M.; Bouzemi, N.; Riant, O.; Aribi-Zouioueche, L. A green route to enantioenriched (*S*)-arylalkyl carbinols by deracemization *via* combined lipase alkaline-hydrolysis/Mitsunobu esterification. *Tetrahedron Asymmetry,* **2013**, *24*, 290-296.
[http://dx.doi.org/10.1016/j.tetasy.2013.01.020]

[53]   Mandal, S.K.; Sigman, M.S. Palladium-catalyzed aerobic oxidative kinetic resolution of alcohols with an achiral exogenous base. *J. Org. Chem.,* **2003**, *68*(19), 7535-7537.
[http://dx.doi.org/10.1021/jo034717r] [PMID: 12968915]

[54]   Sheppard, C.I.; Taylor, J.L.; Wiskur, S.L. Silylation-based kinetic resolution of monofunctional secondary alcohols. *Org. Lett.,* **2011**, *13*(15), 3794-3797.
[http://dx.doi.org/10.1021/ol2012617] [PMID: 21714486]

[55]   Naito, T.; Yoneda, T.; Ito, J.; Nishiyama, H. Enantioselective hydrosilylation of aromatic alkenes catalyzed by chiral bis(oxazolinyl)phenyl-rhodium acetate complexes. *Synlett,* **2012**, *23*, 2957-2960.
[http://dx.doi.org/10.1055/s-0032-1317677]

[56]   Hayward, L.D.; Csizmadia, I.G. The conformations of 1,2-acenaphthene derivatives and steric interactions of contiguous nitroxy groups. *Tetrahedron,* **1963**, *19*, 2111-2121.
[http://dx.doi.org/10.1016/0040-4020(63)85026-7]

[57]   Zimmermann, K.; Haenel, M. W. Novel syntheses of decacyclene by deoxygenating cyclotrimerisation of acenaphthenequinone with zero-valant titanium or phosphorus pentasulfide. *Synlett,* **1997**, 609-611.
[http://dx.doi.org/http://jglobal.jst.go.jp/en/public/20090422/200902160848933081]

[58]   Wang, X.Y.; Cui, J.N.; Ren, W.M.; Li, F.; Lu, C.L.; Qian, X.H. Baker's yeast mediated reduction of substituted acenaphthenequinones: Regio- and enantioselective preparation of mono-hydroxyacenaphthenones. *Chin. Chem. Lett.,* **2007**, *18*, 681-684.
[http://dx.doi.org/10.1016/j.cclet.2007.04.022]

[59]   Yılız, T.; Çanta, N.; Yusufoğlu, A. Synthesis of new chiral keto alcohols by baker's yeast. *Tetrahedron Asymmetry,* **2014**, *25*, 340-347.
[http://dx.doi.org/10.1016/j.tetasy.2014.01.003]

[60]   Wang, L.; Wang, X.; Cui, J.; Ren, W.; Meng, N.; Wang, J.; Qian, X. Preparation of chiral *trans*-5-substistuted-acenaphthene-1,2-diols by baker's yeast-mediated reduction of 5-substistuted-acenaphthylene-1,2-diones. *Tetrahedron Asymmetry,* **2010**, *21*, 825-830.
[http://dx.doi.org/10.1016/j.tetasy.2010.04.048]

[61]   El Ashry, H.E.; Hamid, H.A.; Shoukry, M. Synthesis and reaction of acenaphthenequinones part-1, a review. *Indian J. Heterocycl. Chem.,* **1998**, *7*, 313-332.

[62]   Hashimoto, K.; Kitaichi, Y.; Tanaka, H.; Ikeno, T.; Yamada, T. Nitrous Oxide Oxidation of Secondary and Benzylic Alcohols Using Ruthenium Complex Catalyst. *Chem. Lett.,* **2001**, *30*, 922-923.
[http://dx.doi.org/10.1246/cl.2001.922]

[63]   Zacharias, J.; Koshy, E.P.; Pillai, V.N. Polyvinyl pyrrolidone-supported chlorochromates: preparation and application as solid phase synthetic oxidising reagent. *React. Funct. Polym.,* **2003**, *56*, 159-165.
[http://dx.doi.org/10.1016/S1381-5148(03)00053-1]

[64]   Koshy, E.P.; Zacharias, J.; Pillai, V.N. Poly(*N*-vinylpyrrolidone)-hydrotribromide: A new gel-type resin for alcohol oxidation and alkene dibromination. *React. Funct. Polym.,* **2006**, *66*, 845-850.
[http://dx.doi.org/10.1016/j.reactfunctpolym.2005.11.012]

[65]   Chen, L.; Li, B.D.; Xu, Q.X.; Liu, D.B. A silica supported cobalt(II) Schiff base complex as efficient and recyclable heterogeneous catalyst for the selective aerobic oxidation of alkyl aromatics. *Chin. Chem. Lett.,* **2013**, *24*, 849-852.
[http://dx.doi.org/10.1016/j.cclet.2013.05.017]

[66]   Islam, S.M.; Ghosh, K.; Molla, R.A.; Roy, A.S.; Salam, N.; Iqubal, M.A. Synthesis of a reusable polymer anchored cobalt(II) complex for the aerobic oxidation of alkyl aromatics and unsaturated organic compounds. *J. Organomet. Chem.,* **2014**, *774*, 61-69.
[http://dx.doi.org/10.1016/j.jorganchem.2014.10.010]

[67]   Singh, S.J.; Jayaram, R.V. Oxidation of alkylaromatics to benzylic ketones using TBHP as an oxidant over LaMO₃ (M = Cr, Co, Fe, Mn, Ni) perovskites. *Catal. Commun.,* **2009**, *10*, 2004-2007.
[http://dx.doi.org/10.1016/j.catcom.2009.07.018]

[68]   Burange, A.S.; Kale, S.R.; Jayaram, R.V. Oxidation of alkyl aromatics to ketones by *tert*-butyl hydroperoxide on manganese dioxide catalyst. *Tetrahedron Lett.,* **2012**, *53*, 2989-2992.
[http://dx.doi.org/10.1016/j.tetlet.2012.03.091]

[69]   Maradani, H.R.; Golchoubian, H. Selective and efficient C-H oxidation of alkanes with hydrogen peroxide catalyzed by a manganese(III) Schiff base complex. *J. Mol. Catal. Chem.,* **2006**, *259*, 197-200.
[http://dx.doi.org/10.1016/j.molcata.2006.06.029]

[70]   Wójtowicz, H.; Soroko, G.; Młochowski, J. New recoverable organoselenium catalyst for hydroperoxide oxidation of organic substrates. *Synth. Commun.,* **2008**, *38*, 2000-2010.
[http://dx.doi.org/10.1080/00397910801997793]

[71]   Yang, G.; Zhang, Q.; Miao, H.; Tong, X.; Xu, J. Selective organocatalytic oxygenation of hydrocarbons by dioxygen using anthraquinones and *N*-hydroxyphthalimide. *Org. Lett.,* **2005**, *7*(2), 263-266.
[http://dx.doi.org/10.1021/ol047749p] [PMID: 15646973]

[72]   Zhang, Q.; Chen, C.; Ma, H.; Miao, H.; Zhang, W.; Sun, Z.; Xu, J. Efficient metal-free aerobic oxidation of aromatic hydrocarbons utilizing aryl-tetrahalogenated N-hydroxyphthalimides and 1,4-diamino-2,3-dichloroanthraquinone. *J. Chem. Technol. Biotechnol.,* **2008**, *83*, 1364-1369.
[http://dx.doi.org/10.1002/jctb.1977]

[73]   Yang, G.; Ma, Y.; Xu, J. Biomimetic catalytic system driven by electron transfer for selective oxygenation of hydrocarbon. *J. Am. Chem. Soc.,* **2004**, *126*(34), 10542-10543.
[http://dx.doi.org/10.1021/ja047297b] [PMID: 15327303]

[74]   Zeynizadeh, B.; Yahyaei, S. A mild and convenient method for the reduction of carbonyl compounds with NaBH₄ in the presence of catalytic amounts of MoCl₅. *Bull. Korean Chem. Soc.,* **2003**, *24*, 1664-1670.
[http://dx.doi.org/10.5012/bkcs.2003.24.11.1664]

[75]   Zeynizadeh, B.; Yahyaei, S. Reduction of carbonyl compounds with NaBH₄ under ultrasound irradiation and aprotic condition. *Z. Naturforsch., B. Chem. Sci.,* **2004**, *59*, 704-710.

[76]   Zeynizadeh, B.; Behyar, T. Fast and efficient method for reduction of carbonyl compounds with NaBH₄/Wet SiO₂ under solvent free condition. *J. Braz. Chem. Soc.,* **2005**, *16*, 1200-1209.
[http://dx.doi.org/10.1590/S0103-50532005000700018]

[77]   Zeynizadeh, B.; Behyar, T. Wet THF as a suitable solvent for a mild and convenient reduction of carbonyl compounds with NaBH₄. *Bull. Chem. Soc. Jpn.,* **2005**, *78*, 307-315.
[http://dx.doi.org/10.1246/bcsj.78.307]

[78] Zeynizadeh, B.; Setamdideh, D. Water as a green solvent for fast and efficient reduction of carbonyl compounds with NaBH$_4$ under microwave irradiation. *J. Chil. Chem. Soc.,* **2005**, *52*, 1179-1184. [http://dx.doi.org/10.1002/jccs.200500169]

[79] Zeynizadeh, B.; Zarrin, S.; Ashuri, S. Fast and convenient method for reduction of carbonyl compounds with NaBH$_4$/Cu(dmg)$_2$ in aprotic and protic solvents. *Org. Chem. Indian J,* **2013**, *9*, 469-479.

[80] Kouhkan, M.; Zeynizadeh, B. Wet SiO$_2$ as a suitable media for fast and efficient reduction of carbonyl compounds with NaBH$_3$CN under solvent-free and acid-free conditions. *Bull. Korean Chem. Soc.,* **2010**, *31*, 2961-2966. [http://dx.doi.org/10.5012/bkcs.2010.31.10.2961]

[81] Setamdideh, D.; Khezri, B.; Rahmatollahzadeh, M.; Poramjad, A.A. Mild and efficient reduction of organic carbonyl compounds to their corresponding alcohols with Zn(BH$_4$)$_2$ under protic condition. *Asian J. Chem.,* **2012**, *24*, 3591-3596.

[82] Setamdideh, D.; Khaledi, L. Zn(BH$_4$)$_2$/NaCl: a novel reducing system for efficient reduction of organic carbonyl compounds to their corresponding alcohols. *S. Afr. J. Chem.,* **2013**, *66*, 150-157.

[83] Fanari, S.; Setamdideh, D. Zn(BH$_4$)$_2$/Ultrasonic Irradiation: An Efficient System for Reduction of Carbonyl Compounds to their Corresponding Alcohols. *Orient. J. Chem,* **2014**, *30*, 695-697. [http://dx.doi.org/10.13005/ojc/300240]

[84] Sajiki, H.; Hattori, K.; Hirota, K. Easy and partial hydrogenation of aromatic carbonyls to benzyl alcohols using Pd/C(en)-catalyst. *J. Chem. Soc., Perkin Trans. 1,* **1998**, 4043-4044. [http://dx.doi.org/10.1039/a807486k]

[85] Hattori, K.; Sajiki, H.; Hirota, K. Chemoselective control of hydrogenation among aromatic carbonyl and benzyl alcohol derivatives using Pd/C(en) catalyst. *Tetrahedron,* **2001**, *57*, 4817-4824. [http://dx.doi.org/10.1016/S0040-4020(01)00421-5]

[86] Yu, Z.K.; Zheng, F.L.; Sun, X.J.; Deng, H.X.; Dong, J.H.; Chen, J.Z.; Wang, H.M.; Pei, C.X. Two *pseudo*-N$_3$ ligands and the catalytic activity of their ruthenium(II) complexes in transfer hydrogenation and hydrogenation of ketones. *J. Organomet. Chem.,* **2007**, *692*, 2306-2313. [http://dx.doi.org/10.1016/j.jorganchem.2007.01.058]

[87] Zhao, M.; Yu, Z.; Yan, S.; Li, Y. Ruthenium(II) complex catalysts bearing a pyridyl-supported pyrazolyl-imine ligand for transfer hydrogenation of ketones. *J. Organomet. Chem.,* **2009**, *694*, 3068-3075. [http://dx.doi.org/10.1016/j.jorganchem.2009.05.028]

[88] Ye, W.; Zhao, M.; Du, W.; Jiang, Q.; Wu, K.; Wu, P.; Yu, Z. Highly active ruthenium(II) complex catalysts bearing an unsymmetrical NNN ligand in the (asymmetric) transfer hydrogenation of ketones. *Chemistry,* **2011**, *17*(17), 4737-4741. [http://dx.doi.org/10.1002/chem.201002039] [PMID: 21404339]

[89] Fernández, F.E.; Puerta, M.C.; Valerga, P. Ruthenium(II) Picolyl-NHC complexes: synthesis, characterization, and catalytic activity in amine N-alkylation and transfer hydrogenation reactions. *Organometallics,* **2012**, *31*, 6868-6879. [http://dx.doi.org/10.1021/om300692a]

[90]   Marimuthu, T.; Friedrich, H.B. Microwave-assisted transfer hydrogenation of ketones by Ru(xantphos) arene complexes. *ChemCatChem,* **2012**, *4*, 2090-2095.
[http://dx.doi.org/10.1002/cctc.201200306]

[91]   Cadogan, J.I.; Hickson, C.L.; Hutchison, H.S.; McNab, H. New gas phase reactions of substituted benzyl, phenylaminyl, and phenoxyl radicals. rearrangements to fused 5- and 6-membered heterocyclic systems. *J. Chem. Soc. Chem. Commun.,* **1985**, 643-644.
[http://dx.doi.org/10.1039/c39850000643]

[92]   Cadogan, J.I. Hutchison, H. S.; McNab. H. Gas-phase Reactions of 2-Benzyl- and 2-Benzoyl-phenoxyl Radicals, and of 2-Phenoxybenzyl Radicals: Examples of New Hydrogen-transfer Processes. *J. Chem. Soc., Perkin Trans. 1,* **1991**, 385-393.
[http://dx.doi.org/10.1039/p19910000385]

[93]   Kishner, N. Wolff-Kishner reduction: Huang-Minlon modification. *J. Russ. Phys. Chem. Soc.,* **1911**, *43*, 582-595.

[94]   Wolff, L. Chemischen Institut der Universität Jena: Methode zum Ersatz des Sauerstoffatoms der Ketone und Aldehyde durch Wasserstoff. *Justus Liebigs Ann. Chem.,* **1912**, *394*, 86-108.
[http://dx.doi.org/10.1002/jlac.19123940107]

[95]   Kumar, B.; Kaur, N.; Mehta, R.M.; Thakur, U. A new photoreaction of benzophenone oxime. *Tetrahedron Lett.,* **1978**, *19*, 5031-5032.
[http://dx.doi.org/10.1016/S0040-4039(01)85801-9]

[96]   Kumar, B. Kaur, N.; Kaur, G. One-Pot Photolytic Synthesis of 1-Hydroxyfluorenes. *Synthesis,* **1983**, 115-116.
[http://dx.doi.org/10.1055/s-1983-30241]

[97]   Kumar, B.; Kaur, N. Photochemistry of Diaryl Ketones: A New Photocyclization Reaction. *J. Org. Chem.,* **1983**, *48*, 2281-2285.
[http://dx.doi.org/10.1021/jo00161a030]

[98]   Shi, Y.; Wan, P. Solvolysis and ring closure of quinone methides photogenerated from biaryl systems. *Can. J. Chem.,* **2005**, *83*, 1306-1323.
[http://dx.doi.org/10.1139/v05-138]

[99]   Jones, W.D., Jr; Albrecht, W.L.; Palopoli, F.P. Fluorene Derivatives: Friedel-Crafts Reaction of 2-Fluorenyl Basic Ethers. *J. Org. Chem.,* **1977**, *42*, 4144-4146.
[http://dx.doi.org/10.1021/jo00445a035]

[100] Ador, E.; Crafts, J. Ueber die Einwirkung des Chlorkohlenoxyds auf Toluol in Gegenwart von Chloraluminium. *Ber. Dtsch. Chem. Ges,* **1877**, *10*, 2173-2176.
[http://dx.doi.org/10.1002/cber.187701002241]

[101] Friedel, C.; Crafts, J.M. New general method for the synthesis of hydrocarbons, ketones, *etc. C. R. Hebd. Seances Acad. Sci.,* **1877**, *84*, 1450-1454.

[102] Baeyer, A.; Villiger, V. Einwirkung des Caro'schen Reagens auf Ketone *Ber. Dtsch. Chem. Ges,* **1899**, *32*, 3625-3627.
[http://dx.doi.org/10.1002/cber.189903203151]

[103] Middlemiss, D. A Convenient Synthesis of 4-Oxo-1,2,3,4,4a,9a-hexahydrofluorene and 4-Oxotetrahydrofluorenes. *Synthesis,* **1979**, 987-988.
[http://dx.doi.org/10.1055/s-1979-28904]

[104] Guram, A.S.; Bei, X.; Turner, H.W. Productive utilization of chlorobenzene: palladium-catalyzed selective oxidation of alcohols. *Org. Lett.,* **2003**, *5*(14), 2485-2487.
[http://dx.doi.org/10.1021/ol0347287] [PMID: 12841761]

[105] Bei, X.; Hagemeyer, A.; Volpe, A.; Saxton, R.; Turner, H.; Guram, A.S. Productive chloroarene C-cl bond activation: palladium/phosphine-catalyzed methods for oxidation of alcohols and hydrodechlorination of chloroarenes. *J. Org. Chem.,* **2004**, *69*(25), 8626-8633.
[http://dx.doi.org/10.1021/jo048715y] [PMID: 15575738]

[106] Lim, M.; Oh, S.; Rhee, H. Pd/C catalyzed oxidation of secondary benzylic alcohols using chlorobenzene under an inert condition. *Bull. Korean Chem. Soc.,* **2011**, *32*, 3179-3182.
[http://dx.doi.org/10.5012/bkcs.2011.32.8.3179]

[107] Aït-Mohand, S.; Hénin, F.; Muzart, J. Palladium(II)-Mediated Oxidation of Alcohols using 1,2-Dichloroethane as Pd(0) Reoxidant. *Tetrahedron Lett.,* **1995**, *36*, 2473-2476.
[http://dx.doi.org/10.1016/0040-4039(95)00286-3]

[108] Ajjou, A.N. First example of water-soluble transition-metal catalyst for Oppenauer-type oxidation of secondary alcohols. *Tetrahedron Lett.,* **2001**, *42*, 13-15.
[http://dx.doi.org/10.1016/S0040-4039(00)01882-7]

[109] Ajjou, A.N.; Pinet, J-L. Oppenauer-type oxidation of secondary alcohols catalyzed by homogeneous water-soluble complexes. *Can. J. Chem.,* **2005**, *83*, 702-710.
[http://dx.doi.org/10.1139/v05-037]

[110] An, G.; Lim, M.; Chun, K.S.; Rhee, H. Environmentally benign oxidation reaction of benzylic and allylic alcohols to carbonyl compounds using Pd/C with sodium borohydride. *Synlett,* **2007**, 95-98.

[111] Kim, S.S.; Jung, H.C. An efficient aerobic oxidation of alcohols to aldehydes and ketones with TEMPO/Ceric ammonium nitrate as catalyst. *Synthesis,* **2003**, 2135-2137.
[http://dx.doi.org/10.1055/s-2003-41065]

[112] Kim, S.S.; Rajagopal, G. Efficient aerobic oxidation of alcohols to carbonyl compounds with NHPI/CAN catalytic system. *Synth. Commun.,* **2004**, *34*, 2237-2243.
[http://dx.doi.org/10.1081/SCC-120038507]

[113] Wang, L.; Li, J.; Yang, H.; Lv, Y.; Gao, S. Selective oxidation of unsaturated alcohols catalyzed by sodium nitrite and 2,3-dichloro-5,6-dicyano-1,4-benzoquinone with molecular oxygen under mild conditions. *J. Org. Chem.,* **2012**, *77*(1), 790-794.
[http://dx.doi.org/10.1021/jo202301s] [PMID: 22148499]

[114] Farhadi, S.; Zabardasti, A.; Babazadeh, Z. Aerobic photocatalytic oxidation of activated benzylic and allylic alcohols to carbonyl compounds catalyzed by molecular iodine. *Tetrahedron Lett.,* **2006**, *47*, 8953-8957.
[http://dx.doi.org/10.1016/j.tetlet.2006.10.036]

[115] Prebil, R.; Stavber, G.; Stavber, S. Aerobic oxidation of alcohols by using a completely metal-free catalytic system. *Eur. J. Org. Chem.,* **2014**, 395-402.
[http://dx.doi.org/10.1002/ejoc.201301271]

[116] Zheng, G.; Liu, C.; Wang, Q.; Wang, M.; Yang, G. Metal-free: an efficient and selective catalytic aerobic oxidation of hydrocarbons with oxime and *N*-hydroxyphthalimide. *Adv. Synth. Catal.,* **2009**, *351*, 2638-2642.
[http://dx.doi.org/10.1002/adsc.200900509]

[117] Yang, X.; Wang, Y.; Zhou, L.; Chen, C.; Zhang, W.; Xu, J. Efficient aerobic oxidation of hydrocarbons with O$_2$ catalyzed by DDQ/NHPI. *J. Chem. Technol. Biotechnol.,* **2010**, *85*, 564-568.

[118] Zhang, X.; Ji, X.; Jiang, S.; Liu, L.; Weeks, B.L.; Zhang, Z. Highly efficient synthesis of 9-fluorenones from 9*H*-fluorenes by air oxidation. *Green Chem.,* **2011**, *13*, 1891-1896.
[http://dx.doi.org/10.1039/c1gc15136c]

[119] Marwah, P.; Marwah, A.; Lardy, H.A. An economical and green approach for oxidation of olefins to enones. *Green Chem.,* **2004**, *6*, 570-577.
[http://dx.doi.org/10.1039/b408974j]

[120] Silvestre, S.M.; Salvador, J.A. Allylic and benzylic oxidation reactions with sodium chlorite. *Tetrahedron,* **2007**, *63*, 2439-2445.
[http://dx.doi.org/10.1016/j.tet.2007.01.012]

[121] He, H.; Pei, B.J.; Lee, A.W. Metal free oxidation of alkyl substituted aromatics with aqueous *tert*-butyl hydroperoxide under microwave irradiation. *Green Chem.,* **2009**, *11*, 1857-1861.
[http://dx.doi.org/10.1039/b916265h]

[122] Moriconi, E.J.; Wallenberger, F.T.; O'Conner, W.F. A New Synthesis of 9-Phenanthrol; Absorption Spectra of the Quinhydrone-Type Molecular Compound between 9-Phenanthrol and Phenanthrenequinone. *J. Org. Chem.,* **1959**, *24*, 86-90.
[http://dx.doi.org/10.1021/jo01083a025]

[123] Fujishiro, K.; Mitamura, S. The baeyer-villiger reaction of polycyclic aromatic aldehydes: preparation of polycyclic phenols. *Bull. Chem. Soc. Jpn.,* **1988**, *61*, 4464-4466.
[http://dx.doi.org/10.1246/bcsj.61.4464]

[124] Fu, J-M.; Sharp, M.J.; Snieckus, V. The directed *ortho* metalation connection to aryl-aryl cross coupling. A general regiospecific synthesis of phenanthrons. *Tetrahedron Lett.,* **1988**, *29*, 5459-5462.
[http://dx.doi.org/10.1016/S0040-4039(00)80786-8]

[125] Fu, J-M.; Snieckus, V. The directed *ortho* metalation-palladium catalyzed cross coupling connection. A general regiospecific route to 9-phenanthrols and phenanthrenes. Exploratory further metalation. *Can. J. Chem.,* **2000**, *78*, 905-919.
[http://dx.doi.org/10.1139/v00-055]

[126] Miyaura, N.; Suzuki, A. Stereoselective synthesis of arylated (*E*)-alkenes by the reaction of alk--enylboranes with aryl halides in the presence of palladium catalyst. *J. Chem. Soc. Chem. Commun.,* **1979**, 866-867.
[http://dx.doi.org/10.1039/c39790000866]

[127] Mosettig, E.; Duvall, H.M. Studies in the Phenanthrene Series. XIV. The Preparation of 1- and 4-Phenanthrol. *J. Am. Chem. Soc.,* **1937**, *59*, 367-369.
[http://dx.doi.org/10.1021/ja01281a045]

[128] Paredes, E.; Brasca, R.; Kneeteman, M.; Mancini, P.M. A novel application of the Diels-Alder reaction: nitronaphthalenes as normal electron demand dienophiles. *Tetrahedron,* **2007**, *63*, 3790-3799.
[http://dx.doi.org/10.1016/j.tet.2007.02.054]

[129] Kneeteman, M.N.; Della Rosa, C.D.; Ormachea, C.M.; Giménez, P.; Lopez Baena, A.F.; Mancini, P.M. Synthesis of phenanthrenol derivatives through polar diels-alder reactions employing nitronaphthalenes and (*E*)-1-(Trimethylsilyloxy)-1,3-butadiene. theoretical calculations. *Lett. Org. Chem.,* **2014**, *11*, 333-337.
[http://dx.doi.org/10.2174/1570178611666140123235852]

[130] Samanta, K.; Kar, G.K.; Sarkar, A.K. Intramolecular gold(III) gold catalysed Diels-Alder reaction of 1-(2-furyl)-hex-1-en-5-yn-3-ol derivatives: a short and generalised route for the synthesis of hydroxyphenanthrene derivatives. *Tetrahedron Lett.,* **2012**, *53*, 1376-1379.
[http://dx.doi.org/10.1016/j.tetlet.2012.01.018]

[131] Xia, Y.; Liu, Z.; Xiao, Q.; Qu, P.; Ge, R.; Zhang, Y.; Wang, J. Rhodium(II)-catalyzed cyclization of bis(N-tosylhydrazone)s: an efficient approach towards polycyclic aromatic compounds. *Angew. Chem. Int. Ed. Engl.,* **2012**, *51*(23), 5714-5717.
[http://dx.doi.org/10.1002/anie.201201374] [PMID: 22532502]

[132] Berger, S.; Zeller, K.P. $^{13}C^{13}C$ Spin Coupling Constants in Phenanthrene Derivatives. *Org. Magn. Reson,* **1978**, *11*, 303-307.
[http://dx.doi.org/10.1002/mrc.1270110608]

[133] Tomioka, H.; Kobayashi, N.; Ohtawa, Y.; Murata, S. Multiple Rearrangement of 1-Methoxyfluorenylidene to 1-Phenanthrenol. *J. Org. Chem.,* **1991**, *56*, 2609-2611.
[http://dx.doi.org/10.1021/jo00008a004]

[134] Wu, R.; Silks, L.A.; Olivault-Shiflett, M.; Williams, R.F.; Ortiz, E.G.; Stotter, P.; Kimball, D.B.; Martinez, R.A. A general route for 13C-labeled fluorenols and phenanthrenols *via* palladium-catalyzed cross-coupling and one-carbon homologation. *J. Labelled Comp. Radiopharm.,* **2013**, *56*(11), 581-586.
[http://dx.doi.org/10.1002/jlcr.3066] [PMID: 24285190]

[135] Platt, K.L.; Oesch, F. K-Region *trans*-Dihydrodiols of polycyclic arenes; an efficient and convenient preparation from *o*-quinones or *o*-diphenols by reduction with sodium borohydride in the presence of oxygen. *Synthesis,* **1982**, 459-461.
[http://dx.doi.org/10.1055/s-1982-29834]

[136] Fitting, R. Ueber einige Producte der trockenen Destillation essigsaurer Salze. *Justus Liebigs Ann. Chem.,* **1859**, *110*, 17-23.
[http://dx.doi.org/10.1002/jlac.18591100103]

[137] Li, C.J.; Meng, Y.; Yi, X.H.; Ma, J.; Chan, T.H. Manganese-Mediated Carbon-Carbon Bond Formation in Aqueous Media: Chemoselective Allylation and Pinacol Coupling of Aryl Aldehydes. *J. Org. Chem.,* **1998**, *63*(21), 7498-7504.
[http://dx.doi.org/10.1021/jo980535z] [PMID: 11672403]

[138] Ohmori, K.; Kitamura, M.; Suzuki, K. From axial chirality to central chiralities: pinacol cyclization of 2,2'-biaryldicarbaldehyde to *trans*-9,10-dihydrophenanthrene-9,10-diol. *Angew. Chem. Int. Ed.,* **1999**, *38*, 1226-1229.
[http://dx.doi.org/10.1002/(SICI)1521-3773(19990503)38:9<1226::AID-ANIE1226>3.0.CO;2-T]

[139] Yamamoto, Y.; Hattori, R.; Ito, K. Highly *trans*-selective intramolecular pinacol coupling of dials catalyzed by bulk Cp$_2$TiPh. *Chem. Commun. (Camb.),* **1999**, 825-826.
[http://dx.doi.org/10.1039/a902154j]

[140] Yamamoto, Y.; Hattori, R.; Miwa, T.; Nakagai, Y.I.; Kubota, T.; Yamamoto, C.; Okamoto, Y.; Itoh, K. Diastereoselective inter- and intramolecular pinacol coupling of aldehydes promoted by monomeric titanocene(II) complex Cp(2)TiPh. *J. Org. Chem.,* **2001**, *66*(11), 3865-3870.
[http://dx.doi.org/10.1021/jo001781p] [PMID: 11375008]

[141] Mori, K.; Ohtaka, S.; Uemura, S. InCl$_3$-Catalyzed Reductive Coupling of Aromatic Carbonyl Compounds in the Presence of Magnesium and Chlorotrimethylsilane. *Bull. Chem. Soc. Jpn.,* **2001**, *74*, 1497-1498.
[http://dx.doi.org/10.1246/bcsj.74.1497]

[142] Lin, S.; You, T. An efficient one-pot approach to phenanthrene derivatives using a catalyzed tandem Ullmann-pinacol coupling reaction. *Tetrahedron,* **2008**, *64*, 9906-9910.
[http://dx.doi.org/10.1016/j.tet.2008.08.004]

[143] Shing, T.K.; Tam, E.K.; Tai, V.W; Chung, I. H, F.; Jiang, Q.; Ruthenium-Catalyzed *cis*-Dihydroxylation of Alkenes: Scope and Limitations. *Chemistry,* **1996**, *2*, 50-57.
[http://dx.doi.org/10.1002/chem.19960020111]

[144] Ho, C.M.; Yu, W.Y.; Che, C.M. Ruthenium nanoparticles supported on hydroxyapatite as an efficient and recyclable catalyst for *cis*-dihydroxylation and oxidative cleavage of alkenes. *Angew. Chem. Int. Ed. Engl.,* **2004**, *43*(25), 3303-3307.
[http://dx.doi.org/10.1002/anie.200453703] [PMID: 15213959]

[145] Reisch, H.A.; Enkelmann, V.; Scherf, U. A Novel Nickel(0)-Mediated One-Pot Cascade Reaction to cis-9,10-Dihydroxy-9,10-dihydrophenanthrenes and 9-Phenanthrones. *J. Org. Chem.,* **1999**, *64*, 655-658.
[http://dx.doi.org/10.1021/jo9817523]

[146] Pampanin, D.M.; Kemppainen, E.K.; Skogland, K.; Jørgensen, K.B.; Sydnes, M.O. Investigation of fixed wavelength fluorescence results for biliary metabolites of polycyclic aromatic hydrocarbons formed in Atlantic cod (*Gadus morhua*). *Chemosphere,* **2016**, *144*, 1372-1376.
[http://dx.doi.org/10.1016/j.chemosphere.2015.10.013] [PMID: 26492423]

[147] Tu, Y. Wang, Z. X.; Shi, Y. An efficient asymmetric epoxidation method for trans-Olefins mediated by a fructose-derived ketone. *J. Am. Chem. Soc.,* **1996**, *118*, 9806-9807.
[http://dx.doi.org/10.1021/ja962345g]

[148] Negishi, E.; King, A.O.; Okukado, N. Selective carbon-carbon bond formation *via* transition metal catalysis. 3. A highly selective synthesis of unsymmetrical biaryls and diarylmethanes by the nickel- or palladium-catalyzed reaction of aryl- and benzylzinc derivatives with aryl halides. *J. Org. Chem.,* **1977**, *42*, 1821-1823.
[http://dx.doi.org/10.1021/jo00430a041]

[149] Dansette, P.; Jerina, D.M. A facile synthesis of arene oxides at the K regions of polycyclic hydrocarbons. *J. Am. Chem. Soc.,* **1974**, *96*(4), 1224-1225.
[http://dx.doi.org/10.1021/ja00811a046] [PMID: 4816471]

[150] Cortez, C.; Harvey, R.G. Arene oxide synthesis: Phenanthrene 9,10-oxide. *Org. Synth.,* **1978**, *58*, 12-17.
[http://dx.doi.org/10.15227/orgsyn.058.0012]

[151] Robinson, P.L.; Barry, C.N.; Kelly, J.W.; Evans, S.A., Jr Diethoxytriphenylphosphorane: a mild, regioselective cyclodehydrating reagent for conversion of diols to cyclic ethers. stereochemistry, synthetic utility, and scope. *J. Am. Chem. Soc.,* **1985**, *107*, 5210-5219.
[http://dx.doi.org/10.1021/ja00304a030]

[152] Krishnan, S.; Kuhn, D.G.; Hamilton, G.A. The formation of arene oxides by direct oxidation of arenes. *Tetrahedron Lett.,* **1977**, *16*, 1369-1372.
[http://dx.doi.org/10.1016/S0040-4039(01)93046-1]

[153] Jørgensen, K.B.; Joensen, M. Photochemical synthesis of chrysenols. *Polycycl. Aromat. Compd.,* **2008**, *28*, 362-378.
[http://dx.doi.org/10.1080/10406630802374580]

[154] Wittig, G.; Schollkopf, U. Triphenylphosphinemethylene as an olefin-forming reagent. I. *Ber,* **1954**, *97*, 1318-1330.

[155] Liu, L.; Yang, B.; Katz, T.J.; Poindexter, M.K. Improved methodology for photocyclization reactions. *J. Org. Chem.,* **1991**, *56*, 3769-3775.
[http://dx.doi.org/10.1021/jo00012a005]

[156] Paolobelli, A.B.; Ruzziconi, R.; Lupattelli, P.; Scafato, P.; Spezzacatena, C. A facile access to polycyclic homo- and heteroaromatic hydrocarbons based on the ceric ammonium nitrate-promoted oxidative addition of 3-Aryl-1-[(trimethylsilyl)oxy]cyclohexenes to ethyl vinyl ether. *J. Org. Chem.,* **1999**, *64*(9), 3364-3368.
[http://dx.doi.org/10.1021/jo982195n] [PMID: 11674447]

[157] Johnson, W.S.; Banerjee, D.K.; Schneider, W.P.; Gutsche, C.D.; Shelberg, W.E.; Chinn, L.J. The total synthesis of estrone and three stereoisomers including lumiestrone. *J. Am. Chem. Soc.,* **1952**, *74*, 2832-2849.
[http://dx.doi.org/10.1021/ja01131a037]

[158] Kumar, S. A new efficient route to the phenolic derivatives of chrysene and 5-methylchrysene, precursors to dihydrodiol and diol epoxide metabolites of chrysene and 5-methylchrysene, through Suzuki cross-coupling reaction. *J. Chem. Soc., Perkin Trans. 1,* **1998**, 3157-3161.
[http://dx.doi.org/10.1039/a805502e]

[159] Wilds, A.L.; Shunk, C.H. The preparation of derivatives of chrysene by means of the robinson-mannich base synthesis of unsaturated ketones. *J. Am. Chem. Soc.,* **1943**, *65*, 469-475.
[http://dx.doi.org/10.1021/ja01243a048]

[160] Fu, P. P.; Harvey, R. G. Synthesis of the Chrysene Bay Region anti-Diolepoxide from Chrysene. *J. Chem. Soc. Chem. Comm.,* **1978**, 585-586.

[161] Fu, P.P.; Harvey, R.G. Synthesis of the Dihydro Diols and Diol Epoxides from Chrysene. *J. Org. Chem.,* **1979**, *44*, 3778-3784.
[http://dx.doi.org/10.1021/jo01336a008]

[162] Prévost, C. Iodo-silver benzoate and its use in the oxidation of ethylene derivatives into α-glycols. *Compt. rend.,* **1933**, *196*, 1129-1131.

[163] Prileschajew, N. Oxydation ungesättigter Verbindungen mittels organischer Superoxyde. *Ber. Dtsch. Ges,* **1909**, *42*, 4811-4815.
[http://dx.doi.org/10.1002/cber.190904204100]

[164] Pataki, J.; Lee, H.; Harvey, R.G. Carcinogenic metabolites of 5-methylchrysene. *Carcinogenesis,* **1983**, *4*(4), 399-402.
[http://dx.doi.org/10.1093/carcin/4.4.399] [PMID: 6839413]

[165] Harvey, R.G.; Pataki, J.; Lee, H. Synthesis of the dihydrodiol and epoxide metabolites of chrysene and 5-methylchrysene. *J. Org. Chem.,* **1986**, *51*, 1407-1412.
[http://dx.doi.org/10.1021/jo00359a006]

[166] Amin, S.; Huie, K.; Hecht, S.S.; Harvey, R.G. Synthesis of 6-methylchrysene-1,2-diol-3,4-epoxides and comparison of their mutagenicity to 5-methylchrysene-1,2-diol-3,4-epoxides. *Carcinogenesis,* **1986**, *7*(12), 2067-2070.
[http://dx.doi.org/10.1093/carcin/7.12.2067] [PMID: 3779900]

[167] Amin, S.; Huie, K.; Balanikas, G.; Hecht, S.S. Synthesis and mutagenicity of 5-alkyl-substituted chrysene-1,2-diol-3,4-epoxides. *Carcinogenesis,* **1988**, *9*(12), 2305-2308.
[http://dx.doi.org/10.1093/carcin/9.12.2305] [PMID: 3191576]

[168] Clemmensen, E. Reduktion von Ketonen und Aldehyden zu den entsprechenden Kohlenwasserstoffen unter Anwendung von amalgamiertem Zink und Salzäure. *Ber. Dtsch. Gem. Ges,* **1913**, *46*, 1837-1843.
[http://dx.doi.org/10.1002/cber.19130460292]

[169] Wameling, C.; Glatt, H.R.; Oesch, F.; Seidel, A. Stereospecific synthesis of the diastereomeric revers dihydrodiol epoxides of chrysene picene. *Polycycl. Aromat. Compd.,* **1993**, 191-198.

[170] Yagi, H.; Vyas, K.P.; Tada, M.; Thakker, D.R.; Jerina, D.M. Synthesis of the enantiomeric bay-region diol epoxide of benz[*a*]anthracene and chrysene. *J. Org. Chem.,* **1982**, *47*, 1110-1117.
[http://dx.doi.org/10.1021/jo00345a043]

[171] Lorentzen, M.; Sydnes, M.O.; Jørgensen, K.B. Enantioselective synthesis of (-)-(1*R*,2*R*)-1,2-1-2-dihydrochrysene-1,2-diol. *Tetrahedron,* **2014**, *70*, 9041-9051.
[http://dx.doi.org/10.1016/j.tet.2014.10.016]

CHAPTER 7

Microbial Degradation in the Aquatic Environment

Andrea Bagi*

Faculty of Science and Technology, Department of Mathematics and Natural Science, University of Stavanger, Stavanger, Norway

Abstract: Considering their recalcitrance and toxicity, understanding how PAHs are removed from the marine environment is essential in order to maintain healthy and functional ecosystems. This chapter discusses the major advances in knowledge in the field of polycyclic aromatic hydrocarbon (PAH) biodegradation regarding key players, degradation mechanisms, genetic background, environmental factors influencing biodegradation and the role of interaction among microbial community members. Microbial degradation, or biodegradation, is the natural mechanism by which PAHs are channeled back into the marine carbon cycle and transformed into harmless material. The key players in this process are specialized bacteria, which developed strategies to utilize PAH as carbon and energy source. The rate of biodegradation is one of the major concerns in terms of bioremediation efforts. Factors influencing the rate at which PAHs are converted include temperature, oxygen concentration, availability of nutrients, type and bioavailability of PAHs, and the level of adaptation of the local microbial community.

Keywords: Biodegradation, Marine hydrocarbon degrading bacteria, PAHs.

INTRODUCTION

Polycyclic aromatic hydrocarbons (PAHs) are among those constituents of crude oil, which tend to persist, are highly toxic, accumulate in higher-level organisms and some of them are also well-known carcinogens [1]. Due to these characteristics, PAHs represent high risk to the marine environment. Understanding the processes contributing to their elimination is essential for better

* **Corresponding author Andrea Bagi**: Faculty of Science and Technology, Department of Mathematics and Natural Science, University of Stavanger, Stavanger, Norway; Tel/Fax: +47 51831853; E-mail: andrea.bagi@uis.no

Daniela M. Pampanin and Magne O. Sydnes (Eds.)
All rights reserved-© 2017 Bentham Science Publishers

mitigation of ecosystem damages. The overall fate of PAHs in seawater includes photo-oxidation, uptake and transformation by higher-level organisms, sedimentation (following adsorption to particulate matter) and microbial biodegradation. Each of these processes has a unique outcome and contribution to contaminant fate. Nevertheless, it is only microbial degradation which can completely eliminate PAHs from the environment, by transforming them into harmless products (*e.g.* carbon-dioxide and water) and cell material [2, 3]. Owing to their ability to biodegrade toxic compounds and thereby recycle carbon and energy from pollutants, microbial communities play an essential role in maintaining functional marine ecosystems. Firstly, by reducing the amount of harmful chemicals they directly decrease the environmental impact of anthropogenic pollution. Secondly, the microbial cell material produced from consuming contaminants represents a readily available carbon source for other higher-level members of the marine food web [4 - 7].

Biodegradation in general can be defined as a process in which microorganisms decompose organic molecules into simpler chemicals and thereby, ideally, transforming harmful compounds into non-harmful ones [3]. In this context, complete biodegradation occurs when compounds are transformed into water and carbon dioxide through catabolic (energy producing) processes and into cell material through anabolic (carbon assimilating) processes. The former is often referred to as mineralization. Incomplete or partial biodegradation involves the formation of intermediary metabolites, which can be released back into the environment. Very little research has been done regarding the fate of such intermediary metabolites excreted from microorganisms, despite the fact that some of these compounds are known to be more toxic than the parent molecules [8, 9].

From an environmental risk perspective, both the extent and the speed (kinetics) of biodegradation is important to consider. They depend on the type and abundance of microorganisms involved, the bioavailability of contaminants and various environmental factors (*e.g.* nutrient availability and temperature). The aim of this chapter is to outline current knowledge regarding each of these subtopics. First it will introduce the broader diversity of microorganisms (both eukaryotic and prokaryotic) involved in marine PAH biodegradation, and then the fundamental mechanisms will be discussed with focus on aerob heterotrophic bacterial processes. Finally, environmental factors and the role of interaction within the microbial loop will be discussed.

By definition, microorganisms are single celled, and invisible to the naked eye. Both prokaryotic domains, Bacteria and Archaea, fulfil this requirement, while among Eukaryotes, only Algae, Fungi and Protozoans harbour microbes. Most members of this microbial world have been to a varying extent studied in connection with PAH removal from marine environments. Aerob heterotrophic bacteria are by far the most broadly studied group. While the focus of this chapter will be the aerob heterotroph PAH degrading bacteria, current knowledge regarding the other microbial groups will also be described.

Hydrocarbon Degrading Microorganisms

Crude oil, one of the major sources of marine PAH pollution, is a naturally occurring highly complex mixture of organic components. Its major hydrocarbon groups include alkanes, resins, and asphaltenes besides PAHs. Crude oil has been released into the environment spontaneously throughout historical times. According to estimates by the National Research Council (USA) in 2003, nearly 50% of all the oil entering the marine environment originated from natural oil seeps, amounting to 700 million liters each year globally [10]. Though estimates vary widely, natural seepage is certainly a significant source of crude oil in the oceans [11, 12]. This continuous presence of hydrocarbons had a crucial role in the evolution of specialized microorganisms, which were able to develop mechanisms for utilizing hydrocarbons, mainly alkanes and PAHs, as growth substrates. Already in the 1940s, Claude Zobell published two reviews summarizing rich evidence for marine (and also terrestrial) microorganisms being able to grow on unusual materials, such as asphalt, rubber and wax [13, 14]. Since then, petroleum microbiology as a research area developed and the diversity of hydrocarbon degrading microorganisms has continuously been emerging. It was quickly understood that crude oil in the sea is degraded according to a pattern of sequential depletion defined by the biodegradability of hydrocarbon groups. In terms of biodegradability, PAHs are approximately in the middle of the scale where short, straight-chained alkanes are the most easily, and large asphaltenes are the least easily biodegradable. Hence, it is generally observed that first alkanes, then low molecular weight (LMW) PAHs, and finally high molecular weight (HMW) PAHs are degraded [15 - 17]. As microbial community composition analysis methods clarified later on, this sequential depletion is not only connected to biodegradability, but also to a sequential change in bacterial community composition. Specialist hydrocarbon degraders evolved in response to specific oil components, meaning that different microorganisms prefer different hydrocarbon groups.

Microbial Diversity Involved in PAH Degradation

The role of phototrophic microorganisms, *i.e.*, prokaryotic cyanobacteria, and eukaryotic green algae and diatoms is largely unclear and has proven to be difficult to delineate [4, 18]. Most studies, which found cyanobacteria partially transforming PAHs, are from the 80s and lack direct evidence for PAH assimilation processes. Nevertheless, they showed for example, that *Agmenellum quadruplicatum*, a blue-green algae, can carry out oxidization of naphthalene [19]. The same authors reported that this cyanobacterial strain can activate phenanthrene as well, and form *trans*-9,10-dihydroxy-9,10-dihydrophenanthrene and 1-methoxyphenanthrene [20]. The authors suggested that this activation took place likely *via* monooxygenase enzyme activity, however, no further enzymatic steps were studied. It is hence not possible to conclude about the significance of this transformation regarding the actual degradation (removal) of phenanthrene. Another cyanobacterium, *Microcystis aeruginosa*, was recently shown to grow with shorter generation times when incubated in the presence of a mixture of naphthalene, phenanthrene and pyrene at low concentrations compared to medium without PAHs [21]. Unfortunately, the methodological approach (monitoring only of cell density and chlorophyll-a content) did not allow clarification over whether this can be attributed to actual utilization of the PAHs for growth or some other effect. In some cases, biodegradation was clearly attributed to activities of associated heterotrophic bacteria, rather than the cyanobacteria themselves [22]. Studied cyanobacteria appeared to play an indirect role in biodegradation by supporting the growth and activity of the actual bacterial degraders. In fact, the use of microalgae to produce oxygen necessary for degrading bacteria has already been assessed as a viable strategy for PAH removal in photobioreactors [23]. It is worth noting that while the role of live and active cyanobacteria may remain unclear, it appears that dead cyanobacterial organic matter has a significant impact on PAH fate in sediments [24]. Yan and coworkers found that cyanobacterial biomass added to sediment improved PAHs bioavailability and microbial activity, and enhanced overall degradation of pyrene and benzo[a]pyrene (B[a]P) in sediments.

Similarly to phototrophic microorganisms, very little is known about the role of Archaea in PAH biodegradation. Many speculated that Archaea disappear, once oil pollution occurs, hence their absence was thought to be a good indicator of marine oil pollution. Later on, however, different field results emerged and this hypothesis was discarded [25, 26]. So far, the only Archaea known to be able to degrade aromatic compounds completely is the thermophilic Fe(III)-respiring *Ferroglobus placidus* [27].

Biodegradative potential of terrestrial fungi, such as the well-known white rot fungi, has been studied for a long time, while marine-derived counterparts capable of degrading PAHs were uncharacterized until recently [28 - 30]. A recent review summarized the current knowledge on both environments, focusing mainly on sediments in the marine one [31]. Marine sediments were the source for isolating over 80 strains by only a single study assessing PAH biodegradation [32]. Pyrene degrading *A. sclerotiorum CBMAI849* and benzo[a]pyrene degrading *Mucorracemosus CBMAI847* has also been isolated from marine sediments [28], together with benzo[a]pyrene degrading *Aspergillus* sp. BAP14 [33]. Other groups of marine fungi include pyrene degrading Zygomycetes (*Mucorracemosus, M. racemosus* var. *sphaerosporus*), Deuteromycetes (*Gliocladium virens, Penicillium simplicissimum, P. janthinellum, Phialophoraalba, P. hoffmannii, Trichoderma harzianum*), Dematiaceae (*Scopulariopsis brumptii*) and Sphaeropsidale (*Coniothyrium fuckelii*), with Zygomycetes being one of the most efficient pyrene degrading groups [34]. The significance of fungal PAH biodegradation roots within their widely distributed and applied extracellular lignolytic enzymes, which exhibit very low substrate specificity and therefore can attack a wide range of aromatic structures. Fungi, like other eukaryotic organisms can also transform PAHs intracellularly (*via* reactions carried out by P450 type monooxygenase enzymes). However, these reactions often lead to the production of toxic metabolites, *trans*-diols, which are further transformed into DNA-adduct forming structures [35].

Bacterial PAH Degraders

PAH degraders are widely distributed among many bacterial phyla. This is to say that PAH degraders are not restricted to one particular branch of the bacterial tree of life. As illustrated in Fig. (**1**), they are present in many of the major phyla, including Proteobacteria, Actinobacteria, Bacteroidetes, and Firmicutes [36, 37]. The number of isolated and characterized PAH degraders has been growing continuously since the 80s, with novel strain isolated each year. Mallick and coworkers reported nearly 60 strains with documented LMW PAH degradation pathways and 20 further strains with unknown metabolic pathways by 2011 [38]. With regard to HMW PAHs, 50 bacterial isolates were reported by 2000 [39]. In a review from 2010, sphingomonads and actinomycetes, with special emphasis on mycobacteria, were highlighted as the most versatile groups involved in HMW PAH biodegradation [40]. Besides these, α-Proteobacteria, also members of β- and γ-Proteobacteria were found to be able to grow on various HMW PAHs, with *Pseudomonas* and *Stenotrophomonas* strains being particularly often isolated. Most of the above mentioned marine strains have been isolated from polluted sediments, as these habitats are considered the ultimate pool for PAHs, especially for HMW PAHs. *Novosphingobium pentaromativorans US6-1* is the most recently isolated example from a shallow muddy sediment (Ulsan Bay, Republic of Korea), while the strain *Celeribacter indicus* P73[T], the first fluoranthene-degrading bacterium within the family *Rhodobacteraceae*, is an example of deep-sea sediment bacteria from the Indian Ocean [41, 42].

Possibly the most often reported obligate PAH degrading genus is the *Cycloclasticus* [43 - 46]. *Cycloclasticus* sp., similarly to *Polaromonas* species, are cosmopolitan and considered among the most important obligate PAH degraders in the marine environment, isolated from coastal and also deep-sea sediments all around the world [47].

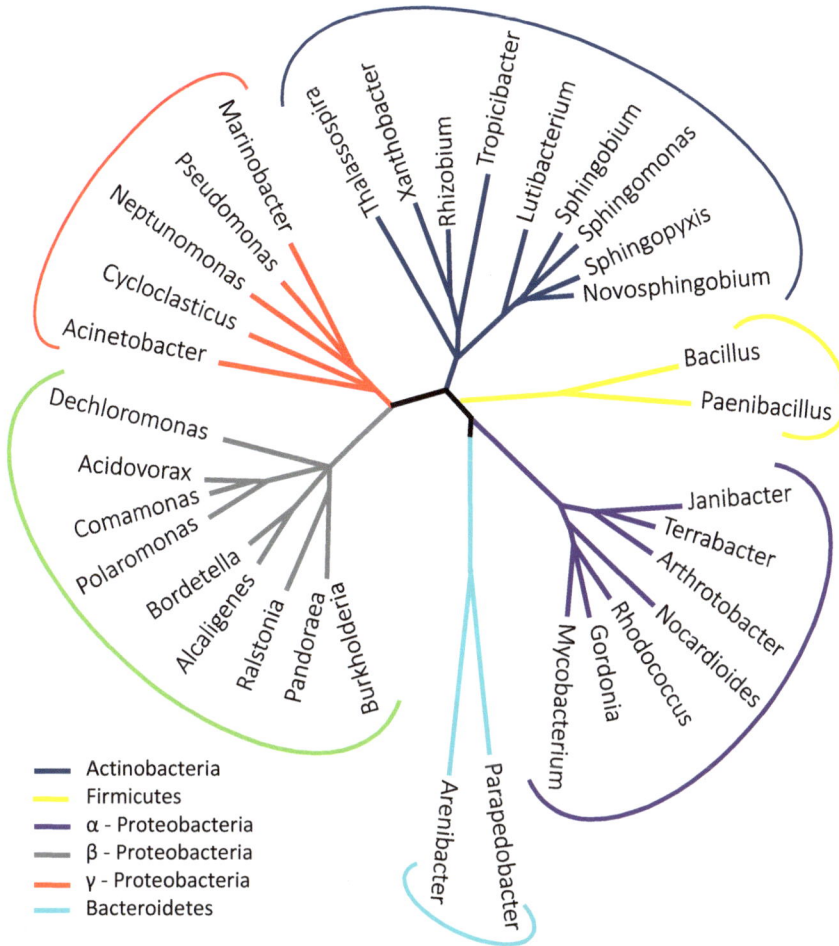

Fig. (1). Phylogenetic tree of genera containing reported PAH-degrader(s). Representative 16S rRNA genes from GenBank were imported into MEGA6. From ClustalW aligned sequences a neighbor-joining tree was built.

Aerob Heterotroph Bacterial Metabolism

Uptake into the Cell

The majority of heterotrophic bacteria transform PAHs in the cytosol with the help of intracellular enzymes, in contrast to fungal counterparts, which mostly employ extracellular enzymes. Hence, PAHs must enter the bacterial cell before biodegradation can take place. This is a challenging process because most PAHs are considered not easily accessible for bacteria. PAHs larger than naphthalene have very low water solubility (below 10 mg/L), tend to adsorb to hydrophobic surfaces (*e.g.* particles) and partition into organic liquid phases (*e.g.* oil in water). The hydrophobic nature of PAHs is measured in their octanol-water partitioning coefficients (k_{ow}). These coefficients are a measurable proxy to evaluate a compound's bioavailability. As microorganisms are believed to take up only water-dissolved compounds, hydrophobic substrate like PAHs with high k_{ow}, are considered as low bioavailability compounds. Bacteria have developed many strategies in order to be able to access these difficult substrates. These include changing the hydrophobicity of their outer cell membrane, attachment to solid or liquid PAHs and biosurfactant production.

Rhodococcus strains are well-known for their ability to increase mycolic acid content of their cell membrane and thereby increasing its hydrophobicity [48]. By doing so these and other Gram-negative bacteria can reduce the distance from solid or liquid phase PAHs. Moreover, in some cases cells can facilitate direct attachment (or adhesion) to PAH crystals and non-aqueous liquid phases. Attachment is then usually followed by biofilm formation on the substrate surface. Direct attachment has been demonstrated using advanced microscopy techniques. For example, a *Mycobacterium* strain was shown to adsorb to solid pyrene prior to biodegradation, while a *Brevibacillus* strain appeared to adhere to solid anthracene according to electron microscopy images [49, 50]. Despite the use of advanced microscopy techniques, it remains unknown exactly how the individual PAH molecules cross the cell membrane [51]. A second strategy, that some bacteria use to facilitate access to PAHs, is the excretion of biosurfactants [52]. Biosurfactants are amphiphilic compounds (*i.e.* compounds with both hydrophilic and hydrophobic structures) belonging to two major chemically distinguished classes, the glycolipids and the lipopeptides. Both groups work according to the same mechanism, usually referred to as pseudo-solubilization [53]. According to one theory, the micelles formed by biosurfactant molecules, containing one PAH in the center, are taken up by the cell *via* engulfment. However, lack of direct evidence renders this mechanism still hypothetical.

Once the accessibility of a PAH is ensured, transport through the cell membrane can take place *via* either passive diffusion or active transport. Passive diffusion of aromatics through cell membrane was demonstrated early on for mono-aromatic compounds benzene and toluene). However, later on many PAHs including naphthalene, phenantrene and even benzo[*a*]pyrene, have been shown to diffuse through the cell envelop without energy requiring processes [54]. It is still believed that active transport through transmembrane proteins by energy consuming reactions is an important mechanism as well [55]. Several experiments using ATP inhibitors, which block the active transport systems at the electron-flow level, confirm that bacteria uses high affinity, active uptake systems in order to accumulate PAHs intracellularly [56]. Considering that PAHs can passively diffuse into the cytosol, there is a risk of high enough concentrations accumulating becoming toxic or even lethal to the cell. In order to control cytosolic PAH concentrations, strictly regulated efflux systems evolved in many bacterial strains. The presence of such active transport sites, which expel excess substrates, has been used as evidence for passive diffusion as uptake mechanism. *Pseudomonas fluorescens* LP6a strain, for example, was shown to actively transport PAHs out from its cytosol, and its efflux proteins were found to be chromosomally encoded. At the same time, the bacterium's biodegradation genes were encoded on a plasmid, implying the major strategy this *Pseudomonas* strain employed to get rid of toxic PAHs was active efflux [57]. Biodegradation genes were likely acquired later *via* horizontal genes transfer.

Intracellular Transformation

The understanding of PAH biodegradation has evolved significantly during the last decade owing to rapid improvements in molecular biological methods, including high-throughput sequencing technologies and advanced proteomics [58 - 60]. A few decades ago, biodegradation was still looked at as a process occurring inside a microbial black box (Fig. **2**). It was known that during biodegradation, substrate is being depleted, cell growth usually occurs and in case of aerobic processes, oxygen is consumed while carbon-dioxide is produced. Mechanisms inside the black box, however, remained unknown.

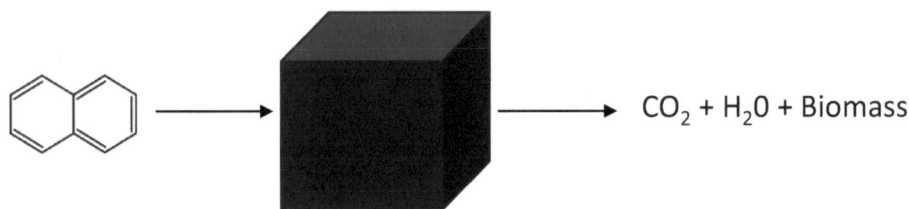

Fig. (2). Illustration of the "black box" approach for naphthalene biodegradation.

Since the 1970s, with improved analytical tools (*e.g.* high performance liquid chromatography and gas chromatography techniques), identification of intermediate metabolic products became possible and the first putative pathways were outlined based on observed intermediates [61 - 63]. After the 1990s, with the development of advanced molecular biological tools researchers began to be able to isolate and characterize the actual enzymes involved in PAH transformation. In parallel, the genetic background (*i.e.* genes coding for degradative enzymes) was discovered and the number of putative degradation pathways increased dramatically for the typically studied PAHs (*i.e.* naphthalene, acenaphthene, fluorene, anthracene, phenantrene, fluoranthene, pyrene, benzo[*a*]pyrene, and benzo[*k*]pyrene). This meant a better understanding of the cellular processes within individual bacterial cells, a crucial and fundamental step in gaining insight into the black box.

Hydrocarbons are energy rich substrates. However, compared to other common organic substrates, PAHs are highly unreactive and apolar, which makes their biological conversion challenging for bacteria [64]. It has been discovered a long time ago that PAHs require an initial, energy consuming activation step to first convert them into more polar metabolites [65]. Afterwards, a number of PAH-specific enzymatic conversions are carried out, before their products can be channeled into common metabolic pathways, such as the tri-carboxylic acid (TCA) cycle. The key feature distinguishing PAH-degraders from non-degraders is the possession of the enzymes involved in activating PAHs and converting them into common TCA cycle-metabolites. Until recently, it was believed that these reactions occur in a linear way [66 - 68]. Although it remains true that the end product of specific PAH degradative pathway reactions is a common intermediary metabolite, the linear concept has been replaced with a concept of a network of reactions occurring in a funneling manner, narrowing down towards the TCA-cycle metabolites.

There are several types of enzymes involved in PAH catabolism. Ring-hydroxylating dioxygenases (RHOs) play a key role as they initiate the entire process. This family of enzymes has been the center of many investigations and genes encoding them have also been targets of studies assessing biodegradative capacity of microbial communities [69]. RHOs are multicomponent enzymes containing a number of α and β subunits (general structure: $\alpha_n\beta_n$), where the α subunit carries the catalytic center [70]. The range of substrates accepted by the enzyme is therefore defined by the α subunit. Besides RHOs, several types of ring-cleavage dioxygenases, dehydrogenases, isomerases, hydroxylases, hydratase-aldolases, aldehyde dehydrogenases, and (oxidative) decarboxylases are also essential to complete the degradation. Fig. (**3**) summarizes the processes involved in the network of reactions that lead to assimilation of carbon from PAHs *via* the TCA cycle. The numbers in brackets refer to the arrows in the Figure, where the arrows represent multistep enzymatic reactions. The scheme described here is general and naturally there are several other possible transformations. Parent PAHs are first converted by dioxygenase enzymes, RHOs, into *cis*-dihydrodiols (structure not shown) which are immediately reduced by dihydrodiol dehydrogenases into PAH-diols (1a). Another possible route PAH parent molecules can take is oxidation *via* Cytochrome P450 type monooxygenases (CYPs). CYPs oxidize the ring resulting in the formation of an epoxide. If a non-enzymatic rearrangement occurs here, then monohydroxy-PAHs are produced (1b). However, if an epoxide hydrolase acts on the epoxide, *trans*-dihydrodiols result. There is a crucial difference between *cis*- and *trans*-dihydrodiols as the latter can be easily converted into reactive metabolites responsible for DNA and protein adduct formation and other kinds of mutagenic DNA damage. Monohydroxy-PAHs on the other hand are usually transformed by another monooxygenase in order to form PAHs-diols (2). These PAH-diols are then converted into either salicylates of phthalates, *via* ring-cleavage reactions. Arrows 3a and 3b represent three consecutive enzyme reactions carried out by ring-cleavage dioxygenases, hydratase-aldolases, and aldehyde dehydrogenases. Salicylates are formed through meta-cleavage with the help of extradiol dioxygenases, while phthalates through ortho-cleavage with the help of intradiol dioxygenases.

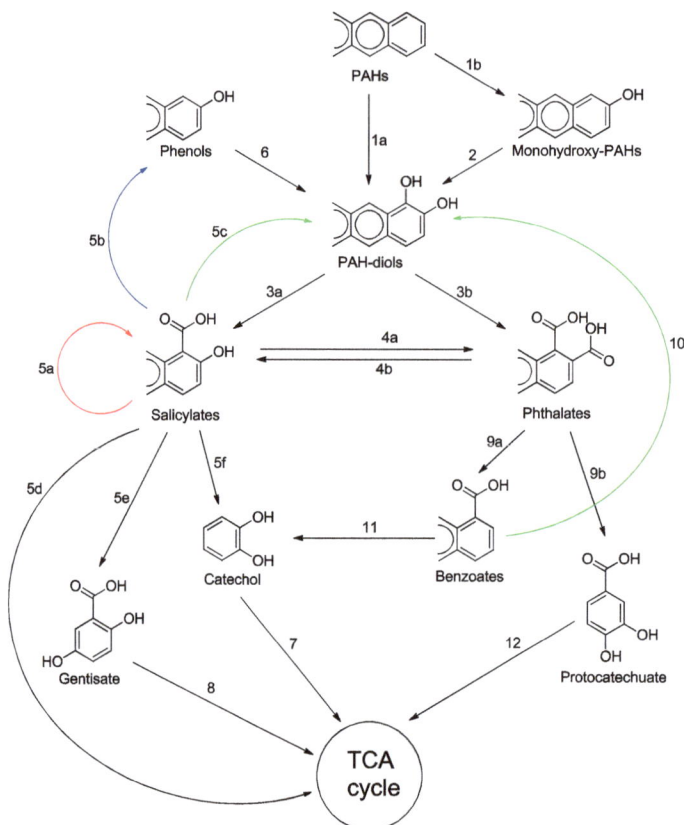

Fig. (3). Schematic overview of the diverse metabolic conversions involved in PAH assimilation by bacteria.

In case of a parent PAH with at least three benzene rings, salicylates can be transformed *via* three possible pathways by different enzymes: (I) ring-cleavage dioxygenases followed by hydratase-aldolase and aldehyde dehydrogenase (5a) until only a salicylate is left, (II) decarboxylases into phenols (5b), (III) oxidative decarboxylases into PAH-diols (5c). Phenols are transformed into PAH-diols *via* a two-step reaction catalysed by aromatic hydroxylases and RHOs (6). A series of these decarboxylations, ring-hydroxylations or ring-cleavage reactions eventually result in the formation of a salicylic acid, which is then converted into either gentisate (5e) or catechol (5f), two of the major intermediary metabolites often used to differentiate the type of PAH catabolic pathways of different bacteria.

Salicylate type ring-cleavage dioxygenases can also channel carbon from salicylate directly into the TCA cycle (5d). Alternatively, the intermediary metabolites, gentisate or catechol, are further cleaved into compounds that can enter the TCA cycle (7, 8).

The alternative route with phthalates as metabolic intermediates, leads to the formation of either benzoates (9a) or protocatechuate (9b). Phthalates can also converted into salicylate (4b) and *vice versa* (4a). Benzoates with several aromatic rings left in their structure are transformed into PAH-diols by RHOs (10) and follow previously outlined reaction until only benzoic acid is left. Benzoic acid needs to be converted into catechol (11) prior entering into the TCA cycle, whereas protocatechuate can directly be channelled there with the catalytic action of ring-cleavage dioxygenases (12).

A detailed description of the degradation pathways for LMW PAHs in bacteria has recently been presented, showing that the diverse metabolic pathways of these degradations are mainly guided by the activities of oxygenase enzymes [32]. This review suggests that each individual oxygenase class is diverse enough to recognize a range of substrates in both a regio- and stereospecific manner. Moreover, it shows that even the simplest PAH (*i.e.* naphthalene) can be degraded in completely different ways in heterotrophy bacteria (*Rhodococcus* sp. strain B4, *Raltsonia* sp. U2, *Geobacillus thermoleovorans*, *Geobacillus* sp. G27, and *Nocardia otitidiscavarium* strain TSH1).

Genetic Background Encoding PAH Degradation Enzymes

The DNA encoded genetic background determines which enzymes a particular bacterium can synthesize. This in turn defines the range of potential substrates and degradation pathways available for that bacterium. Bacteria carry genetic information on circular chromosomes and very often on smaller circular plasmids. These plasmids can be lost during duplication, however, their advantage their transferability not only to "offsprings" but also between different species of bacteria. The variety of enzymes required for complete mineralization of a single PAH are encoded on several genes. Regulatory genes, genes encoding transport proteins are also usually included in the array of genes for biodegradation. All these genes are typically arranged into gene clusters, so called operons, whose expression occurs simultaneously. The first maps of genes involved in degradative pathways suggested that they need to be arranged in a rigid order, preferably within one common open reading frame and regulated by a common promoter. Later studies revealed that this rigid structure is not always the case and the arrangement of genes can occur very flexibly, and in fact superior PAH degraders prefer the flexible structure [71, 72]. The general overall organization of biodegradative genes seems to be well-conserved and biodegradation pathways tend to have genus-specific variations [73]. This means that two PAH degraders can have the same capacity for utilizing aromatics yet possessing different genetic background for it. Genus or species specific regulation of dioxygenase encoding genes can on the other hand determine degradation activity [74]. Some degraders were shown to carry multiple copies of oxygenase genes, and genes encoding a variety of different oxygenases [44, 75 - 77]. Moreover, some *Pseudomonas* and *Rhodococcus* strains can simultaneously carry genes for PAH and alkane degradation [78, 79]. This shows the significance of type of degraders involved in biodegradation. The diversity of oxygenase enzymes involved in degradation can provide important information regarding the kinetics of the process.

Kinetics of Biodegradation

Environmental factors, such as availability of nutrients (mainly nitrogen and phosphor), oxygen concentration, number of degrading cells, and the biologically available concentration of the substrate, are well-known to directly influence biodegradation kinetics [80, 81]. As discussed above briefly, the type of degraders involved in the process has also a significant role in determining kinetics. Although identifying key players has been central to many studies around PAH degradation, the metabolic capacity of those players is rarely determined making it difficult to include such parameter in kinetic models [82].

It is far easier to take into account the number of cells participating in biodegradation. The more biomass is involved, the faster biodegradation can take place. If transforming a particular pollutant also stimulates cell growth, the increasing cell numbers will result in increasing transformation rates. Limits of growth are then set by the other environmental factors. To begin with, the availability of oxygen determines whether aerobic or anaerobic biodegradation can take place. Anaerobic processes tend to occur significantly slower than aerobic ones, hence such conditions are usually suboptimal for biodegradation [83]. Once aerobic conditions prevail, oxygen will act as a limiting resource and its concentration can be included in growth models easily. The situation is similar for nutrients. As cell-growth requires nitrogen and phosphor for synthesis of proteins and DNA, therefore these macronutrients also often become limiting resources [84]. Oceans are generally oligotrophic (*i.e.*, nutrients are scarce), therefore it is extremely important to take into account nutrient concentrations in case of marine biodegradation.

The effect of the above-described factors on single strain cultures utilizing a single PAH, is relatively straightforward to model [3, 85]. The situation becomes more complex once whole communities are degrading mixtures of hydrocarbons [86]. Additionally, the biologically available concentration of substrate (PAHs) is particularly difficult to include in kinetic models. Initial substrate uptake steps can often become the bottleneck of the entire biodegradation process, determining the kinetics. In order to understand how, it is important to go back at the concept of bioavailability. As mentioned before, bioavailability does not have a measurable proxy. A good definition of bioavailability originates from Bosma's work, which was refined by Thullner and co-workers into the following: "bioavailability is the ratio of the actual biodegradation rate of an extant microbial community to its degradation capacity" [87, 88]. By degradation capacity the authors mean the maximum achievable conversion rate of the intracellular enzymes under substrate saturation conditions. In other words, reduced biodegradability resulting from low bioavailability means, that the substrate itself and its interaction with its physicochemical environment are the key factors limiting the degradation process.

Temperature will also influence biodegradation rate through its effect on enzyme activity and through its influence on physicochemical properties of contaminants. Surface seawater temperatures (above the thermocline) range from below 0 °C in the Arctic during winter, up to around 30 °C in tropical areas during the warm season. Nevertheless, when looking at the oceans from bottom to top, the majority of marine habitats are in fact constantly cold, with a temperature lower than 5 °C [89]. Growth rate and metabolic rates vary according to temperature and biodegradation kinetics limiting effect of low temperature was demonstrated already in the 1960s [90 - 92]. Due to thermodynamic laws, an enzyme's activity (rate of conversion) is dependent on environmental temperature and all enzymes have an optimum temperature where maximum conversion rates are achieved [93]. Depending on the origin of the enzyme, whether it evolved under cold conditions or under extremely high temperatures, it will have a correspondingly lower or higher optimum temperature. So-called cold-adapted enzymes operate just as efficiently at low temperatures as do warm-adapted enzymes at higher temperatures [94 - 96].

The question, whether biodegradation occurs with the same rate in colder temperatures as it does in warmer ones, has been challenging to answer. Biodegradation is carried out *via* the activity of populations of degrading bacteria, where each individual uses several enzymes in complex processes. Although, temperature dependence of enzyme kinetics can be extrapolated to metabolic rates of whole communities, such extrapolation is not sufficient to answer the question [97, 98]. Investigators interested in this issue generally designed experiments where a warm-adapted seawater community was exposed to the same amount of contaminants at various low temperatures [99, 100]. Such experiments showed that the lower the incubation temperatures were, the slower biodegradation took place. It did not take many years before a rule of thumb became accepted, *i.e.*, biodegradation rates halve at a 10 °C temperature decrease [101]. However, very few studies compared directly the biodegradation rates of a cold-adapted and a warm-adapted community at a defined temperature (or at their *in situ* ambient temperatures). A recent study showed that arctic seawater degraded naphthalene just as quickly as a temperate seawater at their corresponding *in situ* ambient temperatures [102]. Only this type of experiments could provide real answer to the question, whether one should expect lower or higher biodegradation rates for example in arctic marine environments compared to more temperate ones. According to the current view on hits matter, it is mostly the limited bioavailability of hydrocarbons rather than the slower metabolism of bacteria which causes lower biodegradation rates in permanently cold marine environments [103].

Importance of Interaction

No microbe is an island and no one microorganism can alone metabolize the diversity of PAHs present in a mixture such as crude oil. In order to develop biotechnological applications for tackling oil spills in the future, it is necessary to move beyond the traditional considerations around biodegradation kinetics. Recent findings clearly show that interspecies interactions, among degraders, and also among degraders and non-degraders in the same community, are essential for efficient and fast biodegradation [4]. Fig. (**4**) provides a simplified overview of possible interactions in the microbial loop an oil spill scenario. Until 2009, it was believed that benzo[*a*]pyrene could not be used by any bacteria as sole carbon and energy source. Then Luo and colleagues demonstrated that a strain isolated from seawater collected from the Botan oil port (Xiamen, China) was able to do so to some extent. The most interesting finding in their study was, that the entire enrichment culture community was twice as efficient in removing benzo[*a*]pyrene compared to a single strain or the mixture of three isolated PAH degraders [104]. This confirms the idea that a complex network of microorganism is necessary for efficient biodegradation of HMW PAHs. Moreover, collaboration between bacteria and fungi in a co-culture was shown to enhance benzo[*a*]pyrene biodegradation compared to using either the bacterial or the fungal cultures alone [105]. Fungi are well-known for excreting extracellular lignolytic enzymes, which can carry out the initial attack of the aromatic structure of PAHs. Fungal enzymes therefore facilitate bioaccessibility of PAH as these more polar, activated PAHs can be taken up by bacteria for further conversion.

As mentioned earlier, cyanobacteria and microalgae can form mutually beneficial interactions with PAH degraders as well [22]. The phototrophic microorganisms can most significantly enhance heterotrophic activity by the production of oxygen or biosurfactants. Their biosorptive capabilities may play an important role as well, once biosorbed PAHs are accessible for algae associated bacteria [23, 24]. It is important to remind the reader that besides beneficial interactions, competition for nutrients and other antagonistic behavior is also characteristic of microbial communities, and these often affect the outcome of biodegradation negatively.

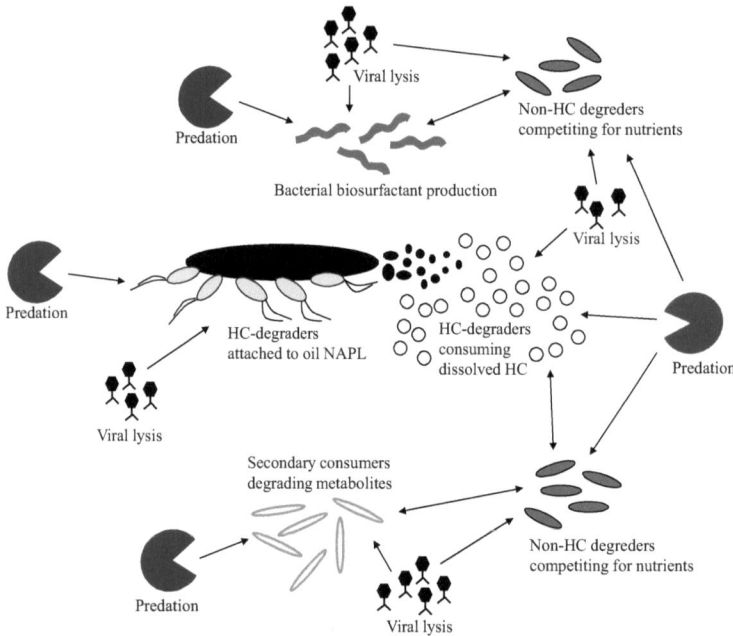

Fig. (4). Interaction between members of hypothetical microbial community during oil degradation.

CONCLUDING REMARKS

The role of environmental biotechnology, with microbial metabolism at its heart, is seen as vital for future sustainable exploitation of marine resources. The marine environment harbors a vast diversity of microorganisms that are well adapted to utilizing PAHs. Bacteria are considered particularly important players in the microbial degradation of PAHs, nevertheless marine fungi are increasingly recognized as major contributors as well. PAH detoxification *via* channeling carbon back into the marine microbial loop and further in the food web is a complex process that requires multiple enzyme reactions within each participating microbe and collaboration between the diverse microbial members. Strategies of individual cells for increasing PAH bioavailability, and collaborative, network-like consumption of intermediary metabolites are both fundamentally important aspects, which are now beginning to be understood with the help of advanced molecular techniques. Today the process of going from the black box approach towards actual environmental biotechnology for taking advantage of bacteria's ability to mitigate pollution is well underway. Findings, such as the detailed cellular responses of *Mycobacterium vanbaalenii* PYR-1 to crude oil, will facilitate development of useful bioremediation tools for the future.

CONFLICT OF INTEREST

The author confirms that the author has no conflict of interest to declare for this publication.

ACKNOWLEDGEMENTS

The author would like to thank the University of Stavanger for providing funding for this work.

REFERENCES

[1]   Pampanin, D.M.; Sydnes, M.O. Polycyclic aromatic hydrocarbons a constituent of petroleum: Presence and influence in the aquatic environment. In: *Hydrocarbons*; Vladimir, K.; Kolesnikov, A., Eds.; InTech: Rijeka, **2013**.

[2]   Atlas, R.M.; Bartha, R. Fate and effects of polluting petroleum in the marine environment. *Residue Rev.,* **1973**, *49*(0), 49-85.
[PMID: 4604473]

[3]   Alexander, M. *Biodegradation and Bioremediation*; Academic Press: London, **1994**.

[4]   McGenity, T.J.; Folwell, B.D.; McKew, B.A.; Sanni, G.O. Marine crude-oil biodegradation: a central role for interspecies interactions. *Aquat. Biosyst.,* **2012**, *8*(1), 10.
[http://dx.doi.org/10.1186/2046-9063-8-10] [PMID: 22591596]

[5]   Head, I.M.; Jones, D.M.; Röling, W.F. Marine microorganisms make a meal of oil. *Nat. Rev. Microbiol.,* **2006**, *4*(3), 173-182.
[http://dx.doi.org/10.1038/nrmicro1348] [PMID: 16489346]

[6]   Azam, F. Microbial control of oceanic carbon flux: The plot thickens. *Science,* **1998**, *280*, 694-696.
[http://dx.doi.org/10.1126/science.280.5364.694]

[7]   Azam, F.; Fenchel, T.; Field, J.G.; Gray, J.S.; Meyer, L.A.; Thingstad, F. The ecological role of water column microbes in the sea. *Mar. Ecol. Prog. Ser.,* **1983**, *10*, 257-263.
[http://dx.doi.org/10.3354/meps010257]

[8]   Lundstedt, S.; White, P.A.; Lemieux, C.L.; Lynes, K.D.; Lambert, I.B.; Oberg, L.; Haglund, P.; Tysklind, M. Sources, fate, and toxic hazards of oxygenated polycyclic aromatic hydrocarbons (PAHs) at PAH-contaminated sites. *Ambio,* **2007**, *36*(6), 475-485.
[http://dx.doi.org/10.1579/0044-7447(2007)36[475:SFATHO]2.0.CO;2] [PMID: 17985702]

[9]   Dasgupta, S.; Cao, A.; Mauer, B.; Yan, B.; Uno, S.; McElroy, A. Genotoxicity of oxy-PAHs to Japanese medaka (*Oryzias latipes*) embryos assessed using the comet assay. *Environ. Sci. Pollut. Res. Int.,* **2014**, *21*(24), 13867-13876.
[http://dx.doi.org/10.1007/s11356-014-2586-4] [PMID: 24510601]

[10] *Committee on Oil in the Sea: Inputs, Fates, and Effects. Oil in the sea III. Inputs, Fates, and Effects*; The National Academies Press: Washington, **2003**.

[11] Hornafius, J.S.; Quigley, D.; Luyendyk, B.P. The world's most spectacular marine hydrocarbon seeps (Coal Oil Point, Santa Barbara channel, California): Quantification of emissions. *J. Geophys. Res. C: Oceans Atmos.,* **1999**, *104*, 20703-20711.

[12] Kvenvolden, K.A.; Cooper, C.K. Natural seepage of crude oil into the marine environment. *Geo-Mar. Lett.,* **2003**, *23*, 140-146.
[http://dx.doi.org/10.1007/s00367-003-0135-0]

[13] Zobell, C.E. Action of Microörganisms on Hydrocarbons. *Bact. Rev.,* **1946**, *10*, 1-49.

[14] Zobell, C.E.; Carroll, W.G.; Haas, H.F. Marine microorganisms which oxidize petroleum hydrocarbons. *Bull. Am. Assoc. Pet. Geol.,* **1943**, *27*, 1175-1193.

[15] Cerniglia, C.E. Biodegradation of polycyclic aromatic hydrocarbons. *Biodegradation,* **1992**, *3*, 351-369.
[http://dx.doi.org/10.1007/BF00129093]

[16] Peng, R-H.; Xiong, A-S.; Xue, Y.; Fu, X-Y.; Gao, F.; Zhao, W.; Tian, Y-S.; Yao, Q-H. Microbial biodegradation of polyaromatic hydrocarbons. *FEMS Microbiol. Rev.,* **2008**, *32*(6), 927-955.
[http://dx.doi.org/10.1111/j.1574-6976.2008.00127.x] [PMID: 18662317]

[17] Haritash, A.K.; Kaushik, C.P. Biodegradation aspects of polycyclic aromatic hydrocarbons (PAHs): a review. *J. Hazard. Mater.,* **2009**, *169*(1-3), 1-15.
[http://dx.doi.org/10.1016/j.jhazmat.2009.03.137] [PMID: 19442441]

[18] Subashchandrabose, S.R.; Ramakrishnan, B.; Megharaj, M.; Venkateswarlu, K.; Naidu, R. Mixotrophic cyanobacteria and microalgae as distinctive biological agents for organic pollutant degradation. *Environ. Int.,* **2013**, *51*, 59-72.
[http://dx.doi.org/10.1016/j.envint.2012.10.007] [PMID: 23201778]

[19] Cerniglia, C.E.; Gibson, D.T.; Van Baalen, C. Algal oxidation of aromatic hydrocarbons: formation of 1-naphthol from naphthalene by *Agmenellum quadruplicatum*, strain PR-6. *Biochem. Biophys. Res. Commun.,* **1979**, *88*(1), 50-58.
[http://dx.doi.org/10.1016/0006-291X(79)91695-4] [PMID: 110329]

[20] Narro, M.L.; Cerniglia, C.E.; Van Baalen, C.; Gibson, D.T. Metabolism of phenanthrene by the marine cyanobacterium *Agmenellum quadruplicatum* PR-6. *Appl. Environ. Microbiol.,* **1992**, *58*(4), 1351-1359.
[PMID: 1599252]

[21] Zhu, X.; Kong, H.; Gao, Y.; Wu, M.; Kong, F. Low concentrations of polycyclic aromatic hydrocarbons promote the growth of *Microcystis aeruginosa. J. Hazard. Mater.,* **2012**, *237-238*, 371-375.
[http://dx.doi.org/10.1016/j.jhazmat.2012.08.029] [PMID: 22954602]

[22] Abed, R.M.; Köster, J. The direct role of aerobic heterotrophic bacteria associated with cyanobacteria in the degradation of oil compounds. *Int. Biodeterior. Biodegradation,* **2005**, *55*, 29-37.
[http://dx.doi.org/10.1016/j.ibiod.2004.07.001]

[23]   Muñoz, R.; Guieysse, B. Algal-bacterial processes for the treatment of hazardous contaminants: a review. *Water Res.,* **2006**, *40*(15), 2799-2815.
[http://dx.doi.org/10.1016/j.watres.2006.06.011] [PMID: 16889814]

[24]   Yan, Z.; Jiang, H.; Li, X.; Shi, Y. Accelerated removal of pyrene and benzo[*a*]pyrene in freshwater sediments with amendment of cyanobacteria-derived organic matter. *J. Hazard. Mater.,* **2014**, *272*, 66-74.
[http://dx.doi.org/10.1016/j.jhazmat.2014.02.042] [PMID: 24681443]

[25]   Röling, W.F.; de Brito Couto, I.R.; Swannell, R.P.; Head, I.M. Response of Archaeal communities in beach sediments to spilled oil and bioremediation. *Appl. Environ. Microbiol.,* **2004**, *70*(5), 2614-2620.
[http://dx.doi.org/10.1128/AEM.70.5.2614-2620.2004] [PMID: 15128510]

[26]   Redmond, M.C.; Valentine, D.L. Natural gas and temperature structured a microbial community response to the deepwater horizon oil spill. *Proc. Natl. Acad. Sci. USA,* **2012**, *109*(50), 20292-20297.
[http://dx.doi.org/10.1073/pnas.1108756108] [PMID: 21969552]

[27]   Schmid, G.; René, S.B.; Boll, M. Enzymes of the benzoyl-coenzyme A degradation pathway in the hyperthermophilic archaeon *Ferroglobus placidus. Environ. Microbiol.,* **2015**, *17*(9), 3289-3300.
[http://dx.doi.org/10.1111/1462-2920.12785] [PMID: 25630364]

[28]   Passarini, M.R.; Rodrigues, M.V.; da Silva, M.; Sette, L.D. Marine-derived filamentous fungi and their potential application for polycyclic aromatic hydrocarbon bioremediation. *Mar. Pollut. Bull.,* **2011**, *62*(2), 364-370.
[http://dx.doi.org/10.1016/j.marpolbul.2010.10.003] [PMID: 21040933]

[29]   Harms, H.; Schlosser, D.; Wick, L.Y. Untapped potential: exploiting fungi in bioremediation of hazardous chemicals. *Nat. Rev. Microbiol.,* **2011**, *9*(3), 177-192.
[http://dx.doi.org/10.1038/nrmicro2519] [PMID: 21297669]

[30]   Bonugli-Santos, R.C.; Dos Santos Vasconcelos, M.R.; Passarini, M.R.; Vieira, G.A.; Lopes, V.C.; Mainardi, P.H.; Dos Santos, J.A.; de Azevedo Duarte, L.; Otero, I.V.; da Silva Yoshida, A.M.; Feitosa, V.A.; Pessoa, A., Jr; Sette, L.D. Marine-derived fungi: diversity of enzymes and biotechnological applications. *Front. Microbiol.,* **2015**, *6*, 269.
[http://dx.doi.org/10.3389/fmicb.2015.00269] [PMID: 25914680]

[31]   Marco-Urrea, E.; García-Romera, I.; Aranda, E. Potential of non-ligninolytic fungi in bioremediation of chlorinated and polycyclic aromatic hydrocarbons. *N. Biotechnol.,* **2015**, *32*(6), 620-628.
[http://dx.doi.org/10.1016/j.nbt.2015.01.005] [PMID: 25681797]

[32]   Salvo, V.S.; Gallizia, I.; Moreno, M.; Fabiano, M. Fungal communities in PAH-impacted sediments of Genoa-Voltri Harbour (NW Mediterranean, Italy). *Mar. Pollut. Bull.,* **2005**, *50*(5), 553-559.
[http://dx.doi.org/10.1016/j.marpolbul.2005.01.001] [PMID: 15907497]

[33]   Wu, Y-R.; He, T-T.; Lun, J-S.; Maskaoui, K.; Huang, T-W.; Hu, Z. Removal of Benzo[a]pyrene by a fungus *Aspergillus* sp. BAP14. *World J. Microbiol. Biotechnol.,* **2009**, *25*, 1395-1401.
[http://dx.doi.org/10.1007/s11274-009-0026-2]

[34]   Ravelet, C.; Krivobok, S.; Sage, L.; Steiman, R. Biodegradation of pyrene by sediment fungi. *Chemosphere,* **2000**, *40*(5), 557-563.
[http://dx.doi.org/10.1016/S0045-6535(99)00320-3] [PMID: 10665394]

[35] Cerniglia, C.E. Fungal metabolism of polycyclic aromatic hydrocarbons: past, present and future applications in bioremediation. *J. Ind. Microbiol. Biotechnol.,* **1997**, *19*(5-6), 324-333.
[http://dx.doi.org/10.1038/sj.jim.2900459] [PMID: 9451829]

[36] Yakimov, M.M.; Timmis, K.N.; Golyshin, P.N. Obligate oil-degrading marine bacteria. *Curr. Opin. Biotechnol.,* **2007**, *18*(3), 257-266.
[http://dx.doi.org/10.1016/j.copbio.2007.04.006] [PMID: 17493798]

[37] Prince, R.C.; Gramain, A.; McGenity, T.J. Prokaryotic Hydrocarbon Degraders. In: *Handbook of Hydrocarbon and Lipid Microbiology*; Timmis, K.N., Ed.; Springer: Berlin, **2010**.
[http://dx.doi.org/10.1007/978-3-540-77587-4_118]

[38] Mallick, S.; Chakraborty, J.; Dutta, T.K. Role of oxygenases in guiding diverse metabolic pathways in the bacterial degradation of low-molecular-weight polycyclic aromatic hydrocarbons: a review. *Crit. Rev. Microbiol.,* **2011**, *37*(1), 64-90.
[http://dx.doi.org/10.3109/1040841X.2010.512268] [PMID: 20846026]

[39] Kanaly, R.A.; Harayama, S. Biodegradation of high-molecular-weight polycyclic aromatic hydrocarbons by bacteria. *J. Bacteriol.,* **2000**, *182*(8), 2059-2067.
[http://dx.doi.org/10.1128/JB.182.8.2059-2067.2000] [PMID: 10735846]

[40] Kanaly, R.A.; Harayama, S. Advances in the field of high-molecular-weight polycyclic aromatic hydrocarbon biodegradation by bacteria. *Microb. Biotechnol.,* **2010**, *3*(2), 136-164.
[http://dx.doi.org/10.1111/j.1751-7915.2009.00130.x] [PMID: 21255317]

[41] Lyu, Y.; Zheng, W.; Zheng, T.; Tian, Y. Biodegradation of polycyclic aromatic hydrocarbons by *Novosphingobium pentaromativorans* US61. *PLoS One,* **2014**, *9*(7), e101438.
[http://dx.doi.org/10.1371/journal.pone.0101438] [PMID: 25007154]

[42] Cao, J.; Lai, Q.; Yuan, J.; Shao, Z. Genomic and metabolic analysis of fluoranthene degradation pathway in *Celeribacter indicus* P73^T. *Sci. Rep.,* **2015**, *5*, 7741.
[http://dx.doi.org/10.1038/srep07741] [PMID: 25582347]

[43] Dyksterhouse, S.E.; Gray, J.P.; Herwig, R.P.; Lara, J.C.; Staley, J.T. *Cycloclasticus pugetii* gen. nov., sp. nov., an aromatic hydrocarbon-degrading bacterium from marine sediments. *Int. J. Syst. Bacteriol.,* **1995**, *45*(1), 116-123.
[http://dx.doi.org/10.1099/00207713-45-1-116] [PMID: 7857792]

[44] Geiselbrecht, A.D.; Hedlund, B.P.; Tichi, M.A.; Staley, J.T. Isolation of marine polycyclic aromatic hydrocarbon (PAH)-degrading *Cycloclasticus* strains from the Gulf of Mexico and comparison of their PAH degradation ability with that of puget sound *Cycloclasticus* strains. *Appl. Environ. Microbiol.,* **1998**, *64*(12), 4703-4710.
[PMID: 9835552]

[45] Kasai, Y.; Kishira, H.; Harayama, S. Bacteria belonging to the genus *cycloclasticus* play a primary role in the degradation of aromatic hydrocarbons released in a marine environment. *Appl. Environ. Microbiol.,* **2002**, *68*(11), 5625-5633.
[http://dx.doi.org/10.1128/AEM.68.11.5625-5633.2002] [PMID: 12406758]

[46] Cui, Z.; Lai, Q.; Dong, C.; Shao, Z. Biodiversity of polycyclic aromatic hydrocarbon-degrading bacteria from deep sea sediments of the Middle Atlantic Ridge. *Environ. Microbiol.,* **2008**, *10*(8), 2138-2149.
[http://dx.doi.org/10.1111/j.1462-2920.2008.01637.x] [PMID: 18445026]

[47] Darcy, J.L.; Lynch, R.C.; King, A.J.; Robeson, M.S.; Schmidt, S.K. Global distribution of Polaromonas phylotypesevidence for a highly successful dispersal capacity. *PLoS One,* **2011**, *6*(8), e23742.
[http://dx.doi.org/10.1371/journal.pone.0023742] [PMID: 21897856]

[48] de Carvalho, C.C.; da Fonseca, M.M. The remarkable *Rhodococcus erythropolis. Appl. Microbiol. Biotechnol.,* **2005**, *67*(6), 715-726.
[http://dx.doi.org/10.1007/s00253-005-1932-3] [PMID: 15711940]

[49] Wick, L.Y.; Ruiz de Munain, A.; Springael, D.; Harms, H.; de, M.A. Responses of *Mycobacterium* sp. LB501T to the low bioavailability of solid anthracene. *Appl. Microbiol. Biotechnol.,* **2002**, *58*(3), 378-385.
[http://dx.doi.org/10.1007/s00253-001-0898-z] [PMID: 11935191]

[50] Liao, L.; Chen, S.; Peng, H.; Yin, H.; Ye, J.; Liu, Z.; Dang, Z.; Liu, Z. Biosorption and biodegradation of pyrene by *Brevibacillus brevis* and cellular responses to pyrene treatment. *Ecotoxicol. Environ. Saf.,* **2015**, *115*, 166-173.
[http://dx.doi.org/10.1016/j.ecoenv.2015.02.015] [PMID: 25700095]

[51] Hua, F.; Wang, H.Q. Uptake and trans-membrane transport of petroleum hydrocarbons by microorganisms. *Biotechnol. Biotechnol. Equip.,* **2014**, *28*(2), 165-175.
[http://dx.doi.org/10.1080/13102818.2014.906136] [PMID: 26740752]

[52] Reis, R.S.; Pacheco, G.J.; Pereira, A.G.; Freire, D.M. Biosurfactants: Production and Applications. In: *Biodegradation - Life of Science*; Chamy, R.; Rosenkrantz, F., Eds.; InTech, **2013**.
[http://dx.doi.org/10.5772/56144]

[53] Iwabuchi, N.; Sunairi, M.; Urai, M.; Itoh, C.; Anzai, H.; Nakajima, M.; Harayama, S. Extracellular polysaccharides of *Rhodococcus rhodochrous* S-2 stimulate the degradation of aromatic components in crude oil by indigenous marine bacteria. *Appl. Environ. Microbiol.,* **2002**, *68*(5), 2337-2343.
[http://dx.doi.org/10.1128/AEM.68.5.2337-2343.2002] [PMID: 11976106]

[54] Sikkema, J.; de Bont, J.A.; Poolman, B. Mechanisms of membrane toxicity of hydrocarbons. *Microbiol. Rev.,* **1995**, *59*(2), 201-222.
[PMID: 7603409]

[55] Li, Y.; Wang, H.; Hua, F.; Su, M.; Zhao, Y. Trans-membrane transport of fluoranthene by *Rhodococcus* sp. BAP-1 and optimization of uptake process. *Bioresour. Technol.,* **2014**, *155*, 213-219.
[http://dx.doi.org/10.1016/j.biortech.2013.12.117] [PMID: 24457306]

[56] Miyata, N.; Iwahori, K.; Foght, J.M.; Gray, M.R. Saturable, energy-dependent uptake of phenanthrene in aqueous phase by *Mycobacterium* sp. strain RJGII-135. *Appl. Environ. Microbiol.,* **2004**, *70*(1), 363-369.
[http://dx.doi.org/10.1128/AEM.70.1.363-369.2004] [PMID: 14711664]

[57] Bugg, T.; Foght, J.M.; Pickard, M.A.; Gray, M.R. Uptake and active efflux of polycyclic aromatic hydrocarbons by *Pseudomonas fluorescens* LP6a. *Appl. Environ. Microbiol.,* **2000**, *66*(12), 5387-5392.
[http://dx.doi.org/10.1128/AEM.66.12.5387-5392.2000] [PMID: 11097918]

[58] Loviso, C.L.; Lozada, M.; Guibert, L.M.; Musumeci, M.A.; Sarango Cardenas, S.; Kuin, R.V.; Marcos, M.S.; Dionisi, H.M. Metagenomics reveals the high polycyclic aromatic hydrocarbon-degradation potential of abundant uncultured bacteria from chronically polluted subantarctic and temperate coastal marine environments. *J. Appl. Microbiol.,* **2015,** *119*(2), 411-424.
[http://dx.doi.org/10.1111/jam.12843] [PMID: 25968322]

[59] Kweon, O.; Kim, S-J.; Blom, J.; Kim, S-K.; Kim, B-S.; Baek, D.H.; Park, S.I.; Sutherland, J.B.; Cerniglia, C.E. Comparative functional pan-genome analyses to build connections between genomic dynamics and phenotypic evolution in polycyclic aromatic hydrocarbon metabolism in the genus *Mycobacterium. BMC Evol. Biol.,* **2015,** *15*, 21.
[http://dx.doi.org/10.1186/s12862-015-0302-8] [PMID: 25880171]

[60] Vandera, E.; Samiotaki, M.; Parapouli, M.; Panayotou, G.; Koukkou, A.I. Comparative proteomic analysis of *Arthrobacter phenanthrenivorans* Sphe3 on phenanthrene, phthalate and glucose. *J. Proteomics,* **2015,** *113*, 73-89.
[http://dx.doi.org/10.1016/j.jprot.2014.08.018] [PMID: 25257624]

[61] Gibson, D.T.; Mahadevan, V.; Jerina, D.M.; Yogi, H.; Yeh, H.J. Oxidation of the carcinogens benzo [*a*] pyrene and benzo [*a*] anthracene to dihydrodiols by a bacterium. *Science,* **1975,** *189*(4199), 295-297.
[http://dx.doi.org/10.1126/science.1145203] [PMID: 1145203]

[62] Heitkamp, M.A.; Freeman, J.P.; Cerniglia, C.E. Naphthalene biodegradation in environmental microcosms: estimates of degradation rates and characterization of metabolites. *Appl. Environ. Microbiol.,* **1987,** *53*(1), 129-136.
[PMID: 3827241]

[63] Mahaffey, W.R.; Gibson, D.T.; Cerniglia, C.E. Bacterial oxidation of chemical carcinogens: formation of polycyclic aromatic acids from benz[a]anthracene. *Appl. Environ. Microbiol.,* **1988,** *54*(10), 2415-2423.
[PMID: 2462407]

[64] Wilkes, H.; Schwarzbauer, J. Hydrocarbons: An Introduction to Structure, Physico-Chemical Properties and Natural Occurrence. In: *Handbook of hydrocarbon and lipid microbiology*; Timmis, K.N., Ed.; Springer-Verlag: Berlin, **2010.**
[http://dx.doi.org/10.1007/978-3-540-77587-4_1]

[65] Widdel, F.; Musat, F. Diversity and Common Principles in Enzymatic Activation of Hydrocarbons. In: *Handbook of hydrocarbon and lipid microbiology*; Timmis, K.N., Ed.; Springer-Verlag: Berlin, **2010.**
[http://dx.doi.org/10.1007/978-3-540-77587-4_70]

[66] Demanèche, S.; Meyer, C.; Micoud, J.; Louwagie, M.; Willison, J.C.; Jouanneau, Y. Identification and functional analysis of two aromatic-ring-hydroxylating dioxygenases from a *sphingomonas* strain that degrades various polycyclic aromatic hydrocarbons. *Appl. Environ. Microbiol.,* **2004,** *70*(11), 6714-6725.
[http://dx.doi.org/10.1128/AEM.70.11.6714-6725.2004] [PMID: 15528538]

[67] Pinyakong, O.; Habe, H.; Yoshida, T.; Nojiri, H.; Omori, T. Identification of three novel salicylate 1-hydroxylases involved in the phenanthrene degradation of *Sphingobium* sp. strain P2. *Biochem.*

Biophys. Res. Commun., **2003**, *301*(2), 350-357.
[http://dx.doi.org/10.1016/S0006-291X(02)03036-X] [PMID: 12565867]

[68]   Schuler, L.; Jouanneau, Y.; Chadhain, S.M.; Meyer, C.; Pouli, M.; Zylstra, G.J.; Hols, P.; Agathos, S.N. Characterization of a ring-hydroxylating dioxygenase from phenanthrene-degrading Sphingomonas sp. strain LH128 able to oxidize benz[a]anthracene. *Appl. Microbiol. Biotechnol.,* **2009**, *83*(3), 465-475.
[http://dx.doi.org/10.1007/s00253-009-1858-2] [PMID: 19172265]

[69]   Wu, P.; Wang, Y.S.; Sun, F.L.; Wu, M.L.; Peng, Y.L. Bacterial polycyclic aromatic hydrocarbon ring-hydroxylating dioxygenases in the sediments from the Pearl River estuary, China. *Appl. Microbiol. Biotechnol.,* **2014**, *98*(2), 875-884.
[http://dx.doi.org/10.1007/s00253-013-4854-5] [PMID: 23558584]

[70]   Peng, R-H.; Xiong, A-S.; Xue, Y.; Fu, X-Y.; Gao, F.; Zhao, W.; Tian, Y-S.; Yao, Q-H. A Profile of Ring-hydroxylating Oxygenases that Degrade Aromatic Pollutants. In: *Reviews of Environmental Contamination and Toxicology. Springer Science+Business Media*; Whitacre, D.M., Ed.; , **2010**.
[http://dx.doi.org/10.1007/978-1-4419-6260-7_4]

[71]   Habe, H.; Omori, T. Genetics of polycyclic aromatic hydrocarbon metabolism in diverse aerobic bacteria. *Biosci. Biotechnol. Biochem.,* **2003**, *67*(2), 225-243.
[http://dx.doi.org/10.1271/bbb.67.225] [PMID: 12728980]

[72]   Tang, H.; Yu, H.; Li, Q.; Wang, X.; Gai, Z.; Yin, G.; Su, F.; Tao, F.; Ma, C.; Xu, P. Genome sequence of *Pseudomonas putida* strain B62, a superdegrader of polycyclic aromatic hydrocarbons and dioxin-like compounds. *J. Bacteriol.,* **2011**, *193*(23), 6789-6790.
[http://dx.doi.org/10.1128/JB.06201-11] [PMID: 22072645]

[73]   Diaz, E. *Microbial Biodegradation: Genomics and Molecular Biology*; Caister Academic Press: Norfolk, **2008**.

[74]   Master, E.R.; Mohn, W.W. Induction of bphA, encoding biphenyl dioxygenase, in two polychlorinated biphenyl-degrading bacteria, psychrotolerant *Pseudomonas* strain Cam-1 and mesophilic *Burkholderia* strain LB400. *Appl. Environ. Microbiol.,* **2001**, *67*(6), 2669-2676.
[http://dx.doi.org/10.1128/AEM.67.6.2669-2676.2001] [PMID: 11375179]

[75]   Romine, M.F.; Stillwell, L.C.; Wong, K-K.; Thurston, S.J.; Sisk, E.C.; Sensen, C.; Gaasterland, T.; Fredrickson, J.K.; Saffer, J.D. Complete sequence of a 184-kilobase catabolic plasmid from Sphingomonas aromaticivorans F199. *J. Bacteriol.,* **1999**, *181*(5), 1585-1602.
[PMID: 10049392]

[76]   Moody, J.D.; Freeman, J.P.; Doerge, D.R.; Cerniglia, C.E. Degradation of phenanthrene and anthracene by cell suspensions of *Mycobacterium* sp. strain PYR-1. *Appl. Environ. Microbiol.,* **2001**, *67*(4), 1476-1483.
[http://dx.doi.org/10.1128/AEM.67.4.1476-1483.2001] [PMID: 11282593]

[77]   Tittabutr, P.; Cho, I.K.; Li, Q.X. Phn and Nag-like dioxygenases metabolize polycyclic aromatic hydrocarbons in *Burkholderia* sp. C3. *Biodegradation,* **2011**, *22*(6), 1119-1133.
[http://dx.doi.org/10.1007/s10532-011-9468-y] [PMID: 21369832]

[78]   Whyte, L.G.; Bourbonniére, L.; Greer, C.W. Biodegradation of petroleum hydrocarbons by psychrotrophic *Pseudomonas* strains possessing both alkane (alk) and naphthalene (nah) catabolic

pathways. *Appl. Environ. Microbiol.,* **1997**, *63*(9), 3719-3723.
[PMID: 9293024]

[79]   Andreoni, V.; Bernasconi, S.; Colombo, M.; van Beilen, J.B.; Cavalca, L. Detection of genes for alkane and naphthalene catabolism in *Rhodococcus* sp. strain 1BN. *Environ. Microbiol.,* **2000**, *2*(5), 572-577.
[http://dx.doi.org/10.1046/j.1462-2920.2000.00134.x] [PMID: 11233165]

[80]   Siron, R.; Pelletier, E.; Brochu, C. Environmental factors influencing the biodegradation of petroleum hydrocarbons in cold seawater. *Arch. Environ. Contam. Toxicol.,* **1995**, *28*, 406-416.
[http://dx.doi.org/10.1007/BF00211621]

[81]   Delille, D.; Pelletier, E.; Rodríguez-Blanco, A.; Ghiglione, J-F. Effects of nutrient and temperature on degradation of petroleum hydrocarbons in sub-Antarctic coastal seawater. *Polar Biol.,* **2009**, *32*, 1521-1528.
[http://dx.doi.org/10.1007/s00300-009-0652-z]

[82]   Castle, D.M.; Montgomery, M.T.; Kirchman, D.L. Effects of naphthalene on microbial community composition in the Delaware estuary. *FEMS Microbiol. Ecol.,* **2006**, *56*(1), 55-63.
[http://dx.doi.org/10.1111/j.1574-6941.2006.00062.x] [PMID: 16542405]

[83]   MacRae, J.D.; Hall, K.J. Biodegradation of polycyclic aromatic hydrocarbons (PAH) in marine sediment under denitrifying conditions. *Water Sci. Technol.,* **1998**, *38*, 177-185.
[http://dx.doi.org/10.1016/S0273-1223(98)00653-2]

[84]   Leahy, J.G.; Colwell, R.R. Microbial degradation of hydrocarbons in the environment. *Microbiol. Rev.,* **1990**, *54*(3), 305-315.
[PMID: 2215423]

[85]   Simkins, S.; Alexander, M. Models for mineralization kinetics with the variables of substrate concentration and population density. *Appl. Environ. Microbiol.,* **1984**, *47*(6), 1299-1306.
[PMID: 6742843]

[86]   Desai, A.M.; Autenrieth, R.L.; Dimitriou-Christidis, P.; McDonald, T.J. Biodegradation kinetics of select polycyclic aromatic hydrocarbon (PAH) mixtures by Sphingomonas paucimobilis EPA505. *Biodegradation,* **2008**, *19*(2), 223-233.
[http://dx.doi.org/10.1007/s10532-007-9129-3] [PMID: 17534722]

[87]   Bosma, T.N.; Middeldorp, P.J.; Schraa, G.; Zehnder, A.J. B mass transfer limitation of biotransformation: quantifying bioavailability. *Environ. Sci. Technol.,* **1996**, *31*, 248-252.
[http://dx.doi.org/10.1021/es960383u]

[88]   Harms, H.; Smith, K.E.; Wick, L.Y. Introduction: Problems of Hydrophobicity/Bioavailability. In: *Handbook of hydrocarbon and lipid microbiology*; Timmis, K.N., Ed.; Springer-Verlag: Berlin, **2010**.
[http://dx.doi.org/10.1007/978-3-540-77587-4_98]

[89]   Morita, R.Y. Psychrophilic bacteria. *Bacteriol. Rev.,* **1975**, *39*(2), 144-167.
[PMID: 1095004]

[90]   Farrell, J.; Rose, A. Temperature effects on microorganisms. *Annu. Rev. Microbiol.,* **1967**, *21*, 101-120.
[http://dx.doi.org/10.1146/annurev.mi.21.100167.000533] [PMID: 4860253]

[91]   Zobell, C.E. The occurrence, effects, and fate of oil polluting the sea. *Air Water Pollut.,* **1963**, *7*, 173-197.
[PMID: 14003851]

[92]   Atlas, R.M.; Bartha, R. Biodegradation of petroleum in seawater at low temperatures. *Can. J. Microbiol.,* **1972**, *18*(12), 1851-1855.
[http://dx.doi.org/10.1139/m72-289] [PMID: 4649739]

[93]   Arrhenius, S. Über die Reaktionsgeschwindigkeit bei der Inversion von Rohrzucker durch Sauren. Zeitschrift für Physik. *Chemique,* **1889**, *4*, 226-248.

[94]   Feller, G. Molecular adaptations to cold in psychrophilic enzymes. *Cell. Mol. Life Sci.,* **2003**, *60*(4), 648-662.
[http://dx.doi.org/10.1007/s00018-003-2155-3] [PMID: 12785714]

[95]   Feller, G.; Gerday, C. Psychrophilic enzymes: hot topics in cold adaptation. *Nat. Rev. Microbiol.,* **2003**, *1*(3), 200-208.
[http://dx.doi.org/10.1038/nrmicro773] [PMID: 15035024]

[96]   Struvay, C.; Feller, G. Optimization to low temperature activity in psychrophilic enzymes. *Int. J. Mol. Sci.,* **2012**, *13*(9), 11643-11665.
[http://dx.doi.org/10.3390/ijms130911643] [PMID: 23109875]

[97]   Price, P.B.; Sowers, T. Temperature dependence of metabolic rates for microbial growth, maintenance, and survival. *Proc. Natl. Acad. Sci. USA,* **2004**, *101*(13), 4631-4636.
[http://dx.doi.org/10.1073/pnas.0400522101] [PMID: 15070769]

[98]   Brauer, V.S.; de Jonge, N.; Buma, A.G.; Weissing, F.J. Does universal temperature dependence apply to communities? An experimental test using natural marine plankton assemblages. *Oikos,* **2009**, *118*, 1102-1108.
[http://dx.doi.org/10.1111/j.1600-0706.2009.17371.x]

[99]   Brakstad, O.G.; Bonaunet, K.; Nordtug, T.; Johansen, O. Biotransformation and dissolution of petroleum hydrocarbons in natural flowing seawater at low temperature. *Biodegradation,* **2004**, *15*(5), 337-346.
[http://dx.doi.org/10.1023/B:BIOD.0000042189.69946.07] [PMID: 15523916]

[100] Brakstad, O.G.; Bonaunet, K. Biodegradation of petroleum hydrocarbons in seawater at low temperatures (05 degrees C) and bacterial communities associated with degradation. *Biodegradation,* **2006**, *17*(1), 71-82.
[http://dx.doi.org/10.1007/s10532-005-3342-8] [PMID: 16453173]

[101] Bagi, A.; Pampanin, D.M.; Brakstad, O.G.; Kommedal, R. Estimation of hydrocarbon biodegradation rates in marine environments: a critical review of the Q_{10} approach. *Mar. Environ. Res.,* **2013**, *89*, 83-90.
[http://dx.doi.org/10.1016/j.marenvres.2013.05.005] [PMID: 23756048]

[102] Bagi, A.; Pampanin, D.M.; Lanzén, A.; Bilstad, T.; Kommedal, R. Naphthalene biodegradation in temperate and arctic marine microcosms. *Biodegradation,* **2014**, *25*(1), 111-125.
[http://dx.doi.org/10.1007/s10532-013-9644-3] [PMID: 23624724]

[103] Brakstad, O.G. Natural and stimulated biodegradation of petroleum in cold marine environments. In:

Psychrophiles: from biodiversity to biotechnology; Margesin, R.; Schinner, F.; Marx, J-C.; Gerday, C., Eds.; Springer-Verlag: Berlin, **2008**.
[http://dx.doi.org/10.1007/978-3-540-74335-4_23]

[104] Luo, Y.R.; Tian, Y.; Huang, X.; Yan, C.L.; Hong, H.S.; Lin, G.H.; Zheng, T.L. Analysis of community structure of a microbial consortium capable of degrading benzo(*a*)pyrene by DGGE. *Mar. Pollut. Bull.,* **2009**, *58*(8), 1159-1163.
[http://dx.doi.org/10.1016/j.marpolbul.2009.03.024] [PMID: 19409577]

[105] Machín-Ramírez, C.; Morales, D.; Martínez-Morales, F.; Okoh, A.I.; Trejo-Hernández, M.R. Benzo[*a*]pyrene removal by axenic- and co-cultures of some bacterial and fungal strains. *Int. Biodet. Biodeg.,* **2010**, *64*, 538-544.
[http://dx.doi.org/10.1016/j.ibiod.2010.05.006]

SUBJECT INDEX

3

32P-postlabelling assay 108

A

Acenaphthene 24, 94, 150, 195, 214
Acenaphthenol 154, 157, 194
Acenaphthenones 153, 154
Acenaphtylene 24
Acetylcholine esterase 10
Acyl CoA oxidase 11
Adriatic Sea 34
Alaskan Beaufort Sea 21, 42
Alkanes ii, 52, 59, 196, 207, 208
Androstenedione 122
Annelid 115, 119, 130
Anthracene 24, 55, 56, 66, 68, 73, 78, 83,
    94, 99, 114, 116, 117, 119, 129, 175,
    204, 212, 214, 227-229
Anthropogenic sources of PAH 20
Anti-cytP450-serum 118
Arctic 3, 9, 16, 21, 42, 109, 219, 220, 231
Aromatics 52, 62, 195, 196, 200, 213, 218
Aryl hydrocarbon hydroxylase 89
Asphaltics 52
Atlantic Coast 45, 46, 57, 63, 108
Atlantic cod 17, 38, 41, 45, 49, 90, 108,
    117, 118, 129, 134, 202
Australian Northwest shelf 34

B

Baltic clam 77
Beluga whales 79, 92, 104, 109
Benzo 6, 11, 17, 24, 33, 55, 56, 78, 93, 94,
    99, 107, 108, 110, 119, 175, 193, 209,
    213, 214, 221, 225, 228, 232

Bioaccumulation 5, 18, 23, 25, 31, 40, 42,
    43, 45, 87, 116, 128
Bioavailability i, 15, 17, 24, 25, 28, 29,
    33, 37, 39, 44, 45, 65, 70, 73, 76, 87,
    93, 97, 100, 205, 206, 209, 212, 219,
    220, 222, 227, 230
Biodegradation 20, 29, 40, 63, 102, 205,
    206, 223-231
Biological effective dose 87, 90
Biological effect monitoring 28, 29, 36,
    107
Biomagnification 8
Biomarker 10, 11, 23, 28, 35, 37, 41, 46,
    47, 49, 60, 92, 93, 107, 108, 110, 112
Biomonitoring 16, 18, 45, 48, 81, 82, 89,
    104, 108, 110
Biota Sediment Accumulation Factor 9
Blue mussels 38, 78, 93

C

California and Gulf of Mexico 34
Carcinogen 38, 48, 73, 74, 81, 103, 193
Catalase 10
Chrysene 13, 24, 55, 56, 68, 82, 119, 128,
    129, 172, 203, 204
Chrysenol 181, 182
Clean Water Act 24
Comet assay 10, 13, 82, 85, 87, 98, 106,
    109, 223
Common mussel 78, 103
Crab urine 114, 115
Cycloalkanes 52
Cytochrome P4501A 10, 36, 87, 90, 107,
    108

Daniela M. Pampanin and Magne O. Sydnes (Eds.)
All rights reserved-© 2017 Bentham Science Publishers

www.ingramcontent.com/pod-product-compliance
Lightning Source LLC
Chambersburg PA
CBHW041726210326
41598CB00008B/789